RADIOISOTOPIC METHODS FOR BIOLOGICAL AND MEDICAL RESEARCH

Radioisotopic Methods for Biological and Medical Research

HERMAN W. KNOCHE

University of Nebraska–Lincoln

New York   Oxford
OXFORD UNIVERSITY PRESS
1991

Oxford University Press

Oxford   New York   Toronto
Delhi   Bombay   Calcutta   Madras   Karachi
Petaling Jaya   Singapore   Hong Kong   Tokyo
Nairobi   Dar es Salaam   Cape Town
Melbourne   Auckland

and associated companies in
Berlin   Ibadan

Library of Congress Cataloging-in-Publication Data
Knoche, Herman W.
Radioisotopic methods for biological and medical research /
by Herman W. Knoche.
p.   cm.   Includes bibliographical references.
Includes index.   ISBN 0-19-505806-2
1. Nuclear medicine   2. Radioactive tracers in biology.
3. Radiation—Safety measures.   4. Radiation dosimetry.
5. Radiometry. I. Title.
[DNLM: 1. Nuclear Physics.   2. Radiation Protection.
3. Radiobiology.   WN 650 K72r]
R895.K544   1991   616.07′575—dc20
DNLM/DLC   90-7824

1 3 5 7 9 8 6 4 2

Printed in the United States of America
on acid-free paper

Dedicated to Some Scintillators in My Life,
Kim, Chris, Connie, and Johnny Knoche

Preface

The usefulness of radioactive materials for scientific investigation was well recognized shortly after World War II, and the Atomic Energy Commission (AEC), which was the forerunner to the Nuclear Regulatory Commission (NRC), promulgated well-conceived regulations for the safe use of radioactive materials before their use became widespread. These regulations required a user of radioactivity to have 40 clock hours of formal or on-the-job training in the safe and effective handling of radioactive materials. Besides regulating the use of radioactivity, the AEC promoted its use by providing training at the laboratories at Oak Ridge, Tennessee, and through educational grants to academic institutions. Consequently, the formal training requirements were readily met by "Radiotracer Courses" offered by most universities as well as those offered at Oak Ridge.

By the early 1970s the first generation of radioactivity users had been trained and licensed, so on-the-job training became widely available at academic institutions. Many laboratory courses that were not specifically designed to teach principles of radiotracer methodology were using radioisotopes to illustrate other scientific principles. Thus, a training alternative to the "radiotracer course" became available and this appears to have caused radiotracer courses to be dropped by many institutions. Although the mystique about radioactive materials is gone, requirements for proper training remain. Public concern about the environment undoubtedly will prevent any relaxation of license requirements.

On-the-job training appears to be an efficient method to provide practical laboratory experience, but the theoretical body of knowledge must be gained as well. It can be gained through bits and pieces of various courses a person may take during collegiate training, or, where available, as a separate course. Another alternative might be self-study with the aid of books, computer programs, and video tapes. Examinations can be used as an instrument to demonstrate knowledge in the theoretical area in lieu of the more traditional method of providing a document showing hours of training in appropriate courses.

This book was designed to serve as a text for either a formal lecture course on the theoretical aspects of radioisotopic methods or for self-study. Particular attention has been given to discussion of the four major topic areas specified in NRC's license application forms: (1) principles and practices of radiation protection; (2) radioactivity measurement, standardization and monitoring techniques, and instruments; (3) mathematics and calculations basic to the use and measurement of radioactivity; and (4) biological effects of radiation.

Mastery of the material should provide the beginning research scientist with the theoretical knowledge needed to use radioactive materials safely and effectively. Hopefully, suitable documentation of such knowledge coupled with practical laboratory experience will satisfy training requirements for radioactive materials licensure.

Many colleagues and students have contributed ideas and suggestions that have shaped the approach taken in this book, but I am particularly indebted to several individuals at the University of Nebraska: L. E. Grimm, R. J. Kittok, R. V. Klucas, R. J. Krueger, A. M. Parkhurst, L. L. Shearman, R. J. Spreitzer, and W. W. Stroup.

I am deeply grateful to my wife Darlene, who not only has supported and encouraged the writing of this book, but also prepared the original art work, and, once again, came to my rescue by doing the word processing when other arrangements did not materialize.

Lincoln, Neb.   H. W. K.
September 1990

Abbreviated Contents

Contents

RADIOISOTOPIC METHODS FOR
BIOLOGICAL AND MEDICAL RESEARCH

CHAPTER 1

An Overview of Regulations and Responsibilities for Users of Radioactive Materials

The devastating effects of high intensities of ionizing radiation on humans was illustrated vividly when the two atomic bombs were dropped on Japan to end World War II. Undoubtedly, the results of these events and the effects of radiation exposure to scientific pioneers caused the world to proceed in the peaceful applications of nuclear technology with considerable caution. Only one peacetime nuclear disaster has occurred in the world, at Russia's Chernobyl reactor site, where 31 persons were killed in 1986, and about 200 more persons probably will have their lives shortened by statistically significant amounts. However, compared with accident and fatality rates in other industries, this is an excellent record; just think of all the coal miners who have been killed during the past 30 years. But the accident at Chernobyl should not have happened, and surely would not have happened if the safety standards for the reactor's design and operation had been as stringent as those in the United States or other Western countries. Thus, the use of radioactive materials and radiation-producing equipment has been safe, although highly regulated, during the course of most of its development. An overview of current regulations concerning the use of radioactive materials in scientific investigations follows.

ATOMIC ENERGY COMMISSION

Although some regulations pertaining to radiation control existed in the United States prior to 1946, it was in that year that the Atomic Energy Act was passed. The Act created the *Atomic Energy Commission* (*AEC*), but only the Federal Government was permitted to control and own radioactive materials. In 1954 the Act was amended to permit private organizations and individuals to own and use radioisotopes.

The AEC was given two general responsibilities:

1. Promote the peaceful use of nuclear technology.
2. Formulate and administer regulations to protect the health and safety of the general public.

Regulations concerning the use of radioactive materials were formulated with apparent conservatism for the times, and great emphasis was placed on providing training programs to persons (such as research scientists and physicians) who might utilize radioactive materials effectively in their work. It appears that among the fields of occupational safety, radiation safety led all others.

NUCLEAR REGULATORY COMMISSION

A congressional amendment and the Energy Reorganization Act was passed in 1974; this divided the responsibilities of the AEC and created two new commissions. The *Nuclear Regulatory Commission* (*NRC*) was charged with the responsibility of writing and enforcing regulations concerning the use of radioactive materials, and the *Department of Energy* (DOE) was given the other responsibilities previously held by the AEC. NRC enforces regulations by controlling the ownership and possession of radionuclides. A license is required for possession of radioactive materials, and license holders are inspected by NRC to determine if regulations are being followed by the licensee. If serious or repeated violations occur, a license may be revoked, and radioactive materials confiscated.

Other Federal agencies have responsibilities that occasionally overlap with NRC. They are the Department of Health, Education, and Welfare (HEW), Food and Drug Administration (FDA), Occupational Safety and Health Administration (OSHA), and the Environmental Protection Agency (EPA).

ICRP AND NCRP

NRC regulations are based on reports and recommendations, primarily from two scientific bodies, the *International Commission on Radiological Protection* (*ICRP*) and, in the United States, the *National Council on Radiation Protection and Measurements* (*NCRP*). These organizations convene committees of scientists, experts in various aspects of radiation, to formulate standards and procedures, which are published as reports. Thus, the ICRP and NCRP are advisory bodies that provide a scientific basis for regulations.

AGREEMENT STATES

Although the NRC is the federal agency responsible for adopting and enforcing rules and regulations that apply to users of radioactive materials, broad administrative responsibilities have been transferred to some state governments. An act in 1959 permitted the NRC (then the AEC) to make agreements with those states that could, and would, operate a suitable radiological health program for the radioactive material users in their states. States that have such agreements with the NRC are called *Agreement States*. Agreement states have their own state regulations and provide personnel to license and inspect users of radioactive materials. However, the NRC must approve a state's regulations and inspect its programs periodically. Effectively, an agreement state simply administers a program that is highly consistent with NRC regulations. Approximately half of the U.S. states are agreement states.

LICENSES

Licenses to use radioactive materials are issued by the NRC or by an agreement state, depending on where the radioactive materials are to be used. There are exceptions; federal agencies, such as the United States Department of Agriculture (USDA) and the Veterans Administration (VA), that have units operating within an agreement state remain under NRC control; also, nuclear power plants are licensed by NRC only.

Licenses for nonmedical uses fall into two classes: general and specific.

General Licenses

Without application, general licenses are granted to individuals and institutions, including profit-making companies. Such licenses permit the purchase and use of small quantities of certain radioisotopes that are deemed harmless to the general public. For example, many smoke alarm detectors contain a small radioactive source, and the general license permits the sale of such smoke alarms to the public.

Exempt Quantities

For small scientific experiments, demonstrations, and some clinical analytical methods, small quantities of appropriate radionuclides may be purchased under the general license. The NRC and agreement states' regula-

TABLE 1-1   Exempt Quantities of Some Radionuclides

Radionuclide	Quantity limit (μCi)
^{3}H	1000
^{14}C	100
^{32}P	10
^{125}I	1
^{90}Sr	0.1

tions provides a list of radionuclides and the maximum amount of each that may be possessed. Thus, *exempt quantities* are the amounts of a particular radionuclide that may be obtained without a specific license. The maximum amount of radioactivity listed as an exempt quantity does vary between different radionuclides as shown by a few examples in Table 1-1. The limits are consistent with the relative radiological health hazards of the radionuclides.

Specific Licenses

Specific licenses permit individuals or institutions to possess larger quantities of various radionuclides than are possible by purchasing exempt quantities. However, the license will specify which radionuclides may be obtained and the maximum quantities of each that can be possessed at any time. Thus, specific licenses are not blanket authorizations, instead they are highly individualized, and the maximum possession limits approved for specified radionuclides will depend on the licensee's needs and capability to use such materials safely and effectively.

Applications for a Specific License

Individuals may apply to either the NRC or the appropriate radiological health division of an agreement state, depending on his or her location. A typical application form is shown in Figure 1-1. Whether issued by an institution, agreement state, or the NRC, all forms are virtually identical in terms of content.

Besides the applicant's name the precise location (1–5) for the proposed use of radioactive materials is required. The specific radionuclides (6a) and maximum possession limits (6b) being requested are listed, along with their proposed use (7). If human subjects are involved additional forms are required. Because of disposal problems of animal carcasses contaminated with radioactive materials, information concerning possible animal use is needed. To determine if the licensee has access to facilities suitable for radioactivity use, diagrams of lab facilities including fume hoods are requested (11). It is understood that radioactive materials will be used only in the facilities described. Information about radiation detection instruments (counters, survey meters) and their maintenance (10) indicate if the applicant has access to proper equipment. Training (8) is required and must be documentable. Finally, experience (9) working with radioactive materials should be listed, including formal laboratory courses and research conducted under the supervision of a licensee.

Institutional or Broad-Scope Licenses

Institutions as well as individuals may obtain Specific Licenses. Such licenses may be called *Institutional* or *Broad-Scope Licenses,* and they are useful in cases in which a company, university, or hospital has a number of individuals using radioactivity. The license would be in the name of the institution, and individuals would be listed as *Users.* The list of radionuclides and their possession limits would be sufficient for the needs of the whole institution, but each *User* would need to have the training and experience required for an individual specific license.

Effectively, a broad-scope or institutional license is an agreement between the institution and licensing agency that delineates responsibilities for user approval, safety programs, training, and disposal of radioactive wastes. The latitude given to an institution depends on the adequacy and qualifications of personnel, or, in more general terms, the institution's capability to provide an adequate radiation safety program. Just like the individual licensees, institutional licensees are inspected by their licensing agency.

Radiation Safety Officer.   Institutional radiation programs generally have a *Radiation Safety Officer* (RSO) and a *Radiation Safety Committee.* Institutions may designate or hire an individual to administer the radiation program, and if the size of the program warrants it, the RSO may be a full-time employee with assistants. Generally, health physicists are selected for RSO positions. An RSO can be a valuable source of help to users in matters such as waste disposal, procurement, contamination, clean-up, and advice for unusual experiments. Occasionally, RSOs must assume police duties if violations occur.

Radiation Safety Committees.   Radiation safety committees usually include several "users," the RSO, and an institutional administrator or officer. The committee may write and approve institutional rules and regulations, act on applications for prospective users, determine what action should be taken if violations occur, and advise the RSO and the administration of the institution.

Form NRH-5
(7-79)

Nebraska Department of Health

APPLICATION FOR RADIOACTIVE MATERIAL LICENSE

INSTRUCTIONS-Complete Items 1 through 16 if this is an initial application. If application is for renewal of a license, complete only Items 1 through 7 and indicate new information or changes in the program as requested in Items 8 through 15. Use supplemental sheets where necessary. Item 16 must be completed on all applications. Mail two copies to: Nebraska Department of Health, Division of Radiological Health, 301 Centennial Mall South, P.O. Box 95007, Lincoln, Nebraska 68509. Upon approval of this application, the applicant will receive a Radioactive Material License, issued in accordance with the requirements contained in Regulations for the Control of Radiation and the Nebraska Radiation Control Act.

1. (a) NAME AND STREET ADDRESS OF APPLICANT. (Institution, firm, hospital, person, ect.) Telephone No: Area Code ( )_____	STREET ADDRESS (ES) AT WHICH RADIOACTIVE MATERIAL WILL BE USED. (If different from 1 (a).)
2. DEPARTMENT TO USE RADIOACTIVE MATERIAL. Person to Contact_____ Telephone No: Area Code ( )_____	3. This is an Application for: (Check appropriate item) a. ☐ New license b. ☐ Amendment to License No._____ c. ☐ Renewal of License No._____
4. INDIVIDUAL USERS (S). (Name and title of individual (s) who will use or directly supervise use of radioactive materials. Give training and experience in Items 8 and 9.)	5. RADIATION SAFETY OFFICER (RSO) (Name of person designated as radiation safety officer if other than individual user. Attach resume of his training and experience as in Items 8 and 9).
6. (a) RADIOACTIVE MATERIAL. (Elements and mass number of each.)	CHEMICAL AND/OR PHYSICAL FORM AND MAXIMUM QUANTITY OF EACH CHEMICAL AND/OR PHYSICAL FORM THAT YOU WILL POSSESS AT ANY ONE TIME. (if sealed source (s), also state name of manufacturer, model number, number of sources and maximum activity per source.)

7. DESCRIBE PURPOSE FOR WHICH RADIOACTIVE MATERIAL WILL BE USED. (If radioactive material is for "human use," FORM NRH-5A must be completed in lieu of this item. If radioactive material is in the form of sealed sources, include the make and model number of the storage and / or device in which the source will be stored and/or used.)

FIG. 1-1   An application form for a radioactive materials license in the state of Nebraska, which is an Agreement State. Forms used by the Nuclear Regulatory Commission and other Agreement States are nearly identical.

Form NRH-5
(7-79)

TRAINING AND EXPERIENCE OF EACH INDIVIDUAL NAMED IN ITEM 4 (Use supplemental sheets if necessary).

8. TYPE OF TRAINING	WHERE TRAINED	DURATION OF TRAINING	ON THE JOB (Circle answer)		FORMAL COURSE (Circle answer)	
a. Principles and practices of radiation protection			Yes	No	Yes	No
b. Radioactivity measurement standardization and monitoring techniques and instruments			Yes	No	Yes	No
c. Mathematics and calculations basic to the use c. and measurement of radioactivity			Yes	No	Yes	No
d. Biological effects of radiation			Yes	No	Yes	No

9. EXPERIENCE WITH RADIATION (Actual use of radioisotopes or equivalent experience).

ISOTOPE	MAXIMUM AMOUNT	WHERE EXPERIENCE WAS GAINED	DURATION OF EXPERIENCE	TYPE OF USE

10. RADIATION DETECTION INSTRUMENTS (Use supplemental sheets if ncessary).

TYPE OF INSTRUMENTS (Include make and model number of each)	NUMBER AVAILABLE	RADIATION DETECTED	SENSITIVITY RANGE (mr/hr)	WINDOW THICKNESS (mg/cm)	USE (Monitoring, Surveying, Measuring)

11. METHOD, FREQUENCY, AND STANDARDS USED IN CALIBRATING INSTRUMENTS LISTED ABOVE.

12. FILM BADGES, DOSIMETERS, AND BIO-ASSAY PROCEDURES USED (For film badges, specify method of calibrating and processing, or name of supplier).

INFORMATION TO BE SUBMITTED ON ADDITIONAL SHEETS

13. FACILITIES AND EQUIPMENT. Describe laboratory facilities and remote handling equipment, storage containers, shielding, fume hoods, etc. Explanatory sketch of facility is attached (Circle answer).   Yes   No

14. RADIATION PROTECTION PROGRAM. Describe the radiation protection program including control measures. If application covers sealed sources, submit leak testing procedures where applicable, name, training, and experience of person to perform leak tests and arrangements for performing initial radiation survey, servicing, maintenance and repair of the source.

15. WASTE DISPOSAL. If a commercial waste disposal service is employed, specify name of company. Otherwise, submit detailed description of methods which will be used for disposing of radioactive wastes and estimates of the type and amount of activity involved.

CERTIFICATE
(This item must be completed by applicant)

16. THE APPLICANT AND ANY OFFICIAL EXECUTING THIS CERTIFICATE ON BEHALF OF THE APPLICANT NAMED IN ITEM 1, CERTIFY THAT THIS APPLICATION IS PREPARED IN CONFORMITY WITH NEBRASKA DEPARTMENT OF HEALTH REGULATIONS FOR THE CONTROL OF RADIATION AND THAT ALL INFORMATION CONTAINED HEREIN, INCLUDING ANY SUPPLEMENTS ATTACHED HERETO, IS TRUE AND CORRECT TO THE BEST OF OUR KNOWLEDGE AND BELIEF.

_____
Applicant name in Item 1

Date_____   By:_____

_____
Title of certifying official authorized to act on behalf of the applicant

User Permits.   In an institution with a broad-scope license, prospective users will submit an application for a *User Permit*. Generally, the application forms are the same as those for individual licenses, and qualifications for becoming a user are the same also. However, the application may be approved by the RSO and/or the radiation safety committee. From the users perspective, a Users Permit effectively is a specific license granted by an institution.

Enforcement of Licenses

A fairly simple regulation prevents persons and institutions from possessing radioactive material without a specific license. It is illegal to transfer (sell or give) radioactive materials to another person or institution unless the recipient has a license to possess the materials. Consequently, radionuclide supply companies require information about a customer's license before they will fill an order, and license numbers must accompany orders. Regulations are violated if a licensee gives radioactive materials to another who does not hold a license that permits the possession of such materials.

License and Permit Conditions

Licenses and user permits are not granted for life, but must be renewed periodically. Renewals usually are granted unless violations indicate that an individual or institution is not performing radiation safety duties properly. Licenses have been revoked or suspended because of continued violations of approved safety practices.

Holders of individual specific licenses and user permits are inspected periodically by the licensing agency or perhaps by an institutional RSO. The purpose of inspections is to ascertain whether the users are following appropriate safety procedures. Practically speaking, the responsibility for radiation safety rests on those persons actually using the radioactive materials. This is one reason why so much emphasis is placed on the training and experience of individual users.

Training and Experience Requirements

The amount of training and experience required for approval of specific licenses or user permits is remarkably consistent. Generally, license applicants will have at least a B.S. degree in a scientific field and special training in using radiation. The required radiation training is divided into four categories (see items 8a, b, c, and d of the application form in Figure 1-1):

8a.   Principles and practices of radiation protection.

8b.   Radioactive measurement standardization and monitoring techniques and instruments.

8c. Mathematics and calculations basic to the use and measurement of radioactivity.

8d. Biological effects of radiation.

These topics may be covered, with sufficient depth, in a formal course of about 40 hours duration. Formal courses are desirable because they are easily documented by transcripts or certificates. On-the-job training is acceptable, but it is difficult to document the content and balance of topics in such training. Self-study, and the subsequent passing of an exam to document knowledge, is another method that is gaining popularity.

Experience working with radionuclides may be gained by formal laboratory courses or performing research utilizing radioisotopes under the supervision of a licensee or permit holder. Documentation of such experience is important for first-time applicants.

Responsibilities of Licensees and Users

Laboratory rules will be discussed more fully in Chapter 16, but in general, a User must

1. Keep records showing the receipt and disposal of all radioactive materials.
2. Use adequate security methods to prevent the general public and untrained individuals from coming in contact with radioactive materials.
3. Dispose radioactive wastes by approved methods only.
4. Prevent unnecessary radiation exposure to anyone, including oneself.

CONCLUSION

The use of radioisotopic materials in biological and medical research has become essential and commonplace in about all areas of laboratory investigations. Although the regulations may appear as impediments, the possibilities of their relaxation seem remote in light of current public attitudes. Consequently, a person embarking on a scientific or medical career is well advised to seek training in the use of radioactivity. Hopefully, this chapter alerts the reader to areas of knowledge needed for licensure, and the remainder of the book provides a suitable theoretical basis for the *effective* and *safe* use of radioactive materials in research.

PART I

Review of Nuclear Physics

CHAPTER 2

Atomic and Nuclear Structure

$$A = Z + N$$

Mass number $\uparrow$ Atomic number.

ATOMIC STRUCTURE

In the Bohr atomic model, an atom is visualized as a positively charged nucleus surrounded by orbiting electrons that balance the nuclear charge. The nucleus consists of neutrons and protons, collectively called *nucleons,* and they constitute most of the mass of an atom. Both types of nucleons have a mass number of one, meaning that a mole of either would weigh about 1 g, and the proton (p^+) carries one unit of positive charge ($1.6021892 \times 10^{-19}$ C), whereas the neutron (n) is electronically neutral. The electrons contribute little to the mass of an atom, but give it most of its volume due to the space they occupy in their orbitals. An electron has a negative charge that in magnitude is equal to the charge on a proton. To clearly identify the usual negatively charged electron, the term *negatron* (e^-) may be used. This distinguishes it from a positively charged particle of the same size, the *positron* (e^+). Both positrons and negatrons, which are antiparticles of each other, play important roles in nuclear phenomena. In our discussions, the terms electron (e^-) and negatron may be considered synonymous.

The energy of an electron in an atom is quantitized, that is, it is restricted to discrete energy levels. The principal energy levels correspond to shells and are identified by quantum numbers that are integers, 1, 2, 3, etc. Instead of quantum numbers, letters may be used to identify principal energy levels; thus electrons in the K shell correspond to those electrons with a quantum number of 1, and successive shells are L = 2, M = 3, N = 4, etc. The K shell represents the orbital closest to the nucleus, and electrons in that shell have lower energies than those in outer shells. Consequently, to free an electron from its atom, more energy is required if the electron is in a lower level than a higher one. The theoretical minimum energy required to remove an electron from a particular shell or orbital is referred to as its *binding energy* or *ionization potential*. If an electron va- *unfilled space* cancy is created in a lower shell, an electron from a higher shell will fall to the lower shell and emit a photon of electromagnetic radiation during the process. The quantum energy of the photon represents the difference be- tween the binding energies of electrons in the shells affected, and the wave- length of the radiation is usually in the X-ray portion of the electromagnetic spectrum when lower shells are involved, but in the UV region when tran- sitions occur between the outermost or valence elecrons.

Additional quantum numbers describe energy levels within the principal levels. The second quantum number, ℓ, determines the number of sublevels within each principal level, and ℓ has values of zero to $n - 1$ where n is the number of the principal level. Thus, when $n = 1$, only one sublevel exists $\ell = 0$. For $n = 2$, $\ell = 0$ or 1, and when $n = 3$, $\ell = 0$, 1, or 2. The ℓ quantum numbers describe the general shape of electron clouds associ- ated with the electrons having such quantum numbers. These sublevels are also given letter designations. The values of ℓ, 0, 1, 2, and 3 correspond to s, p, d, and f so that the principal level and sublevel may be indicated easily; for example, "3p" refers to the second sublevel in the third principal shell. The third quantum number m determines the number of orbitals in a sublevel, and m may have integer values from $+\ell$ to $-\ell$. Thus, if $\ell = 1$, $m = 1, 0,$ or -1. This quantum number describes how electron clouds are directed in space. The fourth quantum number, s, designates the direction an electron spins on its axis, and s has values of either $+\frac{1}{2}$ or $-\frac{1}{2}$.

A generalized view of the energy levels of electrons is one of bands of energies corresponding to the principal quantum numbers or shell number, and within these bands, are closely spaced sublevels, the number of which depends on the shell number. When $n = 1$ only two energy levels are per- mitted, but for $n = 3$, there are 18 allowed energy levels. In the higher shells, there may be some overlap of sublevels between shells.

The chemical activity of an atom is dependent on the configuration of its electrons in its outermost orbitals because these are the ones involved in the bonds that join two or more atoms to form molecules. Those elec- trons responsible for bond formation are called *valence electrons,* and be- cause they have much more energy than the electrons in inner shells they are more easily removed; hence they are the electrons normally removed when ions are created.

MASS AND ENERGY UNITS

Because of uncertainties in measurements, the values of physical constants may vary slightly depending on one's reference sources, but in calculations presented here such differences are of little consequence. Nevertheless, values presented are those given in the 69th edition of the *Handbook of Chemistry and Physics* (Weast, 1988). A table of commonly used physical constants and conversion factors is given in the appendix.

The isotope ^{12}C is assigned a mass of exactly 12.000 g/mol, and Avogadro's number is 6.022045×10^{23} atoms/mol. An *atomic mass unit* (amu) is $\frac{1}{12}$ the mass of one atom of ^{12}C, and corresponds to the mass of one particle (a hypothetical one) having an exact atomic weight of 1 g/mol. Thus

$$1 \text{ amu} = \frac{1 \text{ g/mol}}{6.022045 \times 10^{23} \text{ atoms/mol}} = 1.660566 \times 10^{-24} \text{ g/atom}$$

$$1 \text{ amu} = 1.660566 \times 10^{-27} \text{ kg}$$

The mass of a proton (m_{p+}) at rest (velocity = 0) is 1.00727647 amu, while the corresponding value for a neutron (m_n) is slightly greater, 1.008665012 amu. An electron's rest mass (m_e) is 5.4858026×10^{-4} amu, and a positron has an identical mass.

In classical Newtonian physics, it is assumed that mass cannot be created nor destroyed, and that the mass of a body is constant and equal to its mass when it is at rest (rest mass). With his theory of relativity, Einstein showed that such assumptions were incorrect; however, the mathematical relationships he developed explained why Newtonian laws had appeared to be applicable. When the velocities of bodies or particles are below about one-tenth the speed of light, the masses of bodies are approximately the same as their rest masses; however at velocities approaching the speed of light, masses increase dramatically. The term *relativistic velocity* refers to velocities where significant deviations from Newtonian laws are observed, and these conditions require *relativistic equations* for calculations. Mathematical relationships from Newtonian physics will be used in this book generally, but when dealing with particles with high velocities some of the following relativistic relationships will be utilized.

Einstein's famous mathematical equation that formalizes the concept that energy and mass are continuous is given as equation 2-1.

$$E = mc^2 \tag{2-1}$$

E is the energy, c the velocity of light in a vacuum, and m the relativistic mass, which is the mass of a body or particle in motion. Equation 2-2 shows the relationship between a particle's relativistic mass (m), its rest mass (m_0), and its velocity (v). Again, c is the velocity of light.

$$m = \frac{m_0}{\sqrt{1 - (v/c)^2}} \tag{2-2}$$

Inspection of the Equation 2-2 shows that when $v = 0$, $m = m_0$, which conforms to the definition of rest mass, and m approaches infinity when v approaches c. Because the velocity ratio is squared, the rest mass is a close approximation of the relativistic mass until the body's velocity is within an order of magnitude of the velocity of light. For example, suppose $v = 10^{-1}c$ or about 3×10^7 m/s.

$$m = \frac{m_0}{\sqrt{1 - (10^{-1} c/c)^2}} = 1.005 \, m_0$$

This constitutes about a 0.5% difference between the relativistic and rest masses.

Combining Equations 2-1 and 2-2 provides an energy relationship for a particle in motion expressed in rest mass units.

$$E = \frac{m_0 c^2}{\sqrt{1 - (v/c)^2}} \qquad (2\text{-}3)$$

The energy according to Equation 2-3 consists of two components, the kinetic energy of the particle plus the energy equivalent to its rest mass. Consequently, the kinetic energy (E_k) of a particle with a relativistic velocity is given by Equation 2-4.

$$E_k = \frac{m_0 c^2}{\sqrt{1 - (v/c)^2}} - m_0 c^2 \qquad (2\text{-}4)$$

Although various energy units may be used in nuclear calculations, the more conventional units are based on the electron volt (eV), which is the energy an electron acquires when it drops through a potential of 1 V. An MeV (million electron volts) is equivalent to 10^6 eV and 1.602192×10^{-13} joule (J), while a keV is 10^3 eV.

A convenient conversion factor relating energy (in MeV) to mass (in amu) may be derived from Equation 2-1.

$$E = (1 \text{ amu})(1.660566 \times 10^{-27} \text{ kg/amu})(2.997925 \times 10^8 \text{ m/s})^2$$

$$E = 1.492443 \times 10^{-10} \text{ kg·m}^2/\text{s}^2 = 1.492443 \times 10^{-10} \text{ J}$$

$$E = \frac{1.492443 \times 10^{-10} \text{ J}}{1.602192 \times 10^{-13} \text{ J/MeV}} = 931.5 \text{ MeV}$$

1 amu = 931.5 MeV

The Electromagnetic Spectrum

Figure 2-1 is a diagram of the higher energy portion of the electromagnetic spectrum. General atomic or molecular transitions that involve energies consistent with the various types of radiation are also given. The quantum energy (E) of electromagnetic radiation (the energy of one photon) is related to its frequency (v) or wavelength (λ) as shown in Equation 2-5, where h is Planck's constant (6.626176×10^{-34} J·s or 4.13570×10^{-15} eV·s).

$$E = hv = \frac{hc}{\lambda} \qquad (2\text{-}5)$$

Region	Wavelength	Approximate Energy	Typical Transitions
Cosmic Rays	10^{-3} Å	12.4 MeV	Nuclear
Gamma Rays	1 Å	12.4 keV	
X-Rays	10 Å	1.24 keV	Intershell
Soft X-Rays	100 Å	124 eV	electrons
U. V.	200 nm	6.2 eV	Ionization
Near U. V.	400 nm	3.1 eV	Outershell
Visible	800 nm	1.55 eV	electrons
Near I. R.	2.5 μm	0.5 eV	Molecular
I. R.	25 μm	0.05 eV	vibration
Far I. R.	400 μm		
Microwaves			

FIG. 2-1 A nonscalar diagram of the electromagnetic radiation spectrum.

For wavelengths in meters, Equation 2-6 applies.

$$E = \frac{1.23985 \times 10^{-6}\,\text{eV} \cdot \text{m}}{\lambda} \qquad (2\text{-}6)$$

NUCLEAR NOMENCLATURE

Nuclides

Elements are distinguished by their atomic numbers or the numbers of protons in their nuclei, whereas nuclides have specified numbers of neutrons as well as protons. To denote a particular nuclide, the appropriate element symbol is used and the mass number of the nuclide is written as a superscript to the left of the symbol. Thus, a nuclide of iodine having a mass number of 125 is written ^{125}I. The iodine symbol indicates it contains 53 protons which, when subtracted from the mass number (125 − 53), leaves 72 neutrons. This nuclide might also be written as iodine-125. Although this method is precise and concise, it is not always easy to remember the atomic numbers of elements or subtract numbers, so a more complete notation is also used.

mass # →$^{125}_{53}$I$_{72}$
protons #↗

neutrons = mass # − protons
         125   − 53
         = 72

The superscript area to the right of the symbol is reserved to indicate electronic conditions such as ionic charges or an excited electronic state, for example, $^{125}I^-$ and $^{125}I^*$. An asterisk signifies that an electron has been promoted to an energy level above its normal ground state level. Although the absence of a right superscript suggests that the atom is in an electronically neutral and unexcited state, sometimes a zero is placed there to ensure that the electronic ground state is intended, for example, $^{125}I^0$.

Frequently, Z is used to indicate the *atomic number* and A the *mass number,* and a generalized representation of a nuclide is

$$_Z^A X_{A-Z}$$

Radionuclides, Isotopes, and Isomers

Radionuclides are those nuclides that undergo radioactive decay. Those that do not decay are referred to as stable nuclides.

Isotopes are groups of nuclides that have the same number of protons, but different numbers of neutrons and hence different mass numbers. Isotopes of iodine include $^{125}_{53}I$, $^{126}_{53}I$, $^{127}_{53}I$, $^{128}_{53}I$, $^{129}_{53}I$, $^{130}_{53}I$, and ^{131}I among others. Although the use of the term radioisotope has been discouraged (radionuclide preferred), it refers to an isotope that is radioactive, whereas a stable isotope is one that is not.

Isomers are nuclides that have the same atomic and mass numbers, but differ in the energy levels of their nuclei. The higher energy level is referred to as a *metastable state* if the nucleus can exist in such a state for a reasonable period of time. Isomers in a metastable state are indicated by the addition of an "m" to the mass number. For example, iodine-130 in its metastable energy state is indicated as

$$_{53}^{130m}I$$

Note that the metastable states apply to energy levels of a nucleus and should not be confused with electronic excited states.

Nuclide Charts

Periodic charts provide information about naturally abundant isotopes as elements, but they are not particularly useful for presenting nuclear data. Nuclide charts, such as the one shown in Figure 2-2, show some pertinent characteristics of known nuclides. The information is given in grid boxes formed by increasing proton (Z) and neutron $(A - Z)$ numbers. The shaded boxes indicate stable nuclides whereas the others contain information about radionuclides. For a proton number of 6 (carbon), six nuclides are shown, ^{10}C through ^{15}C. Stable carbon nuclides are ^{12}C and ^{13}C, while ^{14}C is a naturally occurring radionuclide, and its decay characteristics are given in its box. The other carbon isotopes ^{10}C, ^{11}C, and ^{15}C are also radionuclides, but they do not occur naturally.

NUCLEAR STABILITY

In Figure 2-2 a line is formed by the stable nuclides (shaded boxes), and it is referred to as the _line of stability._ The line is close to a 45° angle for nuclides with proton numbers below 20 (calcium). This indicates that within this group the ratio of neutrons to protons is about one for stable nuclides and suggests that pairing of neutrons and protons may be a factor in determining nuclear stability. Above an atomic number of about 20, the slope of the line of stability decreases, hence the neutron/proton ratio increases. For heavier stable nuclei the ratio is about 1.6.

Within a nucleus there are repulsive forces due to the positively charged protons, but these are counteracted by very short range, yet strong, gravitational forces between the nucleons, the protons, and neutrons. Additional neutrons would increase gravitational forces without increasing repulsive forces and hence may increase the overall forces that bind the nuclear particles together.

When radioactive decay processes alter the atomic number of a radionuclide a transmutation of elements results. In such transformations, the radionuclide undergoing decay is called the _parent,_ and the product is called the _daughter_ nuclide. For most decay processes the daughter will lie closer to (or on) the line of stability than the parent. For example, ^{14}C has a neutron/proton ratio of 1.33 and lies slightly to the right of the line of stability (Figure 2-3). When an atom of ^{14}C decays, a neutron is converted to a proton which yields $^{14}_{7}N$. The daughter nuclide has a neutron/proton ratio of one (7/7 = 1) and lies on the line of stability.

Degrees of Stability

Sometimes nuclei are referred to as being very unstable or relatively stable, which implies that degrees of stability exist. Generally, these terms are correlated with relative rates of decay or half-lives (time required for half of a sample to decay) of a radionuclide. Carbon-10, which has a half-life of 19 s, is considered highly unstable whereas ^{14}C (half-life > 5000 yr) is not. It is plausible that some so-called stable nuclides do decay because it is impossible to detect decay rates in which the half-lives are greater than about 10^{20} yr. Consequently, stable nuclides may be more precisely defined as being those nuclides that have undetectable rates of decay. Nuclear stability is related to a nuclear property called the mass defect or binding energy.

Mass Defect and Binding Energy

An atom weighs less than the sum of its components (neutrons, protons, and electrons), and the difference in mass is referred to as the _mass defect_ (MD). The rest mass of an electrically neutral atom of the most abundant

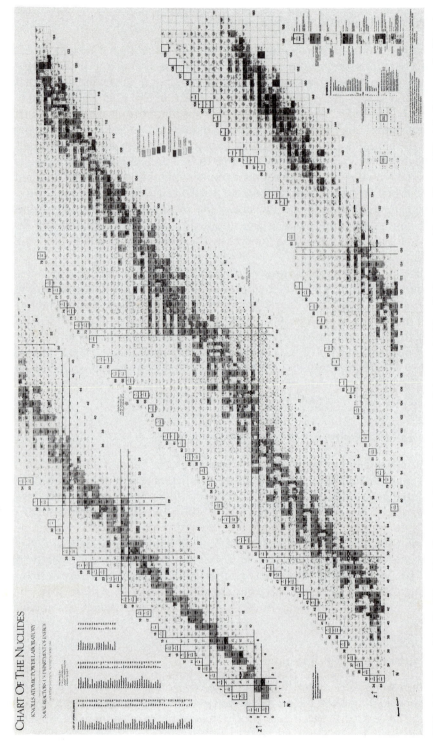

FIG. 2-2 Chart of the Nuclides. (Knolls Atomic Power Laboratory, Schenectady, New York, operated by the General Electric Company for Naval Reactors, the United States Department of Energy. Reprinted by permission)

22

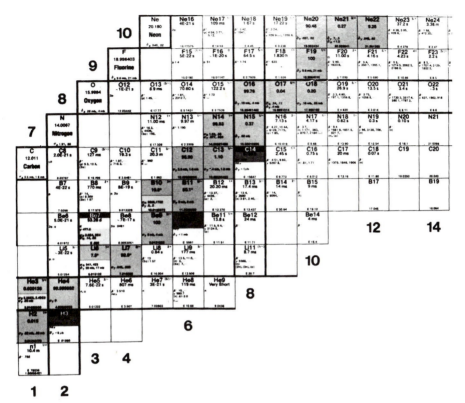

FIG. 2-3 An enlargement of the data for the carbon nuclides shown in Figure 2.2. (Knolls Atomic Power Laboratory, Schenectady, New York, operated by the General Electric Company for Naval Reactors, the United States Department of Energy. Reprinted by permission)

isotope of helium, $_2^4\text{He}^0$ is 4.002603 amu, and it contains two protons, two neutrons, and two electrons. The sum of the masses for the components may be calculated as follows:

$$2\,m_\text{p}{}^+ = 2(1.007276 \text{ amu}) = 2.014552 \text{ amu}$$

$$2\,m_\text{n} = 2(1.008665 \text{ amu}) = 2.017330 \text{ amu}$$

$$2\,m_\text{e} = 2(5.4858 \times 10^{-4} \text{ amu}) = \underline{0.001097} \text{ amu}$$

$$\text{Sum} = 4.032979 \text{ amu}$$

The mass defect is

$$\text{MD} = (4.032979 - 4.002603) \text{ amu} = 0.030376 \text{ amu}$$

This is a significant mass difference and amounts to about 0.76% of the mass of $_2^4\text{He}$. If the mass defect is expressed in energy units, it is referred to as *binding energy* (BE).

$$BE = (0.030376 \text{ amu})(931.5 \text{ MeV/amu})$$

$$BE = 28.295 \text{ MeV}$$

The binding energy represents the amount of energy that would be liberated if the separate atomic components were fused into an atom of $_2^4$He. Another view is that the binding energy is the minimum energy required to separate an atom of $_2^4$He into its components, hence the binding energy represents the energy holding the atom's components together. Theoretically, binding energy should include the binding energy of the nuclear components as well as the binding of the electrons to the nucleus; however, the binding energy of electrons is mathematically insignificant compared to the binding energy of the nucleons. Therefore, the total binding energy is referred to as the nuclear binding energy or nuclear mass defect. The mass contributions of the electrons must be taken into account in the calculations because the mass values for nuclides in most tables are given for neutral atoms, not for bare nuclei.

Nuclear stability is associated with nuclear binding energy or the mass the nucleons lost when fused into a nucleus. It is impossible to determine the masses of individual nucleons within a nucleus, but the average mass can be determined, and this may be expressed as the average binding energy (BE_{av}). The average binding energy is the binding energy per nucleon and may be calculated as illustrated for $_2^4$He.

$$BE_{av} = \frac{28.295 \text{ MeV}}{4} = 7.07 \text{ MeV}$$

Compared to nuclei that are close neighbors in the nuclide chart (Figure 2-2), $_2^4$He has an unusually high average binding energy. The corresponding values for $_2^5$He, $_2^3$He, and $_1^3$H are 5.47, 2.58, and 2.83 MeV, respectively. Neither $_1^4$H nor $_3^4$Li exists; presumably their binding energies are zero or negative. These data suggest that the $_2^4$He nucleus is an extraordinarily stable unit.

Like $_2^4$He, the average binding energies of $_6^{12}$C, $_8^{16}$O, and $_{10}^{20}$Ne are significantly higher than those of their neighboring nuclides, and their nuclei are multiples of $_2^4$He. Thus, it appears that the potential for double pairing of neutrons and protons has an effect on binding energy. Furthermore, many radionuclides decay by emitting an α-particle that is a $_2^4$He nucleus. These observations suggest that the $_2^4$He nucleus may exist as a substructural unit in larger nuclei.

As shown in Figure 2-4, the binding energy per nucleon rises to a maximum at about 50 mass units, which corresponds to Fe. It is fairly constant for nuclides with mass numbers between about 25 to 150 and then decreases slightly with increasing mass numbers. Thus, the effects due to a nucleus' ability to form $_2^4$He subunits seems to be less important for the nuclides with larger atomic numbers.

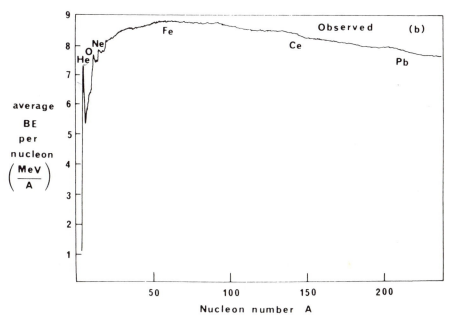

FIG. 2-4 The average binding energy of nuclei as a function of mass number (A). (From *Modern Physics for Applied Science*, by B. C. Robertson. Copyright © by John Wiley and Sons. Reprinted by permission)

Stability will be discussed more when certain aspects of nuclear models are considered.

NUCLEAR MODELS

Liquid Drop Model

In the liquid drop model, the neutrons and protons are visualized as an aggregate with bonding between nucleons. The droplet appears to have surface tension that retains nucleons, but that can be penetrated by energetic particles, either out or in (nuclear emissions or absorptions). The shape of the droplet may represent different energy states and be spherical or assume other solid shapes, such as that of an oblate spheroid (football shape). Oscillations between different orientations of nonspherical shapes are visualized as vibrational excitation modes, and rotational excitation would be possible as well. The average binding energy concept is particularly consistent with this model because it suggests that the binding forces are fairly evenly distributed between the nucleons. Nuclear fission, the splitting of a nucleus into smaller fragments, also fits this model well, but a scheme for multiple nuclear energy levels is difficult to visualize.

Fermi Gas Model

The Fermi gas model assumes the nucleons are loosely bound to one another and free to move independently within the nuclear space. Since nucleons are subject to the Pauli exclusion principle, their movement must not be strictly chaotic. One description that has been applied is that the nucleons behave as a quantitized gaseous particles within a fairly rigid nuclear boundary.

Nuclear Shell Model

The nuclear shell model is analogous to the electronic atomic model where a central attractive force (the nucleus) holds electrons in definite energy shells. Thus, in the core of the nucleus, a similar but stronger force exists that determines the energy shells for the nucleons that surround it. The chemical activity of an atom is dependent on the electronic configuration of the atom, and those that have their electronic shells filled (He, Ne, Ar, Kr, and Xe) are quite stable chemically. Likewise, filled nucleon shells should represent a more stable condition for a nucleus.

If the nuclear shell model has validity, one question needs to be addressed. Are the nucleons separate immutable entities or does the positive charge on the proton and the neutral charge on the neutron represent two different quantum levels for a common particle? As we will see later when radioactive decay processes are discussed, there are specific nuclear reactions in which it appears that neutrons are converted to protons, and other reactions in which the opposite occurs.

Since the nucleons are particles they should have spin quantum levels, $+\frac{1}{2}$ and $-\frac{1}{2}$, if they are analogous to electrons. This yields two charge levels and two spin levels that when filled correspond to the stable $^4_2$He nucleus.

$$\text{SPIN}$$

	n	$p^+ + \frac{1}{2}$
	n	$p^+ - \frac{1}{2}$
charge|	0	+1

Carrying the analogy further, we assume that nucleons fill the lowest unoccupied levels first, and if a lower vacancy exists that condition represents an unstable or excited nuclear state. In $^{12}_6$C$_6$, there is perfect pairing of the nucleon charge levels, 6 protons and 6 neutrons, but in $^{12}_5$B$_7$ the absence of one proton appears as a vacancy in a lower nucleon level. Thus, the nucleus of ^{12}B should be an excited nuclear state, which may undergo de-

12 - 5 = 8
but 7

excitation and conversion to ^{12}C by the shifting one of its nucleons from the "neutron" level to the "proton" level.

^{12}B	^{12}C
n p$^+$	n p$^+$
n p$^+$	n p$^+$
———	———
n p$^+$	n p$^+$
n p$^+$	n p$^+$
———	———
n p$^+$	n p$^+$
n	n p$^+$
n	

[handwritten: $n \longrightarrow p^+ + e^-$]

Indeed, ^{12}B does decay to ^{12}C spontaneously by the emission of an electron; apparently, an $n \rightarrow p^+ + e^-$ reaction occurs in the nucleus! Likewise, $^{12}_{7}N_5$ should represent a nuclear excited state of ^{12}C, and the change of one nucleon from the "proton" level to the "neutron" level would accomplish deexcitation.

[handwritten: $12 - 7 = 6$ but 5]

^{12}N
n p$^+$
n p$^+$
———
n p$^+$
n p$^+$
———
n p$^+$
p$^+$
p$^+$

[handwritten: $p^+ \longrightarrow n + e^+$]

^{12}N does decay to ^{12}C by the emission of a positron and, the apparent reaction is $p^+ \rightarrow n + e^+$. Other emissions are involved in these transitions, but they will be discussed later.

There are no stable nuclides that have odd numbers of neutrons and protons except for a few small nuclei, ^2H, ^6_3Li, $^{10}_5B$, and $^{14}_7N$. As pointed out earlier, neutron–proton pairing appears to be important in nuclear stability for elements with an atomic number below 20, but the ratio of neutrons to protons is considerably greater than one in stable heavy elements. This does not fit the current argument well. Also, there is an anomalous situation involving $^8_3Li_5$ that, as might be predicted, decays to $^8_4Be_4$, but ^8Be subsequently decays to two ^4_2He! The inconsistencies in this theory could be due to an incorrect theory or incomplete knowledge about the quantum nature of nucleons.

Certainly, a consistent system of nucleon energy levels has not been devised to date, but there are certain numbers, so-called _magic numbers,_ that hint of ordered energy levels for nucleons. The magic numbers are 2, 8, 20, 50, 82, and 126.

We have already discussed the stability of the ^4_2He nucleus, which has two neutrons and two protons. Oxygen has eight protons and is the first element in the periodic chart to have three stable isotopes. Also, as noted earlier, $^{16}_8O$ has a higher average binding energy than its neighboring nuclides. Calcium with 20 protons has six stable isotopes, an unusually high number compared to neighboring elements in the periodic chart. Tin (Sn) has 50 protons and 10 stable isotopes, more than any other element. Lead (Pb) has 82 protons and is the end product for three of the four natural decay series that occurs in the very heavy elements. Also there are seven stable nuclei that have 82 neutrons. The most abundant lead isotope is $^{208}_{82}Pb$, which has 126 neutrons, and the end product for the other natural decay series is $^{209}_{83}Bi$, which also has 126 neutrons.

Another feature that fits the nuclear-shell model is that energy changes in nuclei involve discrete quantities of energy. Electromagnetic radiation (γ-rays) emitted by nuclei exhibit discrete, as opposed to continuous, spectra. Also in radioactive decay, the energy liberated in the decay process is constant for a particular radionuclide.

None of the models presented is adequate, but perhaps the model that is eventually accepted will have dual features analogous to the situation with electrons, which sometimes behave as particles, sometimes as waves. Nevertheless, a truly satisfactory model may have to await further developments in our understanding of nuclear forces, and interrelationships between nucleons and strange particles.

ADDITIONAL READING MATERIAL

For greater detail about the theory of relativity, atomic, and nuclear structure the excellent book by Robertson (1981) is recommended. Although mathematics is used, it does not dominate the discussions, and only a knowledge of basic physics is assumed. Another simplified approach to nuclear physics is given by Heckman and Starring (1963).

PROBLEMS

1. Calculate the energy equivalent to the rest mass of an electron in units of MeV or keV. $m_e = 0.91095 \times 10^{-30}$ $p-415$

2. If an electron has a velocity of 6.0×10^7 m/s, what is its relativistic mass?

3. What is the kinetic energy of the electron described in problem 2?

4. Calculate the energy equivalent to the rest mass of a neutron.

5. If the kinetic energy of a neutron is 100 MeV, what is its velocity?

6. Determine the quantum energy of a γ-ray that has a wavelength of 0.15 Å.

7. What is the wavelength of a photon that has a quantum energy of 1.3 MeV?

8. Assuming that the rest mass of an ^{16}O nucleus is 15.994915 amu, calculate the mass defect, binding energy, and binding energy per nucleon for this radionuclide.

REFERENCES

Heckman, H. H., and Starring, P. W. (1963). *Nuclear Physics and the Fundamental Particles*. Holt, Rinehart and Winston, New York.

Robertson, B. C. (1981). *Modern Physics for Applied Science*. Wiley, New York.

Weast, R. C. (1988). *Handbook of Chemistry and Physics*, 69th ed. CRC Press, Boca Raton, FL.

CHAPTER 3

Nuclear Decay Processes

Radioactive decay processes involve nuclear reactions that generally involve specific emissions from the nucleus and result in the tramsmutations of elements. The types of nuclear reactions are classified according to the kinds of nuclear emissions involved and are referred to as *modes of decay.* The principal particles or rays to be considered in our discussion of modes of decay are given in Table 3-1 along with some of their physical characteristics.

α-DECAY

Most α-emitters have a high atomic number ($Z > 82$), and because such elements are not components in most biochemical compounds, they are not useful tracers for biological research, generally. However, from a radiological safety standpoint they are a concern since they are fairly prevalent in our environment, tend to accumulate in bone tissue and cause intense internal biological damage compared to other types of emitters.

The general reaction for α-decay is

$$_{Z}^{A}X^0 \longrightarrow _{Z-2}^{A-4}X^{2-} + \alpha + Q$$

Since the α-particle is a helium nucleus, the reaction could be written as follows:

$$_{Z}^{A}X^0 \longrightarrow _{Z-2}^{A-4}X^{2-} + _{2}^{4}He^{2+} + Q$$

These reactions illustrate that a neutral atom of element X with a mass number of A and an atomic number of Z decays by emitting an α-particle,

TABLE 3-1 Primary Particles and Rays Emitted by Decaying Nuclei

Name	Symbol	Mass (amu)	Electronic charge	Description
Alpha	α	≈ 4.00	$+2$	A $^4_2$He nucleus
Beta (negatron)	β^- or e^-	$\approx 5.49 \times 10^{-4}$	-1	An electron
Beta (positron)	β^+ or e^+	$\approx 5.49 \times 10^{-4}$	$+1$	An antielectron
Gamma	γ	0	0	A photon of electromagnetic radiation
Neutrino	υ	0	0	An unusual, weakly interacting particle
Antineutrino	$\bar{\upsilon}$	0	0	An antimatter counterpart to the neutrino

which is a helium ion with a $+2$ electronic charge. The daughter atom has two less protons and neutrons than the parent atom, and assuming that the parent's electrons were not disturbed when the α-particle was emitted, the product would have two electrons in excess of those needed to balance its nuclear charge. Therefore, the daughter can be considered to be an ion carrying a -2 charge. Of course, the perturbed ion would be expected to lose two electrons readily, perhaps by undergoing a chemical reaction, and when the ejected α-particle (He^{2+}) dissipates its kinetic energy it will pick up two electrons. Thus, a chemical reaction (interactions of valence electrons) would follow and result in the formation of a neutral atom. Although the fate of those two electrons is not of concern for the nuclear reaction, their masses must be accounted for because masses of neutral atoms, not bare nuclei, are used in calculations. Note that the electronic charges and total number of nucleons balance.

Recall that for nuclides with higher atomic numbers, the neutron/proton ratio is about 1.6 for stable nuclides; therefore the elimination of an α-particle (neutron/proton ratio of one) gives the daughter a slightly higher neutron/proton ratio than that of the parent. Consequently, for proton-rich nuclei, α-emission would be a means of moving toward the line of stability.

Calculation of the Decay Energy

The energy released by a nuclear reaction, symbolized as Q, is due to the mass lost in the transformation. With the application of Equation 2-1, the energy can be calcuated after determining the difference in masses between the parent atom and the two product ions (Δm).

$$Q = \Delta m \, c^2 \qquad (3\text{-}1)$$

With the conversion factor derived in Chapter 2 and when Δm is given in units of amu, Equation 3-1 can be simplified:

$$Q = \Delta m \, (931.5 \text{ MeV/amu}) \qquad (3\text{-}2)$$

$$\Delta m = \text{mass of } {}^{A}_{Z}X^0 - (\text{mass of } {}^{A-4}_{Z-2}X^{2-} + \text{mass of } \alpha)$$

The mass of an ion may be calculated from the mass of the corresponding neutral atom with the addition or subtraction of the mass of the appropriate number of electrons.

Let $m_{\text{par}} = $ mass of the neutral parent atom $({}^{A}_{Z}X^0)$

$\qquad m_{\text{dau}} = $ mass of the neutral daughter atom $({}^{A-4}_{Z-2}X^0)$

$\qquad m_{\text{He}} = $ mass of ${}^{4}_{2}He^0$

$\qquad m_{\text{e}} + $ mass of an electron

$$\Delta m = m_{\text{par}} - [(m_{\text{dau}} + 2m_{\text{e}}) + (m_{\text{He}} - 2m_{\text{e}})]$$

$$\Delta m = m_{\text{par}} - [m_{\text{dau}} + m_{\text{He}}]$$

Using ^{214}Po as an example, several features of α-decay will be illustrated.

$$ {}^{214}_{84}Po \longrightarrow {}^{210}_{82}Pb^{2-} + {}^{4}_{2}He^{2+} + Q $$

$\qquad m_{\text{par}} = 213.995176$ amu

$\qquad m_{\text{dau}} = 209.984163$ amu

$\qquad m_{\text{He}} = 4.002603$ amu

$\qquad \Delta m = 213.995176 - (209.984163 + 4.002603)$ amu

$\qquad \Delta m = 0.00841$ amu

Application of Equation 3-2 yields

$$Q \, (0.00841 \text{ amu})(931.5 \text{ MeV/amu})$$

$$Q = 7.834 \text{ MeV}$$

The nuclear reaction liberates 7.834 MeV of energy, and this is called the *decay energy*. The value of the decay energy given in the *Handbook of Chemistry and Physics* (Weast, 1988) is 7.833 MeV, which differs from the calculated value because the significance of figures used in the calculations and accuracy of physical measurements.

The decay energies for natural α-emitter are generally in the range of 4 to 8 MeV, and ^{214}Po falls within the upper portion of this range.

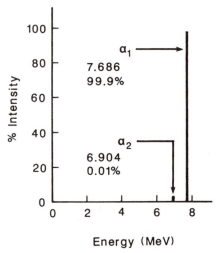

FIG. 3-1   Energy spectrum of α-particles emitted by ^{214}Po.

Dissipation of Q

How might the energy, Q, be released? Obviously the α-particle would
have a significant velocity when ejected, hence some of the decay energy
would go into the separation of the α-particle and the daughter nucleus. An
energy spectrum of α-particles emitted by ^{214}Po reveals that the particles
have discrete energies (Figure 3-1), and close to 100% of the particles have
an energy of 7.686 MeV, but slightly less than 1% have an energy of 6.904
MeV. The kinetic energies of the α-particles account for the majority of the
decay energy. However, the energy separating the α-particle and daughter
nucleus should be distributed between the two particles according to the
law of conservation of momentum; hence, the nucleus should recoil, and
the kinetic energy a daughter nucleus receives is referred to as *recoil energy.*
    The distribution of energy between two bodies when they are separated
by a sudden release of energy can be derived from classical equations, but
are the velocities of α-particles low enough to permit the use of its rest
mass instead of relativistic mass? The velocity for an α-particle with a kinetic energy of 8.0 MeV is about 6.5% the speed of light or 2×10^9 cm/s,
and its relativistic mass is greater than its rest mass by only about 0.22%.
Obviously, the velocity of the recoiling nucleus would be much lower than
that of the α-particle, therefore its rest mass would be very close to its
relativistic mass. For α-decay, rest masses are quite suitable for use in the
following calculations involving momentum and kinetic energy.
    The law of conservation of momentum is expressed in Equation 3-3
where m_1,v_1 and m_2,v_2 are the respective masses and velocities of two particles being forced apart.

$$m_1v_1 = m_2v_2 \tag{3-3}$$

The kinetic energies of the two particles, E_{k1} and E_{k2} are.

$$E_{k1} = \tfrac{1}{2} m_1 v_1{}^2, \qquad E_{k2} = \tfrac{1}{2} m_2 v_2{}^2$$

Let E_t be the total energy used to force the particles apart that should be equal to the sum of the kinetic energies of the two particles ($E_{k1} + E_{k2}$). From these relationships, Equations 3-4 and 3-5 may be derived.

$$E_{k1} = E_t \left(\frac{m_2}{m_1 + m_2} \right) \tag{3-4}$$

$$E_{k2} = E_t \left(\frac{m_1}{m_1 + m_2} \right) \tag{3-5}$$

Assuming all of Q were used to eject the α-particle, the kinetic energies of the particles can be calculated. Let the α-particle be particle 1 and the recoiling nucleus be particle 2.

$$E_t = Q = 7.833 \text{ MeV}$$

$$m_1 = 4 \text{ amu}, \qquad m_2 = 210 \text{ amu}$$

Rounded numbers of masses are acceptable for such calculations. Applying Equations 3-4 and 3-5, the kinetic energies of the particles would be

$$E_{k1} = 7.833 \text{ MeV} \left(\frac{210}{210 + 4} \right) = 7.687 \text{ MeV}$$

$$E_{k2} = 7.833 \text{ MeV} \left(\frac{4}{210 + 4} \right) = 0.146 \text{ MeV} \quad (\text{Recoil energy})$$

The calculated kinetic energy of the α-particle is 7.687 MeV, which corresponds closely to an observed energy of 7.686 MeV. The recoil energy is 0.146 MeV, and the sum of the two particles' kinetic energies is equivalent to the decay energy, Q, when the significance of the figures is considered.

However, the energy spectrum reveals that in a small fraction of the decay events, the α-particle is emitted with an energy of only 6.904 MeV. This indicates that the entire decay energy was not used to separate the particles and that the nucleus should possess some residual energy. This is the case. When an α-particle is emitted with an energy of 6.904 MeV, the nucleus is left in an excited state, and deexcitation occurs by the emission of a photon of electromagnetic radiation, a γ-ray. Thus, $Q = E_{k1} + E_{k2} + E_\gamma$. Equations 3-4 and 3-5 may be used to derive a relationship between recoil energy and α-particle energy.

$$E_{k2} = E_{k1} \frac{m_1}{m_2} \tag{3-6}$$

In this case, the recoil energy, E_{k2} is

$$E_{k2} = 6.904 \text{ MeV } \frac{4}{210} = 0.132 \text{ MeV}$$

$$Q = 7.833 \text{ MeV} = (6.904 + 0.132 + E_\gamma) \text{ MeV}$$

$$E_\gamma = 0.797 \text{ MeV or 797 keV}$$

Gamma ray spectroscopy of decaying $^{214}_{84}$Po atoms reveals that a small fraction of the atoms do emit a γ-ray with a quantum energy of 800 keV, and the intensity of the rays is consistent with the fraction of atoms that decay by emitting α-particles with an energy of 6.904 MeV. Within a reasonable degree of accuracy, Q can be accounted for.

$$Q = 7.833 \text{ MeV} \approx (6.904 + 0.132 + 0.800) \text{ MeV}$$

The discrete energy values for the α-particle and γ-rays, as well as the apparent existence of nuclear excited states, clearly supports the concept that within nuclei discrete energy levels exist, analogous to the electronic levels or shells considered in the chemistry of atoms.

Decay Diagrams

Although decay diagrams or schemes provide basic information about a decay process, they are particularly useful to illustrate nuclear energy levels. Such a diagram for the decay of ^{214}Po is given in Figure 3-2. It shows ^{214}Po decays to ^{210}Pb with a half-life of 1.64×10^{-4}s, and that in 99.99 + % of the cases ^{214}Po decays by emitting an α-particle (α_1) with an energy of 7.686 MeV. An alternative decay route is the emission of a 6.904 MeV α-particle (α_2), which leaves the nucleus in an excited state. The transition between the excited and ground states occurs by the emission of a γ-ray

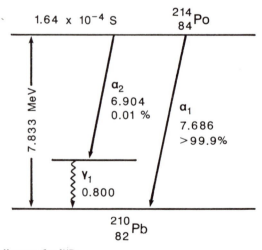

FIG. 3-2 Decay diagram for ^{214}Po.

with an energy of 0.800 MeV. Note that γ-ray transitions are depicted with wavy lines, particle emissions are shown as straight lines, and the transitions involve neutral atoms. The numbering of particles and rays has no significance in regard to routes, it simply is a means of distinguishing the various energies of the particles or rays. Generally, decay diagrams do not have the energies of transitions drawn to scale, and nuclear recoil energies are not specified.

Decay of ^{228}Th

The nuclear reaction for ^{228}Th is similar to ^{214}Po, but the number of nuclear energy levels exhibited by the decaying ^{228}Th nucleus is much greater than that for ^{214}Po.

$$^{228}_{90}\text{Th} \longrightarrow ^{224}_{88}\text{Ra}^{2-} + \alpha + Q$$

Q for the reaction is 5.5200 MeV. Five different energies are observed for α-particles and five for γ-rays as shown in Table 3-2. Calculations similar to those used for ^{214}Po permit the determination of recoil and residual nuclear energies following α-emission, and these values are also included in Table 3-2.

All of the decay energy, Q, is accounted for by the emission of α_1 and the nuclear recoil energy. The emission of an α-particle with 5.3405 MeV of energy (α_2) leaves the nucleus in an excited state (0.0841 MeV), which corresponds to the energy of γ_1. The residual energy for α_3 is accounted for by γ_5, and the energy of γ_5 is essentially equivalent to that for γ_1 plus γ_2. The energies of γ_3 plus γ_1 approximate the residual energy for α_4, and

TABLE 3-2 Energy Values Involved in the Decay of ^{228}Th by α-Emission

Designation	Energy (MeV)		
	α-Particle	Recoil	Residual[a]
α_1	5.4233	0.0967	0
α_2	5.3405	0.0954	0.0841
α_3	5.2114	0.0931	0.2155
α_4	5.1770	0.0924	0.2506
α_5	5.1400	0.0920	0.2881
	γ-Rays		
γ_1	0.0845		
γ_2	0.132		
γ_3	0.167		
γ_4	0.205		
γ_5	0.216		

[a]Residual = Q − (α-particle + recoil). Q = 5.5200 MeV.

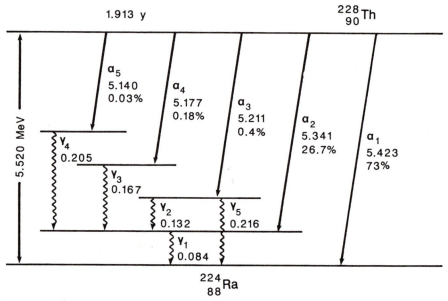

FIG. 3-3   Decay diagram for ^{228}Th.

γ_4 plus γ_1 account for the residual energy for α_5. These relationships permit the construction of a decay diagram for ^{228}Th as shown in Figure 3-3.

Summary of α-Decay Characteristics

1. The primary nuclear emission is an α-particle, which is a helium nucleus ($^{4}_{2}He^{2+}$).

2. Gamma rays may or may not accompany α-emissions depending on the characteristics of the radionuclide in question.

3. Both types of emissions are easily detected.

4. The energy spectra of α-particles and γ-rays are discrete.

5. The decay energy is accounted for by the energies of the α-particle, recoiling daughter nucleus, and γ-ray(s) if applicable.

6. Decay energies are typically in the range of 4 to 8 MeV and are much higher than most other modes of decay.

β-DECAY

Three types of processes are included in the broad classification called β-decay: (1) negatron emission (β⁻-emission), (2) positron emission (β⁺-emission), and (3) orbital electron capture, or more simply, electron capture

(EC). Although there are unifying characteristics for these processes, each will be discussed separately.

NEGATRON DECAY

Many of the useful radiotracers in biological and medical research are negatron emitters, but radionuclides decaying by this mode may be found throughout the chart of the nuclides.

In negatron decay (β^--decay), the decaying nucleus simultaneously emits a β^--particle and an antineutrino ($\bar{\nu}$) (antiparticles are indicated by a bar above the symbol for the particle). A β^--particle is identical to the usual negatively charged electron (e^-), but it possesses kinetic energy as well; thus, it can be envisioned as a high speed electron. The antineutrino is a more unusual particle whose existence was not proven until the late 1950s. It is chargeless and massless, but does possess energy and angular momentum. In the usual sense, it is neither a particle nor a photon of electromagnetic radiation, and because it interacts with matter so weakly, it is very difficult to detect. In Heckman and Starring's (1963) intriguing account of the postulated existence and subsequent detection of neutrinos and antineutrinos, it was suggested that such particles can pass through the mass of the earth 100 billion times with a 50% chance of survival. For comparison, a few millimeter thickness of paper will absorb most negatrons.

The general reaction for β^--decay is

$$^A_Z X^0 \longrightarrow \ ^A_{Z+1} X^+ + \beta^- + \bar{\nu} + Q$$

The significant nuclear reaction is the conversion of one neutron to a proton by the emission of an electron (β^--particle) and an antineutrino.

$$n \longrightarrow p^+ + \beta^- + \bar{\nu}$$

Incidentally, bare neutrons are unstable and do decay with a half-life of about 12 min.

Negatron decay lowers the neutron/proton ratio in a nucleus, and for a neutron-rich radionuclide, causes the daughter nuclide to be closer to the line of stability than was the parent.

Since the antineutrino is massless, the β^--particle has the mass of an electron and the daughter would have to gain one electron to become a neutral atom, the change in mass (Δm) for the reaction is

$$\Delta m = m_{par} - [(m_{dau} - m_e) + m_e]$$

$$\Delta m = m_{par} - m_{dau}$$

From the change in mass (in amu), Q may be calculated using Equation 3-2. Note that for Q to be positive (meaning a release of energy), Δm must be positive, and hence the mass of the daughter must be less than that of the parent. Thus, β^--decay is not a possible spontaneous mode of decay unless $m_{par} > m_{dau}$.

The range of decay energies for β^--emitters extends over a couple of orders of magnitude. A very weak emitter, ^{3}H, has a decay energy of 0.01861 MeV while that for ^{32}P, a hard β^--emitter, is 1.710 MeV. In general, Q for β^--emitters is significantly less than for α-emitters.

Decay of ^{14}C

Carbon-14 is an example of a radionuclide that decays by negatron emission.

$$^{14}_{6}C \longrightarrow ^{14}_{7}N^+ + \beta^- + \bar{\nu} + Q$$

$$Q = 0.156 \text{ MeV}$$

The energy spectrum of the β^--particles emitted by ^{14}C is shown in Figure 3-4. Unlike the spectra for α-particles (Figure 3-1), β^--spectra exhibit continuous ranges of energies from zero to some maximum value, called E_{max} (Figure 3-4). In the case of ^{14}C, E_{max} is equal to Q, which indicates that a few of the β^--particles have kinetic energies that approach the value of Q. However, the *average energy, $\bar{E}$,* is only 0.045 MeV which is 29% of E_{max}. Although the shape of β^--ray spectra for other β^--emitters differ somewhat from ^{14}C, the general features are the same. There will be an upper limit for energy, E_{max}, and $\bar{E}$, which is a function related to the shape of the curve, is usually in the range of 30 to 40% of E_{max}. All β^--ray spectra extend from zero energy.

An electron with a kinetic energy of 0.156 MeV has a velocity of about 2×10^{10} cm/s, about two-thirds the speed of light, and its relativistic mass is about 1.8 times greater than its rest mass. A comparison of velocities of

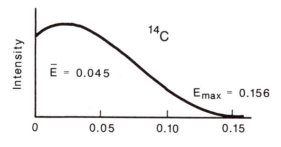

Particle energy (MeV)

FIG. 3-4 Energy spectrum of β^--particles emitted by ^{14}C.

typical α- and β⁻-particles show those of β⁻-particles to be much higher even though their kinetic energies are considerably less. Of course, this is due to the differences in masses of the two types of particles.

If nuclei have discrete energy levels, why are the energies of the β⁻-particles continuous? No γ-rays are emitted by ^{14}C so we can eliminate that aspect in this example. The answer lies in the energies of the antineutrinos. When an atom of ^{14}C decays, the decay energy, Q, is distributed between the negatron and the antineutrino; therefore, an energy spectrum of the antineutrinos would appear as the inverse of the β⁻-ray spectrum. That is, in the decay of an individual nucleus, the sum of the energies of the antineutrino and β⁻-particle is equal to E_{max}. The decay diagram for ^{14}C is simple, and E_{max} is given as the transition energy (Figure 3-5).

In our discussion of the distribution of energy in negatron decay, recoil by the daughter nucleus was not addressed. The recoil energy varies depending upon the energies of the β⁻ and $\bar{\nu}$ and their emission directions. However, their small masses cause recoil energies to be insignificantly small compared to Q in most cases. For ^{14}C, recoil energies will vary between zero to a maximum of about 7eV. While the maximum recoil energy is insignificant compared to Q (0.0045%), it does exceed most bond energies. Consequently, molecular rearrangements in compounds that experience the decay of one of their atoms occurs frequently because the recoil energy may be redistributed, and of course, an elemental transmutation has occurred to induce the process.

Some other useful β⁻-emitters are ^{3}H, ^{32}P, and ^{35}S, and for each, $Q = E_{max}$. Gamma rays are not associated with their decay. However, for other radionuclides, Q does not necessarily equal E_{max}, and γ-rays may be emitted concomitantly.

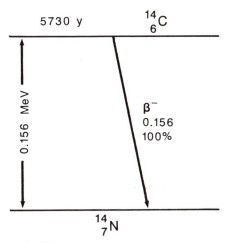

FIG. 3-5   Decay diagram for ^{14}C.

Decay of ³⁹Cl

An example of a β⁻-emitter that also emits γ-rays is ³⁹Cl. It emits β⁻-particles with three different E_{max} values, one of which corresponds to the decay energy, 3.44 MeV (Figure 3-6). In 85% of the decays, 1.91 MeV of energy is distributed between the β⁻-particle and antineutrino, which leaves the nucleus in an excited state. Deexcitation occurs primarily by the emission of γ_3 (1.52 MeV) or by the sequential emission of γ_1 (0.246 MeV) and γ_2 (1.27 MeV). About 2.18 MeV of energy is involved in the emissions of the β⁻-particle and antineutrino in 8% of the decays, and the resultant excited nucleus loses its energy primarily by the emission of γ_2 (1.27 MeV).

Decay of ¹³¹I

The decay diagram for ¹³¹I is quite complex because it involves β⁻-emissions with five different E_{max} values and 11 γ-ray energies. In some of the decays a metastable state is created in the daughter nucleus, ¹³¹Xe. Metastable states will be discussed more in the isomeric transition mode of decay section of this chapter. The γ-rays of ¹³¹I are generally utilized more frequently than the β⁻-particles for detection purposes (counting).

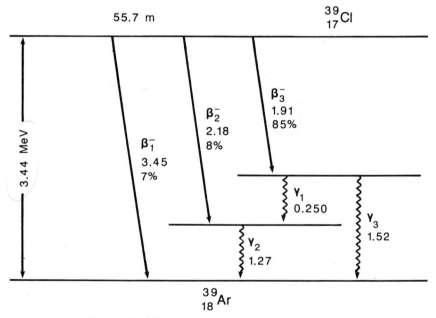

FIG. 3-6 Decay diagram for ³⁹Cl.

Summary of β^--Decay Characteristics

1. The primary nuclear emissions are a negatron, which is a high velocity electron, and a chargeless massless particle called an antineutrino.

2. Gamma rays may or may not accompany the primary emissions depending upon the particular radionuclide.

3. Negatrons and γ-rays are easily detected, but antineutrinos are not detectable by commercially available instruments.

4. The energy spectra of negatrons and antineutrinos are continuous, but those of γ-rays, if emitted, are discrete.

5. The maximum energy of the β^--particles for a particular transition is called E_{max}, and for individual nuclei, the sum of the energies of the antineutrino and negatron equals E_{max}.

6. The decay energy is accounted for by E_{max} plus the energy of γ rays, if applicable.

7. Decay energies are typically in the range of a few keV to several MeV.

8. For spontaneous β^--decay, $m_{par} > m_{dau}$.

POSITRON DECAY

Positron emitters do not occur naturally, but may be produced with nuclear reactors and accelerators. Unlike β^--emitters, most β^+-emitters are found among elements with lower atomic numbers ($Z < 85$) and several are useful tracers in biological materials. The general reaction for β^+-decay is

$$^A_Z X^0 \longrightarrow {}^A_{Z-1} X^- + \beta^+ + \nu + Q_n$$

and the apparent nuclear transition is:

$$p^+ \longrightarrow n + \beta^+ + \nu$$

The conversion of a proton to a neutron increases the neutron/proton ratio in a nucleus; consequently, such a reaction should be expected for unstable nuclei that are proton rich. Of course, this alteration in the neutron/proton ratio is opposite to that in β^--decay. However, the interconversion of neutrons and protons is not a classical reversible reaction because other quite different particles are involved. The positron has the same mass as an electron, but has a positive rather than a negative charge and is the antiparticle of a normal electron (negatron). Since an interaction between matter and antimatter causes the annihilation of both, positrons are particularly vulnerable in our world full of electrons. As will be discussed in Chapter 5, when positrons and negatrons interact, their combined masses are converted to energy in the form of two photons of electromagnetic radiation.

The neutrino (v) is the antiparticle of the unusual particle discussed in β^--decay, the antineutrino, and their physical properties are identical in magnitude, but opposite in character. Thus, the primary emissions of β^+-decay are antimatter relative to the primary emissions in β^--decay, and, in one case, a neutron is converted to a proton while the other is vice versa. Since a neutron is heavier than a proton and the decay energy is derived from the loss of mass, it seems that the decay of neutrons to protons (β^--decay) could be spontaneous, but the conversion of a proton to a neutron would require an input of energy. As far as individual nucleons are concerned, that is indeed the case. However, recall the discussion in Chapter 2 concerning mass defect and binding energy. This phenomenon clearly indicates that the masses of nucleons vary depending on the particular kind of nucleus in which the nucleon is found. A consideration of the change of mass involved in β^+-decay may shed some light on the conditions needed for the conversion of a proton to a neutron. The daughter atom has one excess electron since the proton number is reduced by one, the positron has the mass of an electron and the neutrino is massless. Thus, for the general reaction shown above:

$$\Delta m = m_{par} - [(m_{dau} + m_e) + m_e]$$

$$\Delta m = m_{par} - [m_{dau} + 2m_e]$$

For the spontaneous decay of a radionuclide by β^+-emission, the mass of the parent must exceed the mass of the daughter by at least the mass of two electrons. That is, the mass of the neutral parent must be greater than the neutral daughter by about 0.0011 amu (equivalent to 1.022 MeV). Apparently, the conversion of a proton to a neutron occurs at the expense of the nucleus' binding energy.

There may be some confusion about the way decay energies for β^+-emitters are reported. The energy (Q_n) released by the *nuclear reaction* shown above is

$$Q_n = [m_{par} - (m_{dau} + 2\, m_e)\text{ amu}](931.5 \text{ MeV/amu})$$

However, the positron is annihilated subsequently by an electron, and this releases energy equivalent to the masses of the positron and electron, or $2m_e$ (1.022 MeV). Consequently, when the annihilation reaction is included with the nuclear reaction, the total energy released is

$$Q = [(m_{par} - m_{dau})\text{ amu}](931.5 \text{ MeV/amu})$$

In the "Table of the Isotopes" (Weast, 1988) the decay energies of β^+-emitters include annihilation energy. Nevertheless, it should be noted that the maximum kinetic energy (E_{max}) a positron may possess cannot exceed the energy released by the nuclear reaction (Q_n).

The energy spectra for the positron and neutrino are completely analogous to those of the negatron and antineutrino. Positrons exhibit a continuous spectrum with an E_{max} that is a characteristic of the particular radionuclide, and in individual decays, the energy, E_{max}, is distributed between the positron and neutrino.

Decay of ^{22}Na

This radionuclide decays by two modes, one of which is β^+-decay, the other is a mode called electron capture that will be discussed later. As shown in Figure 3-7, about 90% of the atoms decay by β^+-emission with an E_{max} of 0.544 MeV. Analogous to negatron decay, this energy is distributed between the positron and the neutrino. The nucleus is left in an excited state, and deexcitation occurs by the emission of a γ-ray (1.277 MeV). A small fraction of the β^+-transitions has an E_{max} value equal to Q minus the energy equivalent to the mass of two electrons (1.022 MeV). About 10% of the atoms decay by electron capture, which leaves the nucleus in the same state of excitation as when β^+-emission occurs with an E_{max} of 0.544 MeV.

Summary of β^+-Decay Characteristics

1. The primary nuclear emissions are a positron and a neutrino, both of which are the antiparticles of those emitted in β^--decay.

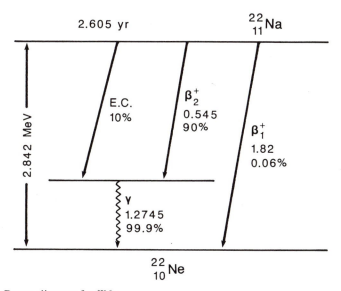

FIG. 3-7 Decay diagram for ^{22}Na.

2. Gamma rays may or may not accompany the primary emissions, depending on the characteristics of the radionuclide.

3. Positrons and γ-rays are easily detected, but neutrinos are not.

4. The energy spectra of the positrons and neutrinos are continuous, while those of the γ-rays, if emitted, are discrete.

5. The maximum energy of the β^+-particle, observed for a particular transition, is called E_{max}, and for individual decay events, the sum of the energies of the neutrino and β^+-particle equals E_{max}.

6. Decay energies are accounted for by E_{max} plus 1.022 MeV (the energy equivalent to the mass of two electrons) and the energies of γ-rays, if applicable.

7. Decay energies are typically in the range of a few MeV.

8. For spontaneous β^+-decay, $m_{par} > m_{dau} + 2m_e$.

9. Another mode of decay, electron capture, is a competing decay mode for some β^+-emitters.

ORBITAL ELECTRON CAPTURE

Orbital electron capture (electron capture or EC) may be considered to be a variation of β^+-decay because the conversion of a proton to a neutron occurs in both modes, and in many radionuclides the two modes are alternate decay processes. Electron capture is observed in isotopes of most elements, and iron-55 and ^{125}I are useful tracers that decay exclusively by this mode.

In electron capture, the nucleus absorbs one of the atom's orbital electrons which reacts with a proton to form a neutron.

$$e^- + p^+ \longrightarrow n + \nu$$

Usually, the electron is captured from the first principal quantum level or K shell, consequently *K capture* is an alternative name for this process. Recall the fundamental reaction in β^+-decay:

$$p^+ \longrightarrow n + \beta^+ + \nu$$

The conversion of nucleons is the same, but the capture of an electron appears as the equivalent to the emission of a positron, an electron's antiparticle. The neutrino is the same in both cases, and the alteration of the neutron/proton ratio is identical to that of β^+-decay. The similarities of these two modes of decay help in understanding why they may be competing processes in the decay of some radionuclides. One significant difference in the two decay modes concerns the change in mass (Δm) for the nuclear reaction. The general reaction for electron capture is

$$^A_Z X^0 \longrightarrow \,_{Z-1}^{A} X^0 + \nu + Q$$

Since one of the parents' orbital electrons is absorbed by its nucleus and causes the neutralization of one proton, both the parent and daughter are neutral atoms. The neutrino is the only nuclear emission and is massless, therefore

$$\Delta m = m_{par} - m_{dau}$$

$$Q = \Delta m \ (931.5 \ \text{MeV/amu})$$

Thus, for electron capture to occur, the mass of the parent must be only slightly greater than the mass of the daughter, whereas, for the β^+-decay, the mass of the parent must exceed the mass of the daughter plus two electrons.

A second significant difference between electron capture and β^+-decay is the distribution of the decay energy. Since there is only one particle emitted (ν) by the nucleus, it must carry all of the energy of the transition instead of having it distributed between two particles. Therefore, the neutrinos emitted by electron capture are monoenergetic, that is, they exhibit a discrete energy spectrum.

In electron capture, it is quite probable that the nucleus may be left in an excited state after the emission of the neutrino, and if so, deexcitation by the emission of a γ-ray(s) occurs (see Figure 3-7). This situation is similar to those described for other modes of decay.

Extranuclear Emissions

Since neutrinos are not detected easily, and if the decay process does not yield an excited nuclear state, how could the decay be detected? This leads to another significant difference between electron capture and β^+-decay. When a lower electronic level (shell or orbital) of an atom is vacant, an electron from a higher level will fall to fill the vacant level with the concomitant emission of a photon of electromagnetic radiation. The energy of the photon is equivalent to the difference between the electron binding energies of the levels or shells involved, and for most of the transitions involving the lower electronic levels, the photons' energies (hence wavelengths) will be in that portion of the electromagnetic spectrum referred to as X-rays. The X-rays will be characteristic of the daughter element because the daughter is formed by the electron capture event that, of course, precedes the rearrangement of electrons. Effectively, the outermost electron of the parent atom falls to the vacant level in the newly created daughter, but this usually occurs in several steps involving transitions between only one or two electronic levels at a time. Therefore, one or more photons of X-rays (each with a characteristic wavelength) plus photons of UV light may be emitted before the daughter atom achieves a stable electronic configuration.

Auger (o' - zha') electron emission is another type of extranuclear event that occurs with some radionuclides undergoing electron capture. X-Rays, arising from transitions among the lower electronic levels, may interact with outer orbital electrons. Effectively, the energy of the X-ray is absorbed by the electron, which gives it sufficient energy to overcome its binding energy and attain a low velocity. Thus, the sum of the electron's binding energy and kinetic energy is equal to the energy of the absorbed X-ray. The general process for such interactions between photons (from any source) and electrons (in any type of absorber) is called the photoelectric effect and will be discussed in Chapter 5. The production of auger electrons is a special case of the photoelectric effect because it occurs within a single atom.

The extranuclear manifestation of the electron capture process is the emission of X-rays and possibly auger electrons, which are monoenergetic and have low velocities. Both types of emissions arise from the orbital electrons, not the decaying atom's nucleus.

Distinction Between X- and γ-Rays

In the electromagnetic spectrum diagram shown in Figure 2-1, γ-rays have shorter wavelengths (< 1 Å) and higher quantum energies than X-rays (1–100 Å). However, the electromagnetic spectrum is a continuum so precise dividing lines are not particularly useful for many purposes. In our discussions, X- and γ-rays will be distinguished on the basis of their source; γ-rays arise from nuclear transitions while X-rays arise from electronic transitions, usually involving the inner shell electrons. The wavelengths of electromagnetic radiation are generally consistent with the typical classification scheme, but some X-rays may have shorter wavelengths than γ-rays according to our definition.

Summary of Electron Capture Decay Characteristics

1. The primary nuclear emission is a monoenergetic neutrino.
2. Gamma rays may or may not accompany the primary emissions, depending on the characteristics of the radionuclide.
3. Secondary emissions arise from orbital electron transitions and are X-rays and possibly monoenergetic low velocity auger electrons.
4. The primary emission is not detectable, but γ-rays, X-rays, and auger electrons are.

5. The decay energy is accounted for by the energies of the neutrino, and, if applicable, the γ-rays, X-rays, and auger electrons.

6. For spontaneous decay, $m_{par} > m_{dau}$.

7. Electron capture may be a competing mode of decay for β⁺-emitters.

ISOMERIC TRANSITIONS

As defined in Chapter 2, isomers are nuclides that differ only in the energy levels of their nuclei; hence isomeric transitions refer to nuclear deexcitation processes. Excited nuclear states may arise by various nuclear reactions including all of the types of radioactive decay modes discussed in this chapter. Also, nuclei may be excited by bombarding them with particles or rays.

Deexcitation may occur immediately after excitation, but if the excited state persists for a measurable half-life, it is referred to as a metastable state. Recall that isomers are indicated by adding a lower case "m" to the mass number. Thus, ^{69m}Zn refers to the metastable state or isometric state of ^{69}Zn. Isomeric transitions are deexcitation processes for metastable nuclei, and the ones to be discussed here are not all inclusive, instead they are the processes that may be detected easily with common types of counters.

Gamma Emission

A common deexcitation process is the emission of a γ-ray or rays.

$$^{Am}_{Z}X \longrightarrow {}^{A}_{Z}X + \gamma \ (Q)$$

In this process there is no transmutation of elements, and the emission spectrum exhibits discrete energy values. One or more γ-rays of specific energies may be emitted in the course of the deexcitation process. The decay energy, Q, is simply the difference in energy levels between the metastable and ground states and may be determined from the energy of the γ-rays emitted during the process. The half-lives of isomers varies from about 10^{-14}s to many years.

Internal Conversion

In the internal conversion process, the excitation energy of a nucleus is transferred quantitatively to an orbital electron causing its ejection from the atom. For many isomers, internal conversion and γ-emission are alternative decay modes, and in internal conversion it appears as though a γ-ray, emitted by the nucleus, interacts with an orbital electron by the photoelectric effect (see Chapter 5). The kinetic energy of the electron is equal to the excitation energy (γ-ray energy) minus the binding energy of the electron. Consequently, such electrons, called *conversion electrons,* have discrete energies. However, a particular isomer may emit conversion electrons with slightly different kinetic energies because the γ-rays may interact with electrons in different electronic orbitals, which, of course, have different binding energies. In our discussion of extranuclear emissions arising from electron capture, the consequences of an electron vacancy were described—the emission of auger electrons and X-rays. Similar electronic transitions occur when a vacancy is created by the ejection of a conversion electron. Thus, in addition to a conversion electron, a decaying atom may emit either or both auger electrons and X-rays.

As mentioned above, internal conversion may be an alternative to γ-emission in some isomers, and the proportion of atoms that yield conversion electrons relative to those that emit γ-rays is a characteristic of the isomer. For example, 11% of the atoms of ^{137m}Ba emit conversion electrons and 91% emit γ-rays during the deexcitation process.

Emission of a Positron–Negatron Pair

Deexcitation by the emission of a positron–negatron pair is not a common process and is not possible unless the excitation energy exceeds 1.022 MeV, which is the energy equivalent to the mass of the two particles. Both particles, e^+ and e^-, arise from the nucleus and possess kinetic energy. Unless the emission of the pair lowers the energy level of the nucleus to the ground state, γ-rays may be emitted as well to complete the deexcitation process.

Summary of Isomeric Transitions

1. Isomeric transitions are nuclear deexcitation processes.
2. Among useful radiotracers, γ-ray emission is the most common deexcitation process.
3. Internal conversion is an intraatomic photoelectric process involving orbital electrons. It yields conversion electrons and X-rays, which may result in the emission of auger electrons as well.
4. Some radionuclides may emit a negatron–positron pair if the energy of the nuclear excitation state is greater than 1.022 MeV.

SOURCES OF NUCLEAR AND ATOMIC DATA

A convenient source of information concerning nuclides is the "Table of the Isotopes" given in *The Handbook of Chemistry and Physics* (Weast, 1988). For radionuclides, their atomic masses, half-lives, decay modes, decay energies, particle energies, accompanying γ-ray energies, and other data are given. The same reference source also contains a table of "X-ray Wavelengths" that gives wavelengths and energies for the various electronic transitions possible for each element.

The "Chart of the Nuclides" (Figure 2-2) is also a source of nuclear data.

PROBLEMS

1. The mass of $^{211}_{84}$Po is 210.986627 amu, and it decays by α-emission to $^{207}_{82}$Pb which has a mass of 206.975872 amu. Calculate the decay energy (Q). Assuming all of the decay energy is distributed between the α-particle and daughter nucleus, calculate the kinetic energies of the α-particle and daughter nucleus.

2. The mass of $^{42}_{21}$Sc is 41.965514 amu, and by positron emission it decays to $^{42}_{20}$Ca, which has a mass of 41.958618 amu. Show the nuclear reaction, calculate the decay energy (Q), and the maximum value for E_{max}.

3. By negatron emission $^{24}_{11}$Na decays to $^{24}_{12}$Mg. The mass of $^{24}_{12}$Mg is given as 23.985042 amu, and Q for the nuclear reaction is listed as 5.514 MeV. Calculate the mass of $^{24}_{11}$Na and compare it with tabular data given by Weast (1988).

4. Construct a decay scheme for

$$^{95}_{40}\text{Zr} \longrightarrow {}^{95}_{41}\text{Nb} + \beta^- + \bar{\nu} + \gamma$$

$Q = 1.124$ MeV, $t_{1/2} = 64.03$ da, and the energies of the emissions are

E_{max} (MeV)	β-Intensity (%)	γ-Energy (MeV)	γ-Intensity (%)
0.396	44	0.236	0.9
0.360	55	0.724	44
0.890	1	0.757	55

5. Construct a decay diagram for $^{224}_{90}$Th, which decays by α-emission to $^{220}_{88}$Ra. The decay energy (Q) is 7.305 MeV, the half-life is 1.045 s, and the energies of emissions are given on the following page.

Particle	Energy (MeV)	Abundance (%)
α_1	7.170	79
α_2	6.997	19
α_3	6.768	1.2
α_4	6.702	0.6
γ_1	0.177	—
γ_2	0.235	—
γ_3	0.297	—
γ_4	0.410	—

REFERENCES

Heckman, H. H., and Starring, P. W. (1963). *Nuclear Physics and the Fundamental Particles*. Holt, Rinehart and Winston, New York.

Slack, L., and Way, K. (1959). *Radiations from Radioactive Atoms in Frequent Use*. U.S. Atomic Energy Commission Report.

Walker, F. W., Miller, D. G., and Feiner, F. (1983). *Chart of the Nuclides*. Knolls Atomic Power Lab., General Electric Co.

Weast, R. C. (1988). *Handbook of Chemistry and Physics*, 69th ed. CRC Press, Boca Raton, FL.

CHAPTER 4

Radioactive Decay and Labeled Compounds

TERMS AND MATHEMATICS OF RADIOACTIVE DECAY

Disintegrations

When an atom of a radionuclide has undergone one of the nuclear decay processes described in Chapter 3, we say the atom has *disintegrated*. The term disintegration is an overstatement; it simply means that a significant spontaneous nuclear transition has occurred, and some prefer the term transition to disintegration. Here we will symbolize a disintegration as "d."

For an individual atom, we cannot predict when it may disintegrate, regardless of knowledge about the radionuclide's half-life. In fact, it may disintegrate anytime between $t = 0$ and $t = \infty$. However, if we observe a large number of atoms of the same radionuclide and know its half-life, we can predict with reasonable accuracy the fraction of atoms that will disintegrate in a given period of time.

Lifetimes

The *half-life* of a radionuclide is the time required for half of a large number of atoms to decay. Thus, after one half-life, half of the original number of atoms remain, after two half-lives, one-fourth remain, after three half-lives one-eighth remain, and so on. Here, half-life will be denoted by the symbol, $t_{1/2}$.

The *average-life* of a radionuclide is the average length of time atoms exist when a large number of atoms are observed, and the symbol for average-life is $\bar{t}$. In calculations, half-lives are used more commonly than average-lives.

Disintegration Rate

The *disintegration rate* or *rate of decay* for a large number of atoms of a particular radionuclide follows first-order reaction kinetics, which means that the number of atoms disintegrating in a small interval of time is directly proportional to the number of atoms present. The corresponding mathematical relationship is shown by Equation 4-1.

$$\frac{-\Delta N}{\Delta t} = \lambda N \qquad (4\text{-}1)$$

In Equation 4-1, N is the number of atoms present at any time, and λ is a constant of proportionality called the *decay constant*. The decay constant is a characteristic of the radionuclide and is related to its half-life mathematically. The fraction, $\Delta N/\Delta t$ represents the change in the number of atoms during a small interval of time. Since atoms are lost, $-\Delta N/\Delta t$ is the disintegration rate or rate of decay, which may be expressed as disintegrations per second (dps), becquerels (Bq), or disintegrations per minute (dpm).

Of course, N changes each time an atom disintegrates. Consequently, Equation 4-1 is valid only when Δt is very small compared to the half-life of the radionuclide. More properly, the equation can be expressed as a differential equation by making Δt infinitesimally small, hence dt, and ΔN becomes dN.

$$\frac{-dN}{dt} = \lambda N \qquad (4\text{-}2)$$

Derivation of Decay Equations

Those familiar with calculus recognize that after rearrangement, Equation 4-2 may be integrated.

$$\int \frac{dN}{N} = \int -\lambda dt$$

$$\ln N = -\lambda t + C \qquad (4\text{-}3)$$

In Equation 4-3, ln N is the natural logarithm (base e) of N, and C is a constant of integration that must be evaluated. Let N_0 equal the number of atoms present when time is zero ($t = 0$), then it may be substituted for N in Equation 4-3 when $t = 0$.

$$\ln N_0 = -\lambda(0) + C$$

$$C = \ln N_0$$

Thus, the constant of integration is the natural logarithm of the number of atoms present at time zero. Now, $\ln N_0$ can be substituted for C in Equation 4-3 to yield Equation 4-4.

$$\ln N = -\lambda t + \ln N_0$$

$$\ln N = \ln N_0 - \lambda t \tag{4-4}$$

Common logarithms, with a base of 10, may be employed, and here common logarithms will be indicated by log rather than ln, which denotes natural logarithms. Since $\ln X = 2.303 \log X$ Equation 4-4 may be transformed to Equations 4-5 and 4-6.

$$2.303 \log N = 2.303 \log N_0 - \lambda t \tag{4-5}$$

$$\log N = \log N_0 - 0.4342 \lambda t \tag{4-6}$$

Equation 4-4 may be rearranged and converted to an exponential form.

$$\ln N = \ln N_0 - \lambda t \tag{4-4}$$

$$\ln \frac{N}{N_0} = -\lambda t$$

$$\boxed{N = N_0 e^{-\lambda t}} \tag{4-7}$$

Decay Constant, Half-Life, and Average-Life Relationships

Handbooks containing physical characteristics of radionuclides rarely list decay constants; instead half-lives are given. The *Handbook of Chemistry and Physics* (Weast, 1988) is a convenient source for such data, and a table for a few radionuclides is given in Appendix A. The decay constant for a particular radionuclide may be calculated easily if its half-life is known. According to the definition of half-life, when $t = t_{1/2}$, $N = \frac{1}{2}N_0$. Substitutions in Equation 4-4 yields:

$$\ln \tfrac{1}{2}N_0 = \ln N_0 - \lambda t_{1/2}$$

$$\ln \frac{\tfrac{1}{2}N_0}{N_0} = -\lambda t_{1/2}$$

$$\ln 2 = \lambda t_{1/2}$$

$$\lambda = \frac{\ln 2}{t_{1/2}} \quad \text{or} \quad \lambda = \frac{0.693}{t_{1/2}} \tag{4-8}$$

Rearrangement of Equation 4-8 permits calculation of half-life from the decay constant.

$$t_{1/2} = \frac{\ln 2}{\lambda} \quad \text{or} \quad t_{1/2} = \frac{0.693}{\lambda} \tag{4-9}$$

From Equation 4-8 it is apparent that the units for the decay constant must be reciprocal units of time, such as seconds, minutes, days, or years. The units employed may depend on the magnitude of the half-life, but the units of time (t) used in the decay equations must cancel those in the decay constant.

The average-life of a radionuclide is related to its half-life and decay constant. Those relationships may be derived by calculus. First, the sum of the individual lifetimes for a number of atoms is found, then division by the number of atoms originally present yields the average-life.

Let S_t = the sum of lifetimes

N_0 = the number of atoms originally present

$-dN$ = the number of atoms disintegrating during the time dt.

Thus, when $(-dN)t$ is integrated from the time when $N = N_0$ to $N = 0$, the sum of the lifetimes would be obtained.

$$S_t = \int_{N = N_0}^{N = 0} (-dN)(t)$$

Unfortunately, there are two variables in this equation, but by the use of Equation 4-4 the variable, t, may be eliminated.

$$t = \frac{\ln N_0 - \ln N}{\lambda}$$

Substituting the function of N for t

$$S_t = \int_{N_0}^{0} (-dN) \frac{(\ln N_0 - \ln N)}{\lambda}$$

$$S_t = \frac{-\ln N_0}{\lambda} \int_{N_0}^{0} dN + \frac{1}{\lambda} \int_{N_0}^{0} \ln N (dN)$$

$$S_t = \frac{-\ln N_0 (N)}{\lambda} \Big|_{N_0}^{0} + \frac{1}{\lambda} [(N)(\ln N) - N] \Big|_{N_0}^{0}$$

$$S_t = \frac{\ln N_0 (N)_0}{\lambda} - \frac{N_0 \ln N_0}{\lambda} + \frac{N_0}{\lambda}$$

$$S_t = \frac{N_0}{\lambda}$$

$$\bar{t} = \frac{S_t}{N_0} = \frac{N_0}{N_0 \lambda}$$

$$\bar{t} = \frac{1}{\lambda} \tag{4-10}$$

Since

$$\lambda = \frac{\ln 2}{t_{1/2}}$$

$$\bar{t} = \frac{t_{1/2}}{\ln 2}$$

$$\bar{t} = 1.443 \ t_{1/2} \tag{4-11}$$

CALCULATIONS

Hand-held calculators with logarithmic and exponential function keys are quite convenient devices for solving decay equations, but there is no advantage in using common as opposed to natural logarithms. Therefore, equations involving common logarithms will be presented, but not used in example calculations.

Example Problem 4-1. If 1.5×10^4 atoms of ^{32}P were present in a sample on May 10, how many atoms would be left on June 12? The use of Equation 4-4 follows. First, one would consult appropriate tables to ascertain the half-life for ^{32}P, which is 14.28 days (da).

$\ln N = \ln N_0 - \lambda t$ $t_{1/2} = 14.28$ da

$\ln N = \ln (1.5 \times 10^4) - \dfrac{\ln 2 \ (32 \ \text{da})}{14.28 \ \text{da}}$ $\lambda = \dfrac{\ln 2}{14.28 \ \text{da}} = \dfrac{0.693}{14.28 \ \text{da}}$

$\ln N = 9.6158 - 1.55327$ $t = 32$ da

$\ln N = 8.0625$ $N_0 = 1.5 \times 10^4$ atoms

$\quad N = 3.173 \times 10^3$ atoms $N = ?$ $\qquad$ (4-4)

Using Equation 4-7 to solve the same problem follows.

$$N = N_0 \ e^{-\lambda t}$$

$$N = (1.5 \times 10^4) \ e^{-\frac{\ln 2 \ (32 \ \text{da})}{14.28 \ \text{da}}}$$

$$N = (1.5 \times 10^4) \ e^{-1.55327}$$

$$N = (1.5 \times 10^4)(0.21156)$$

$$N = 3.173 \times 10^3 \text{ atoms}$$

Example Problem 4-2. After 23 days of decay it was determined that there were 2.5×10^6 atoms of ^{35}S left. How many atoms of ^{35}S were present initially? Tables show the half-life of ^{35}S to be 87.2 days. The use of Equation 4-7 follows.

$$N = N_0 e^{-\lambda t} \qquad t_{\frac{1}{2}} = 87.2 \text{ da}$$

$$t = 23 \text{ da}$$

$$\lambda = \frac{\ln 2}{87.2 \text{ da}}$$

$$N = 2.5 \times 10^6 \text{ atoms}$$

$$N_0 = ?$$

$$2.5 \times 10^6 = (N_0) e^{-\frac{\ln 2 \,(23 \text{ da})}{87.2 \text{ da}}}$$

$$2.5 \times 10^6 = (N_0) e^{-0.18283}$$

$$2.5 \times 10^6 = (N_0)(0.83291)$$

$$N_0 = 3.002 \times 10^6 \text{ atoms}$$

Another method of solving Problem 4-2 using Equation 4-7 follows.

$$N = N_0 e^{-\lambda t}$$

$$\frac{N}{e^{-\lambda t}} = N_0$$

$$N_0 = N \, e^{\lambda t}$$

$$N_0 = (2.5 \times 10^6) \, e^{\frac{\ln 2 \,(23 \text{ da})}{87.2 \text{ da}}}$$

$$N_0 = (2.5 \times 10^6) \, e^{0.18283}$$

$$N_0 = (2.5 \times 10^6)(1.20061)$$

$$N_0 = 3.002 \times 10^6 \text{ atoms}$$

Example Problem 4-3.   How long would it take to reduce the number of atoms in an ^{125}I sample to 10% of its original number? The half-life for ^{125}I is 59.9 days.

Use Equation 4-4 or 4-6 since Equation 4-7 must be converted to Equation 4-4 to solve the problem.

$$\ln N = \ln N_0 - \lambda t \qquad t_{\frac{1}{2}} = 59.9 \text{ da}$$

$$\lambda = \frac{\ln 2}{59.9 \text{ da}}$$

$$N_0 = 100$$

$$N = 10$$
$$t = ?$$
$$t = \frac{\ln N_0 - \ln N}{\lambda}$$
$$t = \frac{\ln 100 - \ln 10}{\ln 2 \,/\, 59.9 \text{da}}$$
$$t = \frac{(4.60517 - 2.30259) \text{ da}}{0.0115}$$
$$t = \frac{2.30258 \text{ da}}{0.1157}$$
$$t = 198.98 \text{ da}$$

GRAPHS OF DECAY EQUATIONS

Equation 4-7 states that the number of atoms, N, decreases exponentially with time, and this is shown by Figure 4-1. Generally, linear plots are more useful for analyzing and presenting data; therefore, semilogarithmic plots are used commonly. Inspection of Equation 4-4 reveals that it is a linear equation if $\ln N$ is taken as the dependent variable and t as the independent variable since $y = ax + b$ describes a straight line.

$$\ln N = -\lambda t + \ln N_0 \tag{4-4}$$

Consequently, a plot of $\ln N$ versus t yields a straight line as shown in Figure 4-2. The slope of the line is $-\lambda$ and the intercept on the $\ln N$ axis is $\ln N_0$. Plotting common logarithms yields a similar graph (Figure 4-3) except the slope is -0.4342λ and the intercept $\log N_0$. Equation 4-6 shows these relationships.

RADIOACTIVITY OR ACTIVITY

Frequently, the term radioactivity is shortened to activity. However, confusion can result; for example, suppose a person is investigating enzyme kinetics by the use of radioactive tracers. The term activity might indicate enzyme activity or radioactivity. Because of possible ambiguities, some scientific journals do not accept activity as a substitute for radioactivity; nevertheless, the term activity will be used here.

The activity of a sample is defined as the number of atoms disintegrating per unit of time. Comparison of this definition to that given earlier for dis-

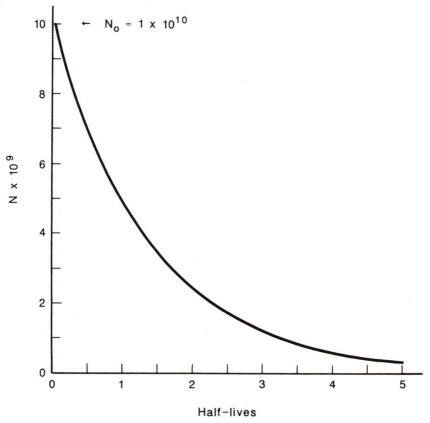

FIG. 4-1   Radioactive decay. The decrease in the number of atoms (N) as affected by time (t).

integration rate or rate of decay reveals that the definitions are equivalent. Activity is denoted by the symbol A so,

$$A = \frac{-\Delta N}{\Delta t} = \lambda N$$

$$A = \lambda N \qquad (4\text{-}12)$$

The SI unit for activity is the Becquerel (Bq), which is equivalent to 1 dps. If the activity is expressed in becquerels or dps, then $-\Delta N$ represents the number of disintegrations in 1 s, and $\Delta t = 1$ s. However, these units result only when the decay constant has units of reciprocal seconds. Earlier it was stated that Equation 4-1 is valid only if Δt is small compared to the half-life of the radionuclide in question. If Δt is 1/100 of $t_{1/2}$, the activity calculated by Equation 4-1 will be in error by less than 0.5%, an amount generally lower than measurement errors.

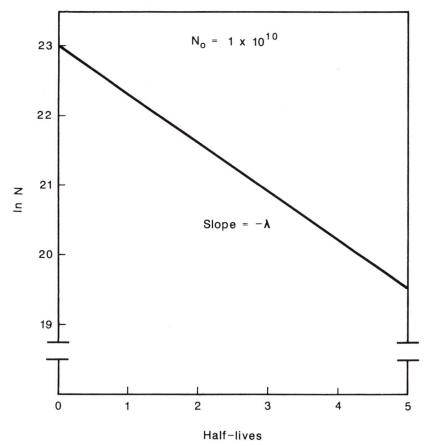

FIG. 4-2 Radioactive decay. A semilogarithmic plot of the natural logarithm of the number of atoms (ln N) versus time (t).

It is not practical to use radionuclides with half-lives less than a day or so, unless they can be used at the point of their production. Consequently, a practical rule is to always use minutes or seconds for time units when using Equation 4-12. When minutes are used, the activity units will be disintegrations per minute or dpm.

Example Problem 4-4.   What is the activity of a sample containing 5×10^{-9} mol of ^{45}Ca atoms? The half-life of ^{45}Ca is 163.8 days.

$$A = \lambda N \qquad t_{1/2} = (163.8 \text{ da})(1440 \text{ min/da})$$

$$\lambda = \frac{\ln 2}{(1.638)(1.44)(10^5) \text{ min}}$$

$$N = (5 \times 10^{-9} \text{ mol})(6.022 \times 10^{23}) \text{ atoms/mol}$$

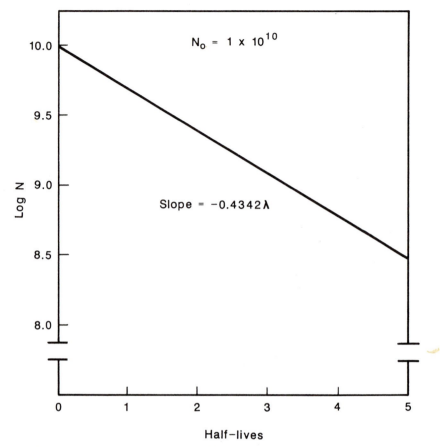

FIG. 4-3   Radioactive decay. A semilogarithmic plot of the common logarithm of the number of atoms (log N) versus time (t).

$$A = ?$$

$$A = \frac{(\ln 2)(5)(6.022)(10^{14}) \text{ atoms}}{(1.638)(1.44)(10^5) \text{ min}}$$

$$A = 8.848 \times 10^9 \text{ dpm}$$

Example of Problem 4-5.   A sample containing ^{59}Fe exhibits an activity of 7.33×10^5 dps. How many atoms of ^{59}Fe must be present in the sample? The half-life of ^{59}Fe is 44.51 days, and there are 8.64×10^4 s/day according to the table of Constants and Conversion Factors in Appendix A.

$$t_{1/2} = 44.51 \text{ da}$$

$$\lambda = \frac{\ln 2}{(44.51 \text{ da})(8.64 \times 10^4 \text{ s/da})}$$

$$A = 7.33 \times 10^5 \text{ dps (atoms/s)}$$

$$N = ?$$

$$N = \frac{A}{\lambda}$$

$$N = \frac{7.33 \times 10^5 \text{ atoms/s}}{\dfrac{\ln 2}{[(4.451)(8.64)(10^5)\text{s}]}}$$

$$N = 4.067 \times 10^{12} \text{ atoms}$$

Units for Activity

Becquerel.   As mentioned earlier, the becquerel (Bq) is the SI unit for activity, and it is equivalent to 1 dps. The becquerel has not achieved overwhelming popularity, probably because the dps is more descriptive, but some scientific journals demand usage of SI units.

Curie.   Originally, the *curie* (Ci) was defined as the amount of radon in equilibrium with 1 g of radium. Later, it was redefined as the disintegration rate for any radioactive sample that undergoes the same number of disintegrations per unit time as that for 1 g of pure ^{226}Ra. The actual number changed periodically because of improvements in measuring activity, the atomic weight of ^{226}Ra, and its decay constant. In 1950 the International Union of Pure and Applied Physics defined the curie as the quantity of any radionuclide in which the number of disintegrations per second is 3.7×10^{10}. Consequently,

$$1 \text{ Ci} = 3.7 \times 10^{10} \text{ Bq}$$
$$1 \text{ Ci} = 3.7 \times 10^{10} \text{ dps}$$

or

$$1 \text{ Ci} = 2.22 \times 10^{12} \text{ dpm}$$

Millicuries (mCi), 3.7×10^7 dps or 2.22×10^9 dpm, and microcuries (μCi), 3.7×10^4 dps or 2.22×10^6 dpm, are used frequently as units in scientific research. Originally, a lower case c was used as the symbol for a curie and may be found in the literature prior to the mid-1960s.

Example Problem 4-6.   How many becquerels and how many curies of ^{3}H would there be in one mole of pure ^{3}H$_2$ gas? We must use Equation 4-12 to calculate activity in units of dps to obtain becquerels, then becquerels or

dps can be converted to curies. The half-life of ^{3}H is 12.26 years and there are 3.1536×10^7 s/yr according to the tables in Appendix A.

$$A = \lambda N \qquad t_{1/2} = (12.26 \text{ yr})(3.1536 \times 10^7 \text{ s/yr})$$

$$\lambda = \frac{\ln 2}{(1.226)(3.1536)(10^8) \text{ s}}$$

$$N = (2)(6.022 \times 10^{23}) \text{ atoms}$$

$$A = \frac{(\ln 2)(2)(6.022)(10^{23}) \text{ atoms}}{(1.226)(3.1536)(10^8) \text{ s}}$$

$$A = 2.159 \times 10^{15} \text{ atoms/s}$$

$$A = 2.159 \times 10^{15} \text{ dps or } 2.159 \times 10^{15} \text{ Bq}$$

$$A = \frac{2.159 \times 10^{15} \text{ dps}}{3.7 \times 10^{10} \text{ dps/Ci}}$$

$$A = 5.835 \times 10^4 \text{ Ci}$$

The units defined in this section are called absolute units, meaning that they are based on actual disintegration rates. Relative activity units are useful and are described in the following section.

ACTIVITY MEASUREMENTS

Activity may be measured by counting emissions from the radioactive sample and such data are expressed as a counting rate, either *counts per minute* (cpm) or *counts per second* (cps). An emission from a radioactive source is detected (produces one count) when it interacts with the detector of the counting instrument, but usually not all emissions interact with the detector. The fraction of emissions detected depends on geometric relationships between the detector and source, the nature of the emissions, the chemical and physical properties of the source, and the properties of the detector.

Background and Net Counting Rates

Since there is considerable natural radiation in our environment, counting instruments will yield appreciable counting rates without a radioactive sample being present. Spurious electronic pulses contribute to the background of instruments also. This "blank counting rate" is called the *background counting rate,* or simply *background,* and it must be subtracted from the gross counting rate of the sample to obtain the *net counting rate, net cpm,* or *net cps.* The net counting rate is a measure of sample radioactivity.

Comparisons of Counting Rates

Since the proportion of emissions that interact with the detector of the counting instrument is constant for a given sample, net cpm or net cps are relative, yet suitable, units of activity. Usually, however, the proportion of emissions interacting with the detector will be different for different radionuclides because the emissions may not be of the same type or energy. Changes in the physical and chemical properties of the sample may alter the proportion of emissions detected even though the same radionuclide is being counted. Thus, it is not legitimate to compare the activities of two samples in units of net cpm or net cps unless the proportion of emissions detected is equal or is known for both samples so that appropriate corrections can be applied.

The absolute units of radioactivity are defined in terms of disintegrations of the radionuclide per minute or second and therefore are not the same, necessarily, as the rate of emission of radiation. When a ^{32}P atom disintegrates only one β^--particle and one $\bar{\nu}$ are emitted. Of course, the $\bar{\nu}$ cannot be detected by present counting instruments while the β^- is easily detected, and the disintegration rate for ^{32}P is equal to the β^--emission rate. Approximately 51% of ^{59}Fe atoms disintegrate by emitting a β^--particle ($E_{max} = 0.475$ MeV) and an $\bar{\nu}$, followed by a 1.099 MeV γ-ray, and 48% emit a β^--particle ($E_{max} = 0.273$ MeV) and an $\bar{\nu}$ followed by a 1.292 MeV γ-ray. Thus, for most ^{59}Fe atoms there are two emissions that are capable of being detected, a β^- and a γ, for each disintegration. Obviously, the decay scheme for a particular radionuclide must be known before the disintegration rate and emission rate can be correlated.

Counting Efficiency

Both problems mentioned above can be circumvented by determining counting efficiency (CE), which is simply the quotient, *net* counting rate divided by the disintegration rate of the sample.

$$CE = \frac{cpm}{dpm} \quad \text{or} \quad \frac{cps}{dps} \quad\quad (4\text{-}13)$$

For a given radionuclide, the counting efficiency may be obtained by determining net counting rate of a standard sample for which the activity (disintegration rate) is known. To determine the activity of a sample containing the same nuclide, the sample, prepared in the same manner as the standard, is assayed with the same instrument operating under the same conditions. The counting rate obtained divided by the counting efficiency will yield the activity of the sample in absolute units. Counting efficiencies are frequently expressed as percentages, although when employed in calculations the quotient, cpm/dpm, is used directly. From statistical considerations, to be developed in Chapter 13, it is advantageous to assay radio-active samples with instruments and sample preparation techniques that yield high counting efficiencies.

Because the emission rate may exceed the disintegration rate for some nuclides, counting efficiencies greater than 100% are feasible. However, for most radionuclides this is not the case, and counting efficiencies considerably below 100% will usually be encountered.

ACTIVITY DECAY

Since the number of atoms of a pure sample of a radionuclide decreases exponentially with time, and the activity of a sample is directly proportional to the number of atoms present, the activity of a sample decreases at the same exponential rate. This is shown mathematically by letting A_0 represent the activity of a sample when $t = 0$. Thus, Equation 4-12 becomes

$$A_0 = \lambda N_0 \tag{4-14}$$

Recall Equation 4-7

$$N = N_0 e^{-\lambda t}$$

Multiplying both sides of the equation by λ yields

$$\lambda N = \lambda N_0 e^{-\lambda t}$$

Since $A = \lambda N$ and $A_0 = \lambda N_0$, the exponential relationship of activity with time is

$$A = A_0 e^{-\lambda t} \tag{4-15}$$

Logarithmic forms of the equation are

$$\ln A = \ln A_0 - \lambda t \tag{4-16}$$

$$\log A = \log A_0 - 0.4342 \lambda t \tag{4-17}$$

Because Equations 4-15, 4-16, and 4-17 show the same relationships between activity and time as Equations 4-7, 4-4, and 4-6 show for the number of atoms and time, their plots have identical forms. Compare Figure 4-2 to Figure 4-4.

The activity decay equations are used commonly in radiotracer research because the experimental measurements are activity measurements, almost invariably. Frequently, one of these Equations 4-15, 4-16, or 4-17 must be used to correct for decay during the course of an experiment.

Example Problem 4-7. Suppose a person received 5 mCi of ^{131}I on a Monday morning at 9:00 A.M., but did not have the other things ready for an experiment until the following Thursday at 9:00 P.M. Would the person have enough ^{131}I for the experiment on Thursday? If the experiment required 5 mCi, then the answer is definitely no, but let us calculate how much activity would have been present at 9:00 P.M. on Thursday. Use Equation 4-15 and let $t = 0$ at 9:00 A.M. Monday morning, then $A_0 = 5$ mCi. Time for decay (t) is 3.5 days, and the half-life of ^{131}I is 8.04 days.

$$A = A_0\, e^{-\lambda t}$$

$$A = (5\ \text{mCi})\, e^{-\frac{(\ln 2)(3.5\ \text{da})}{(8.04\ \text{da})}}$$

$$A = (5\ \text{mCi})\, e^{-(0.30174)}$$

$$A = (5\ \text{mCi})(0.73953)$$

$$A = 3.698\ \text{mCi}$$

The use of a logarithmic equation, such as 4-16, follows.

$$\ln A = \ln A_0 - \lambda t$$

$$\ln A = \ln(5) - \frac{(\ln 2)(3.5\ \text{da})}{8.04\ \text{da}}$$

$$\ln A = 1.60944 - 0.30174$$

$$\ln A = 1.3077$$

$$A = 3.698\ \text{mCi}$$

Example of Problem 4-8. A radiotracer supply company received an order for ^{131}I. The customer specified the need to receive at least 1 mCi ^{131}I by 10:00 A.M. March 15. The company responds by planning to prepare the order at 10:00 A.M. on March 13 and immediately shipping the package by overnight express. How much ^{131}I should be placed in the container to ensure that 1 mCi of activity remains at the specified time?

$$t = 2\ \text{da} \qquad A = A_0\, e^{-\lambda t}$$

$$\lambda = \frac{\ln 2}{8.04\ \text{da}} \qquad A_0 = A\, e^{\frac{(\ln 2)(2\ \text{da})}{(8.04\ \text{da})}}$$

$$A = 1\ \text{mCi} \qquad A_0 = (1\ \text{mCi})\,(e^{0.17242})$$

$$A_0 = ? \qquad A_0 = (1\ \text{mCi})(1.1882)$$

$$A_0 = 1.882\ \text{mCi}$$

Example Problem 4-9. A scientist wishes to determine the uptake of phosphorus by tomato plants growing hydroponically. Although oversimplified, the experiment could be conducted by placing a known amount of ^{33}P-labeled phosphate in the hydroponic solution and then determining the amount of ^{32}P activity after a period of time. Suppose a 7-day uptake period was selected and that the hydroponic solution contained 1 μCi of ^{32}P activity initially. After 7 days, the total ^{32}P activity in the plant was found to be 6.78×10^4 dpm. What percentage of the phosphate was taken up by the plant?

The half-life of ^{32}P is 14.28 days. Let us calculate the activity of the original ^{32}P tracer after 7 days of decay.

$$A = A_0 \, e^{-\lambda t} \qquad t = 7 \text{ da}$$

$$\lambda = \frac{(\ln 2)}{14.28 \text{ da}}$$

$$1 \, \mu\text{Ci} = 2.22 \times 10^6 \text{ dpm}$$

$$A_0 = 2.22 \times 10^6 \text{ dpm}$$

$$A = \,?$$

$$A = (2.22 \times 10^6 \text{ dpm}) \, e^{-\frac{(\ln 2)(7 \text{ da})}{14.28 \text{ da}}}$$

$$A = (2.22 \times 10^6 \text{ dpm})(e^{-0.33978})$$

$$A = (2.22 \times 10^6 \text{ dpm})(0.71193)$$

$$A = 1.580 \times 10^6 \text{ dpm}$$

The percentage in the plant is

$$\frac{6.78 \times 10^4 \text{ dpm}}{1.580 \times 10^6 \text{ dpm}} \times 100 = 4.29\%$$

If the correction for decay had not been made, the calculated percentage would have been

$$\frac{6.78 \times 10^4 \text{ dpm}}{2.22 \times 10^6 \text{ dpm}} \times 100 = 3.05\%$$

Consequently, the later calculation is low by 1.24%, which represents an error of 28.9%. Unless the half-life of the radiotracer is much greater than the term of an experiment, corrections for decay must be made.

HALF-LIFE DETERMINATIONS

Activity decay equations are useful in determining the half-life of a radio-nuclide, even if absolute activities are not determined. The counting efficiency of the instrument does not need to be known so long as operating conditions are held constant to ensure that the counting efficiency does not vary during the length of the experiment.

A sample is prepared, counted, and stored for an interval of time, then recounted. The alternate counting and storing process continues for a period of time sufficient to yield valid data. Measurements over a period equivalent to one-half of the radionuclide's half-life may be sufficient, but observations over one or more half-lives are better. The reduction in the counting rate provides a rough estimate of the half-life.

Example Problem 4-10. From the set of laboratory data given in Table 4-1, determine the half-life for the hypothetical radionuclide. A plot of the data is shown in Figure 4-4.

The slope of the plot is -0.0574 da^{-1}

$$-\lambda = -0.0574 \text{ da}^{-1}$$

$$\lambda = \frac{\ln 2}{t_{1/2}} = 0.0574 \text{ da}^{-1}$$

$$t_{1/2} = \frac{\ln 2}{0.0574 \text{ da}^{-1}}$$

$$t_{1/2} = 12.08 \text{ da}$$

The data in Table 4-1 were analyzed by linear regression, and that analysis yielded the following data.

$$\text{Correlation coefficient:} \quad -0.99895$$
$$\text{slope:} \quad -0.05720$$
$$y \text{ intercept:} \quad 9.67080$$

According to this analysis, the data do fit a straight line well as shown by the closeness of the correlation coefficient to the value of 1.0. The line defined by this analysis is

$$\ln A = -0.0572 \text{ da}^{-1} (t) + 9.6708$$

TABLE 4-1 Data for a Half-Life Determination

Date	Net cpm	ln (net cpm)	Time (days)
March 1	15,936	9.676	0
March 3	14,198	9.561	2
March 5	12,332	9.420	4
March 7	11,070	9.312	6
March 9	10,353	9.245	8
March 11	8,920	9.096	10
March 13	8,103	9.000	12
March 15	7,009	8.855	14
March 17	6,342	8.755	16
March 19	5,653	8.640	18

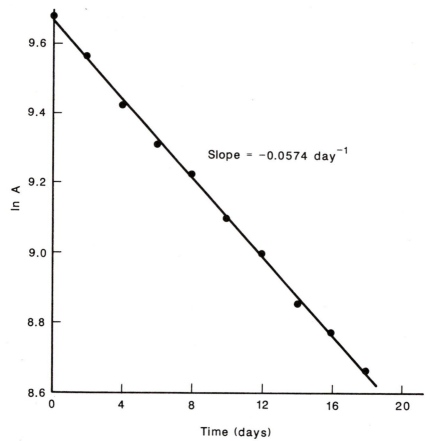

FIG. 4-4 A plot for the determination of a half-life. Data for the plot are given in Table 4-1.

Such a line would be almost indistinguishable from the one shown in Figure 4-4, which was drawn without analysis, and the slopes determined by both methods are similar. Nevertheless, linear regression analysis is preferred for accuracy. Using the slope determined by linear regression the half-life may be recalculated.

$$t_{1/2} = \frac{\ln 2}{0.0572 \text{ da}^{-1}}$$

$$t_{1/2} = 12.12 \text{ da}$$

The graphic method of determining half-life is not practical for radionuclides with very long half-lives. With a half-life of 5730 years, a decrease in activity for a sample of ^{14}C over a period of years cannot be detected without rigorous statistical methods and advanced instrumentation. However, Equation 4-12 may be used to determine the decay constant if absolute activities and radionuclide quantities can be measured accurately.

Example Problem 4-11. Exactly 7.000 ng of a pure radionuclide was found to have an activity of 17,340 dpm. If the atomic weight of this radionuclide is 169.00 g/mol, what is its half-life?

$$A = \lambda N = \frac{(\ln 2)(N)}{t_{1/2}}$$

$$t_{1/2} = \frac{(\ln 2)(N)}{A}$$

$$N = \frac{(7 \times 10^{-9} \text{ g})(6.022 \times 10^{23} \text{ atoms/mol})}{1.69 \times 10^2 \text{ g/mol}} = 2.4943 \times 10^{13} \text{ atoms}$$

$$A = 1.734 \times 10^4 \text{ atoms/min}$$

$$t_{1/2} = \frac{(\ln 2) \, 2.494 \times 10^{13} \text{ atoms}}{1.734 \times 10^4 \text{ atoms/min}}$$

$$t_{1/2} = 9.971 \times 10^8 \text{ min}$$

$$t_{1/2} = \frac{9.971 \times 10^8 \text{ min}}{5.256 \times 10^5 \text{ min/yr}}$$

$$t_{1/2} = 1{,}897 \text{ yr}$$

LABELED COMPOUNDS

Many radioisotopic experiments involve labeled compounds in which a radioactive isotope has been incorporated. For example, acetic acid, CH_3-CO_2H, contains C, H, and O. Theoretically, radioisotopes of any of these three elements could be utilized to produce labeled acetic acid; however, because of their short half-lives, radioisotopes of oxygen are not very useful. Consequently, ^{14}C or ^3H might be selected for labeling acetic acid. Consider the use of ^{14}C. There are three choices in labeling the molecule as shown below.

1. $^{14}CH_3-^{12}CO_2H$
2. $^{12}CH_3-^{14}CO_2H$
3. $^{14}CH_3-^{14}CO_2H$

The position(s) of the label is indicated, using normal organic nomenclature. The first example would be [2-^{14}C]acetic acid, the second [1-^{14}C]acetic acid, and the third [1,2-^{14}C]- or [U-^{14}C]acetic acid. The designation "U" stands for _uniformly labeled,_ meaning that the ^{14}C atoms are uniformly distributed between all available positions. The terms "r," "G," and "N" indicate _random, general,_ and _nominal_ labeling patterns where the distribution of labeled positions is uncertain.

Specific Radioactivity

Specific radioactivity or, more simply, _specific activity_ of a sample is the amount of radioactivity per quantity unit. Many different units may be encountered because there are several acceptable units for activity, such as dpm, Bq, and Ci, and quantity units may be in terms of weight, volume, or number of atoms (moles). Thus, Bq/mg, mCi/mol, and dpm/ml are acceptable examples.

Radioisotopic Carriers

In most cases, the mass of a labeled compound needed for manipulation of the experimental system is much greater than the quantity of labeled compound needed for accurate radioactive measurements. Excess radioactivity would be a needless expense and quite possibly an unnecessary radiation safety hazard. Therefore, nonlabeled acetic acid may be added to labeled acetic acid to reduce its specific activity to an appropriate level. The nonlabeled acetic acid is called the _carrier,_ and generally, the carrier represents a large majority of the mass of the sample. It is the nonlabeled compound that is manipulated, and the labeled compound is "carried along." Consequently, a sample of [1-^{14}C]acetic acid might be mostly $^{12}CH_3-^{12}CO_2H$ with a little $^{12}CH_3-^{14}CO_2H$. Consider [U-^{14}C]acetic acid. Such a sample might contain $^{12}CH_3-^{12}CO_2H$ and a small amount of $^{14}CH_3-^{14}CO_2H$, and/or equal amounts of $^{14}CH_3-^{12}CO_2H$ and $^{12}CH_3-^{14}CO_2H$.

Maximum Specific Activities

Maximum specific activities are obtained by eliminating the carrier. Thus, if a particular labeled compound is _carrier-free,_ it has the maximum specific activity, and that specific activity may be calculated readily since

$$A = \lambda N = \frac{\ln 2}{t_{1/2}} N$$

Knowing the half-life of ^{14}C, we can calculate the activity for [1-^{14}C]acetic acid starting with 1 mol of the labeled compound.

$$A = \lambda N \quad N = 6.022 \times 10^{23} \text{ atoms}$$

$$t_{1/2} = 5730 \text{ yr}$$

$$t_{1/2} = (5730 \text{ yr}) \, 3.1536 \times 10^7 \text{ s/yr}$$

$$t_{1/2} = 1.807 \times 10^{11} \text{ s}$$

$$\lambda = \frac{\ln 2}{1.807 \times 10^{11} \text{ s}}$$

$$A = \frac{(\ln 2)(6.022 \times 10^{23} \text{ atoms/mol})}{1.807 \times 10^{11} \text{ s}}$$

$$A = 2.31 \times 10^{12} \text{ atom/s·mol}$$

$$A = 2.31 \times 10^{12} \text{ dps/mol}$$

The specific activity could be expressed in Ci/mol.

$$\frac{2.31 \times 10^{12} \text{ dps/mol}}{3.7 \times 10^{10} \text{ dps/Ci}} = 62.43 \text{ Ci/mol}$$

Obviously, the maximum specific activity of [1,2-^{14}C]- or [U-^{14}C]acetic acid, on a mole basis, would be twice that of [1-^{14}C]- or [2-^{14}C]acetic acid since two instead of one ^{14}C atoms are incorporated into the molecule. If it is desired to have the specific activity expressed in activity per unit weight, the molecular weight of the compound must be used in the calculation. Should one use the atomic weight of ^{14}C or ^{12}C in such calculations? It depends on the relative proportion of the labeled compound and nonlabeled compound. Carrier-free [U-^{14}C]acetic acid would have a molecular weight of

$$^{14}\text{C} - 14.0031 \times 2 = 28.0062$$

$$\text{H} - 1.0079 \times 4 = 4.0316$$

$$\text{O } 15.9994 \times 2 = \frac{31.9988}{64.0366 \text{ g/mol}}$$

and a specific activity of

$$\frac{124.86 \text{ Ci/mol}}{64.0366 \text{ g/mol}} = 1.9498 \text{ Ci/g}$$

If the nonlabeled molecular weight had been used, the calculated value would have been

$$\frac{124.86 \text{ Ci/mol}}{60.0524 \text{ g/mol}} = 2.079 \text{ Ci/g}$$

The difference represents an error greater than 6%.

The average molecular weight of a non-carrier-free sample of [U-^{14}C]acetic acid could be calculated if the relative proportions of unlabeled and labeled molecules were known. However, in most cases the ratio of labeled to unlabeled is quite small, which yields an insigificant error in calculations when normal (nonradioisotopic) molecular weights are used. Therefore, in most laboratory experiments that do not involve carrier-free tracers, the normal molecular weights are employed.

The maximum activity per mole of a radionuclide, or its maximum specific activity, is inversely proportional to the radionuclide's half-life. This is shown by Equation 4-12.

$$A = \lambda N = \frac{\ln 2}{t_{1/2}} N$$

Tritium has a half-life of 12.26 yr, and its maximum specific activity, dps/mol, is

$$A = \frac{(\ln 2)(6.022 \times 10^{23} \text{ atom/mol})}{(12.26 \text{ yr})(3.1536 \times 10^7 \text{ s/yr})} = 1.08 \times 10^{15} \text{ dps/mol}$$

Earlier in this section we calculated the maximum specific activity of ^{14}C to be about 2.3×10^{12} dps/mol ($t_{1/2} = 5730$ yr), whereas ^{32}P, with a half-life of 14.28 days, is

$$A = \frac{\ln 2(6.022 \times 10^{23} \text{ atom/mol})}{(14.28 \text{ da})(8.64 \times 10^4 \text{ s/da})} = 3.383 \times 10^{17} \text{ dps/mol}$$

Other factors affect efficiency of detection, but generally detection sensitivity increases with specific activity and hence with decreasing half-lives as well.

PROBLEMS

The tables given in Appendix A, as well as the "Table of the Isotopes" (Weast, 1988) may be useful for solving the following problems.

1. If a sample possesses an activity of 1.11×10^7 dpm, what is its activity in units of Bq, Ci, and μCi?

2. What would the disintegration rate be for a sample that contained 1×10^{14} atoms of ^{125}I? Express your answer in units of Bq, MBq, and μCi.

3. A sample of $Na^{125}I$ has a specific activity of 3.852×10^{18} dpm/mol. The sample is not carrier-free since it contains some $Na^{127}I$, which is not radioactive. What is the isotopic abundance of ^{125}I in the sample? That is, what percentage of the iodine atoms are ^{125}I atoms?

4. What is the half-life of a radionuclide if 1×10^{-9} mol of it yields a disintegration rate of 2.4154×10^5 dps?

5. If a sample contained 1×10^{12} atoms of ^{35}S on July 4, how many atoms of ^{35}S should be present the following September 2?

6. The initial activity of a sample of [^{3}H]thymidine was 1.260×10^4 dpm. What was its activity exactly 1 year later?

7. A sample containing ^{32}P was prepared August 4, 1990. If the sample's activity was 4.72×10^3 dpm on September 1, 1990, what was its activity when the sample was first prepared?

8. A gross counting rate of 1432 cpm was obtained for an experimental sample containing ^{32}P. At the same time a 0.1-ml aliquot of a ^{32}P standard solution was counted under identical conditions. The standard yielded 9291 counts in 3 min, and the specific activity of the standard was 1×10^5 dpm/ml at the time it was counted. A background measurement gave 640 counts in 20 min for the instrument used. What was the activity of ^{32}P in the experimental sample in units of dpm?

9. If another 0.1-ml aliquot of the ^{32}P standard described in problem 8 was counted 17 days after the time the measurements in problem 8 were performed, what should the net counting rate be for the 0.1-ml aliquot? Assume sample preparation, counter's performance, and background were identical on both days.

10. A solution containing 10 mg/ml of $Na_2{}^{14}CO_3$ was assayed by counting a 0.1-ml aliquot with a liquid scintillation spectrometer. The data obtained are

Sample: 17,281 cts/10 min

Background: 280 cts/10 min

The counting efficiency of the instrument was determined to be 76.6% (0.766 cpm/dpm). What is the specific activity of $Na_2{}^{14}CO_3$ in units of dpm/mol?

11. The following data were obtained for the determination of the half-life for an unknown radionuclide.

Time of assay (t) (days)	Corrected activity (A) (net cpm)
0	2612
2	1955
4	1511
6	1249
8	986
10	814
12	718
14	421

Determine the decay constant (λ), half-life ($t_{1/2}$), and average-life ($\bar{t}$) for the unknown radionuclide.

12. A β-ionization detector for a gas chromatograph contains a sealed source of ^{90}Sr that had an activity of 20.0 mCi at the time the detector was manufactured. How old will the detector be when the ^{90}Sr activity has been reduced to 16.0 mCi?

13. Yeast was cultured in a medium containing a ^{32}P-labeled phosphate buffer that had a specific activity of 1 μCi/liter at the time of inoculation. After growth, the yeast was harvested by centrifugation and the specific activity of the supernatant fluid was determined to be 500 dpm/ml. If the radioactivity measurements were made exactly 3 days apart and evaporation of the medium was negligible, what fraction of the ^{32}P was taken up by the yeast?

14. The label on a bottle of tritiated toluene states

[^{3}H]Toluene
Specific activity: 3.02 × 10^5 dpm/g
Date: November 15, 1989
Volume: 10 ml

   a. Calculate the specific activity of the [^{3}H]toluene as of November 15, 1991, and give your answer in units of dpm/ml. The density of toluene is 0.8669 g/ml.

   b. On November 15, 1991, exactly 0.1 ml of the [^{3}H]toluene was counted and yielded 75,831 counts in 10 min; the background for the counting instrument was 23 cpm. Determine the counting efficiency for the instrument.

15. You are planning an experiment requiring ^{131}I that will be performed in two phases that, within a few minutes, are 7 days apart. The first phase will require 10 μCi, and 20 μCi is needed for the second phase. Making only one order and assuming that the amount of activity needed will be received at the beginning of the first phase, determine the minimum number of microcuries of ^{131}I that must be purchased for this experiment.

16. What is the maximum specific activity that could be obtained for [1-^{14}C]phenylalanine? Express your answer in dpm/mol, MBq/mol, and μCi/mg.

17. If the specific activity of a D-[2-^{14}C]-glucose sample is 7.9 mCi/mmol, what fraction of the glucose molecules is labeled?

18. After a 1 g sample of carrier free [^{32}P]-phosphoric acid had been allowed to decay for five half-lives, the sample was repurified, and pure phosphoric acid was isolated. What would the specific activity of the repurified phosphoric acid be at that time?

19. Describe how to prepare 1 liter of an acetate buffer solution that has a sodium acetate concentration of 1 × 10^{-3} M and a specific activity of

6.66 × 10^5 dpm/ml. The sodium [1-^{14}C]acetate to be used for this purpose has a specific activity of 60 μCi/mg.

20. a. A benzene solution of [^{14}C]cholesterol (15 ml) was assayed by counting a 0.1-ml aliquot with a liquid scintillation spectrometer. A counting rate of 84,535 counts/5 min, background of 2389 counts/10 min, and counting efficiency of 86% were obtained. Determine the total activity of the cholesterol present in the original solution in units of μCi.

   b. If the specific activity of the [^{14}C]cholesterol is 20 mCi/mmol and the molecular weight of the cholesterol is 386.66 g/mol, what is the concentration of cholesterol in the benzene solutions in units of μg/ml?

REFERENCE

Weast, R. C. (1988). *Handbook of Chemistry and Physics,* 69th ed. CRC Press, Boca Raton, FL.

CHAPTER 5

Interactions of Radiation with Matter

OVERVIEW OF RADIATION ABSORPTION

Interactions of radiation and matter involve specific processes whereby a particle or ray's energy is transferred to the molecules, atoms, or atomic components of matter. These represent energy absorption processes with matter being the *absorber*. Several different processes may occur with each type of radiation, but our discussion will be directed toward processes important for radiation detection and radiological health considerations.

The major types of radiation emitted by radionuclides (α-, β-, γ-, and X-rays) have an important common property—their propensity to create ions in matter, either directly or indirectly. Hence, they belong to the somewhat broader category of radiation called *ionizing radiation.* Although ionization is involved in many detection methods and is a major factor in causing radiobiological damage, the absorption of radiation by other processes is important quantitatively.

From a particle's point of view, an absorber (any type of matter) must appear as a maze of electronic charges including many low mass negatively charged bodies (electrons) and fewer large mass bodies carrying large positive charges (nuclei). Since electrons are more numerous and more widely dispersed than nuclei, the probability of a particle "hitting" an electron is greater. Actually, particles or rays may interact with electrons or nuclei without physical collisions since forces, particularly coulombic forces, extend a distance from the body. Thus, a collision simply means that the ray or particle comes close enough to an absorber component for the two to interact.

A particle's charge and velocity (energy) have a significant effect on its probability of interacting with electrons or nuclei. Thus, the predominant mode of interaction is different for different types of radiation and is dependent on the energy of the ray or particle. In general, particles transfer energy through a series of interactions, while γ- or X-rays tend to interact by "one hit" processes. The distance a particle or ray travels in an absorber is governed by the cumulative probabilities of interactions as a particle or ray traverses the absorber's electronic maze.

α-PARTICLES

The primary interactions between α-particles and matter yield electronically excited molecules (M*) and ions (M$^+$ or M$^-$), hence the principal processes are called *excitation* and *ionization*. Both processes occur as an individual particle's energy is dissipated in an absorber.

Excitation

As a positively charged α-particle traverses an absorber, it may come close enough to an electron (bound to an atom or molecule) so that the attractive forces "pull" the electron from its molecular orbital to a vacant higher energy orbital. The molecule would then exist in an electronically excited state (M*), and the α-particle would have lost some of its kinetic energy in the process. The energy lost by the particle is equivalent to the energy difference between the molecular excited state (M*) and its ground state (M^0). After the interaction, the α-particle will continue its flight at a slightly lower velocity, and it will undergo many subsequent interactions until its kinetic energy is dissipated.

Ionization

The physical picture for ionization is similar to that described for excitation, except that the attractive forces between the α-particle and electron are strong enough to "pull" the electron from the absorber molecule or atom completely. Relative to the electron, the α-particle is moving fast so the electron cannot combine with the particle; instead it is left behind. However, the pull may have given the electron a low velocity. This process creates an *ion pair*, the molecule minus one electron (M$^+$) and the electron (e$^-$) itself. It also reduces the kinetic energy of the α-particle by an amount equal to the energy necessary to remove the electron (ionization potential), plus the kinetic energy imparted to the electron.

The average energy imparted to electrons is about 100–200 eV, however, some electrons may have energies approaching a few keV. As will be discussed later, electrons with such kinetic energies will cause additional ionization and excitation in the absorber. Such electrons are called *secondary electrons,* and the ionization they cause is referred to as *secondary ionization,* whereas *primary ionization* is the ionization directly caused by the incident particle. For α-radiation, about 60–80% of the ionization is due to secondary ionization. In air, the average amount of energy expended in the creation of an ion pair is about 34 eV, which is about twice the amount of energy for the first ionization potential of oxygen and nitrogen.

As an α-particle traverses an absorber, it loses energy by interactions, both excitation and ionization, and hence loses velocity. Near the end of an α-particle's path its velocity will be reduced to a level comparable to that of the orbital electrons of the absorber molecules, and, at that point, the α-particle may alternately gain and lose electrons until it comes to rest.

Specific Ionization

Specific ionization refers to the density of ions produced in an absorber and is generally expressed as the number of ion pairs per unit path length. Mathematically, it is related to the rate of energy loss by a particle, and two important factors in the rate of energy loss are a particle's charge and velocity.

As a particle's velocity (hence kinetic energy) decreases, its specific ionization increases. Although mathematical equations exist to describe this relationship, intuitively it is reasonable to expect a greater probability of an interaction occurring the longer a molecule or atom is exposed to an ionizing particle.

Figure 5-1 shows the specific ionization of α-particles as a function of distance traveled in air. Initially, the particles have an energy of about 14 MeV and a velocity of 2.6×10^9 cm/s. They lose this energy over a distance of about 20 cm. The specific ionization rises slowly until the energy is reduced to about 5 MeV (14.7 cm) then rises rapidly as the particle's velocity decreases further. The maximum occurs when the particle's energy and velocity have been reduced to about 0.37 MeV and 4.21×10^8 cm/s, respectively. At this point, the particle is about 3–4 mm from the end of its path. Further energy loss causes the specific ionization to fall precipitously because the particle starts picking up and losing electrons and hence has a reduced charge.

Cloud chambers provide a method for visualizing ionization because ions serve as centers for condensation of vapor. The tracks produced by α-particles are straight and relatively wide (1–2 mm) compared to β-particles, with the ends of the tracks being much broader than the beginnings.

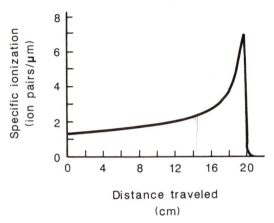

FIG. 5-1   Specific ionization curve for monoenergetic α-particles as a function of distance traveled in air. (Redrawn with permission from *Nuclear and Radiochemistry,* by G. Frielander, J. W. Kennedy, E. S. Macias, and J. M. Miller. Copyright © 1981 by John Wiley and Sons)

Track thickness is a function of specific ionization. The lengths of tracks depend on initial energies, but usually are several centimeters long. For example, a radionuclide emitting 7.0 MeV α-particles will produce tracks about 6 cm long.

Range

The *range* of a particle is the maximum distance it will travel in a given medium. Ranges may be determined from absorption curves that are derived by counting a source repeatedly with different thicknesses of the absorber between the source and detector. Figure 5-2 shows an absorption curve for monoenergetic α-particles with air as the absorber. The plot is flat for most of the curve because an α-source emits particles with discrete energies, hence all particles will penetrate approximately the same absorber thicknesses. Therefore, one might expect the curve to become perfectly vertical when the range is reached. Although the curve drops rapidly, it has a measurable slope, which indicates that there is slight variation in the ranges of individual particles. This is due to collision probability factors and scattering (particles having their paths deflected somewhat). Because the range is not definite, two terms are used to report data. The *mean range* ($\bar{R}$) is the thickness required to stop 50% of the particles, and the **extrapolated range** (R_{ex}), determined by extending the curve to the range axis, reflects the thickness required to stop all particles. Mean ranges are reported most commonly.

In air, approximate ranges of α-particles may be calculated using Equa-

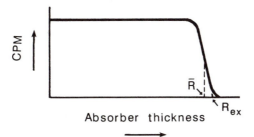

FIG. 5-2 Range curve for a source emitting monoenergetic α-particles in air.

tions 5-1 or 5-2. In these equations, R is the range in centimeters and E is energy in MeV.

$$R = 0.56\,E \qquad \text{for } E < 4\text{ MeV} \qquad (5\text{-}1)$$

$$R = 1.24\,E - 2.62 \qquad \text{for } 4 < E < 8\text{ MeV} \qquad (5\text{-}2)$$

Absorber Thicknesses

At this point, ranges and absorber thicknesses have been expressed as *linear thicknesses* (X_ℓ), but the stopping power of an absorber is more closely related to the mass of material a particle must pass through rather than a linear distance. As shown in Table 5-1, the linear thickness range of 7.0 MeV α-particles is about 60 mm in air, but it is only 74 μm in water and 2 μm in lead. However, the masses that such thicknesses represent do not show nearly as wide a variation. For that reason, *mass thickness* (X_m) is frequently used to express ranges and absorber thicknesses. Mass thickness is the product of the linear thickness and density of the absorber (ρ), as shown by Equation 5-3. Appropriate units for mass thickness are mg/cm² or g/cm².

$$X_m = X_\ell \cdot \rho \qquad (5\text{-}3)$$

TABLE 5-1 Ranges of 7.0 MeV α-Particles in Various Absorbers

| | Range | |
Absorber	(mm)	(mg/cm²)
Air	60.6	7.4
Water	0.074	7.4
Aluminum	0.034	9.2
Copper	0.014	12.5
Lead	0.002	2.3

NEGATRONS

High-speed electrons, including β^--particles, cause ionization and excitation by processes very similar to those described for α-particles. Additional processes called bremsstrahlung and Cerenkov radiation are important types of interactions for energetic negatrons.

Excitation and Ionization

The charge on negatrons is opposite that on α-particles; therefore repulsive rather than attractive forces are involved in the displacement of electrons of the absorber. Otherwise, the processes at the atomic level are very similar. In fact, the amount of energy required to produce an ion pair in air is essentially the same for negatrons as it is for α-particles, 34 eV. However, more secondary electrons, sometimes called *delta rays* (δ-rays), are produced by incident electrons than α-particles. About 70–80% of the total ionization is due to secondary electrons. Other features, such as particle paths, specific ionization, and range, are different, primarily because of the large differences in mass and velocity between the two types of particles.

Due to conservation of momentum, quantitative differences in energy transfers occur in collisions involving particles of equal mass (two electrons) compared to particles of greatly different masses (an α-particle is about 7000 times heavier than an electron). Collisions between α-particles and electrons do not cause appreciable deflections in the α-particle's path, but a collision of a primary electron with an absorber electron, which is also in motion, may change the directions of both significantly. Thus, a primary electron (incident electron) will change its direction somewhat with each interaction, and the extent of the change is dependent on the angle of incidence and particle velocities. Interactions with nuclei generally cause very large deflections of electrons or nuclear absorption.

Because electrons have equal masses, some of their collisions result in the secondary electrons having more energy than the residual energy of the primary electron. Of course, an energetic secondary electron would dissipate its energy in a manner indistinguishable from that of a primary electron. These effects cause ionization tracks of β^- particles to be very tenuous with many direction changes and some branching due to the production of energetic secondary electrons or δ-rays. The irregular branching path and production of energetic secondary electrons is referred to as *scattering*. In some interactions an electron, either primary or secondary, may be defected backward and this phenomenon is called *backscattering*.

For equivalent energies, electrons have much higher velocities than α-particles. The respective velocities of 1.0 MeV α-particles and electrons are 2.3 and 94.1% of the speed of light. Consequently, the distance between interactions is much greater for electrons than α-particles, which means that the specific ionization is lower for electrons, and electrons have a greater range in a given absorber.

Specific Ionization

The specific ionization of electrons in air is dependent on the electron energies (velocities), but is fairly constant over a broad range at higher energies. For example, the specific ionizations of 100 MeV and 50 keV electrons are 5.4 and 4.1 ion pairs/mm, respectively. However, specific ionization increases when electrons are slowed to the eV range of energies, and maximum specific ionization occurs at about 146 eV (velocity, 7.2×10^8 cm/s), but then it is only 485 ion pairs/mm. For comparison, the maximum for α-particles is about 7000 ion pairs/mm and occurs at a particle energy of 370 keV (velocity, 4.2×10^8 cm/s).

Ionization stops when a particle's energy is reduced to about 12.5 eV, which is the ionization potential of oxygen.

Bremsstrahlung

Bremsstrahlung is a German word for "braking radiation" and refers to the electromagnetic radiation emitted by a high-velocity charged particle when it penetrates a dense absorber. At very high velocities, interactions with absorber electrons (ionization and excitation) are not favored; instead the incident particle winds through the coulombic maze of the absorber. Charged particles, such as electrons, emit radiation when subjected to acceleration or deceleration. Since velocity is a vector quantity, a change in direction represents acceleration or deceleration. The wavelength of the radiation emitted by electrons is characteristic of X-rays, and in fact the radiation of X-ray machines is produced by bombarding a heavy metal (tungsten) absorber with a beam of electrons.

Energy loss by bremsstrahlung is dependent on the velocity of the electron and the atomic number of the absorber. Above 1000 MeV, bremsstrahlung is almost exclusively the mode of energy loss, but ionization becomes the predominate mode for energies of about 1.0 MeV and below. The energies encountered with most β⁻-emitters are not high enough for bremsstrahlung to occur appreciably in low mass absorbers such as air, plastic, and aluminum. However in dense absorbers, hard β⁻ emitters ($E_{max} > 1.0$ MeV) will produce bremsstrahlung. As examples, ^{32}P ($E_{max} = 1.71$ MeV) causes bremsstrahlung in lead, but not in plastic, and neither ^{14}C ($E_{max} = 0.156$ MeV) nor ^{3}H ($E_{max} = 0.0186$ MeV) produces appreciable bremsstrahlung in any type of absorber.

In practice, ^{32}P is usually stored in plastic-lined lead containers. The thickness of the lead (a few millimeters) is more than sufficient to absorb

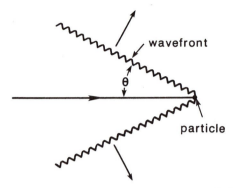

FIG. 5-3 Schematic diagram for the generation of Cerenkov radiation.

all β^--particles, but X-rays are capable of penetrating much greater thicknesses. Therefore, to prevent the production of X-rays by the bremsstrahlung process, a plastic liner is used to reduce the velocities of the more energetic β^--particles so that when they enter the lead, bremsstrahlung will not occur.

Cerenkov Radiation

Although particles cannot have velocities that exceed the velocity of light in a vacuum (c), their velocities may exceed the velocity of light in a transparent medium, and if that condition exists, *Cerenkov radiation* is emitted. At high velocities particles create a path of polarized molecules or atoms in a medium, and when the particle's velocity exceeds the velocity of light in that medium, an electromagnetic shock wave is generated, which is analogous to the sonic shock wave created when an airplane flies faster than the speed of sound. Cerenkov radiation has a directional attribute because the shock wave describes a cone about the particle's path as indicated in Figure 5-3, and the half-conical angle, θ, is related to the index of refraction (n) of the medium and the particle's velocity (v) according to Equation 5-4.

$$\cos \theta = \frac{c}{nv} \tag{5-4}$$

If c_m is the velocity of light in a transparent medium, the index of refraction for the medium is c/c_m and

$$\cos \theta = \frac{c_m}{v} \tag{5-5}$$

Obviously, c_m must be less than v for a meaningful relationship, which restates the original assertion that the velocity of the particle exceeds that of light in the medium.

A continuous spectrum of electromagnetic radiation is emitted, but it is concentrated in the ultraviolet region. The spectrum extends down into the shorter wavelength portion of the visible region, and this is responsible for the eerie blue-white glow observed in the water of water-cooled reactors. Cerenkov radiation production is not limited to negatrons; other charged particles may dissipate energy similarly, but because relativistic velocities are involved, the phenomenon is observed more commonly with lighter particles such as negatrons and positrons. Relativistic equations for energy calculations must be used for obvious reasons.

Utilization of Cerenkov radiation for assay of strong β-emitters will be discussed in Chapter 10.

Range and Absorption

The ranges of β^--particles are not nearly as distinct as those of α-particles because β^--emitters exhibit continuous, rather than discrete energy spectra. Also, scattering is much more pronounced with light particles (electrons) than with α-particles.

An absorption curve for a particular β^--emitter approximates a straight line when the logarithm of the counting rate (a measure of particles not absorbed) is plotted against absorber thickness as shown in Figure 5-4. The range is determined from such an absorption curve by extrapolation of the curve to the background counting level. Absorption curves for hard β^--emitters ($E_{max} > 1.0$ MeV) and dense absorbers will "tail out" above background counting rates beyond the true range of the β^--particles. This is due to bremsstrahlung. The thickness of the absorber is great enough to stop the β^--particles, but not thick enough to stop the X-rays produced when some of the more energetic particles undergo bremsstrahlung.

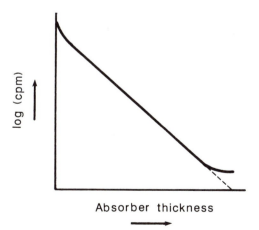

FIG. 5-4   Idealized absorption curve for a β-emitter with a metal absorber.

Although no theoretical basis has been derived, the ranges of β^--particles are pseudoexponential functions of their E_{max} values. Equations that fit experimental data reasonably well have been developed to provide the mathematical relationships shown below. In Equations 5-6 and 5-7, the ranges are in units of mg/cm^2 and E_{max} is in MeV.

for $0.01 < E_{max} < 2.5$ MeV, $R = 412\ E_{max}^{(1.265\ -\ 0.0954\ \ln\ E_{max})}$ (5-6)

for $E_{max} > 2.5$ Mev, $R = 530\ E_{max} - 106$ (5-7)

Self-absorption

Because β^--particles, particularly the low energy ones, are readily absorbed by thin absorbers, a solid sample is subject to self-absorption. Upper layers of a thick source will absorb a portion of the β^--particles emitted from the lower layers. For this reason, solid samples are rarely used for quantitative measurements because corrections for self-absorption would be required.

POSITRONS

Although positrons and negatrons have opposite charges, the kinds of interactions they undergo, ionization, excitation, bremsstrahlung, and Cerenkov radiation, are essentially the same except when the particles come to rest.

Annihilation

Because positrons and negatrons represent antimatter, they will annihilate each other, and this occurs after a positron loses most of its kinetic energy. Ordinary matter has an abundance of negatrons with which a positron may combine so positrons at rest do not exist very long.

In the usual annihilation process, the combined masses of the particles are converted to two photons of electromagnetic radiation (X-rays) that are emitted at an angle of 180° from one another. The energies of the photons are equal and easily calculated since the energy of one photon should be equivalent to the rest mass of one electron.

$E = (5.4858 \times 10^{-4}$ amu$) (931.5$ MeV/amu$)$

$E = 0.511$ MeV

The emission of 0.511 MeV X-rays from matter is a reliable signal that positrons are being annihilated.

A much rarer mode of annihilation involves the emission of three photons, and this occurs when the electron and positron have parallel spins.

γ- AND X-RAYS

As a class, γ-rays penetrate much greater thicknesses of absorbers than either α- or β-particles, and the kinds of interactions they undergo are much different than those of the charged particles. Instead of losing energy in many steps, a photon is consumed in one interaction. All of the interaction processes to be discussed yield electrons with appreciable energies, and in one type of interaction, positrons are produced. The energies of these secondary particles are absorbed through ionization and excitation, which accounts for the ionizing property of γ-radiation. Although quantitative differences exist due to differences in quantum energies, X-rays interact by the same processes as γ-rays.

For photon energies generally encountered in radioisotopic work, the absorption of γ- and X-rays occurs by three primary processes.

Photoelectric Effect

The *photoelectric effect* may be visualized as the absorption of a γ-ray by an orbital electron in an atom or molecule (Figure 5-5). As a result, the electron is expelled with a kinetic energy equal to the energy of the γ-ray minus the binding energy of the electron. Electrons arising by this kind of interaction are called *photoelectrons,* have discrete energies, and dissipate their kinetic energies by ionization and excitation primarily. If an orbital vacancy is created by the loss of the photoelectron, the vacancy is filled by an electron from a higher orbital, hence a low energy photon (X-ray) is emitted, which in turn is likely to interact by the photoelectric process in the same or a neighboring atom. A majority of the photoelectric interactions take place with electrons in the first energy level (K shell). Therefore, the rearrangement of electrons within the interacting atom is similar to the process that occurs after electron capture, which was discussed in Chapter 3.

The photoelectric effect is more prevalent in absorbers of high atomic

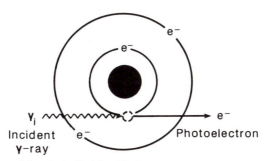

FIG. 5-5 Diagram of the photoelectric effect.

number and with γ-rays of low energy. For photons with energies above 1.0 MeV, the photoelectric effect is not a quantitatively important mode of absorption. Although other processes predominate, it still occurs to a limited extent with γ-rays of much higher energy. This is fortunate because the photoelectric effect yields electrons with discrete energy levels, and this property is used in spectrometers to measure γ-ray energies and hence identify γ-emitting radionuclides.

Compton Effect

In the *Compton effect,* a γ-ray interacts with an electron, but unlike the photoelectric effect, only a portion of the γ-ray's energy is absorbed by the electron and a secondary (or degraded) γ-ray is produced. This process is shown diagrammatically in Figure 5-6. Electrons ejected from absorber atoms by the Compton effect are called *Compton electrons.* The proportion of the primary γ-ray's energy absorbed by the Compton electron is variable and depends on the angle (θ) between the directions of the incident and degraded rays. Because θ may vary between 0 and 180°, the degraded rays may be referred to as scattered rays and the process as *Compton scattering.*

Equation 5-8 shows that the ratio of energies (in MeV) for the scattered (E_s) and incident (E_i) rays is related to the angle θ and the rest mass of an electron (0.511 MeV).

$$\frac{E_s}{E_i} = \frac{0.511}{E_i(1 - \cos \theta) + 0.511} \qquad (5\text{-}8)$$

When $\theta = 0$, the situation represents no interaction ($E_s = E_i$), but as θ increases, E_s decreases to a minimum at $\theta = 180°$. Thus, the kinetic ener-

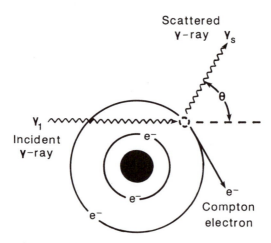

FIG. 5-6 Diagram of the Compton effect.

gies of Compton electrons from a continuum between near zero to some maximum value that corresponds to the condition when a γ-ray is scattered backward. Note that E_s cannot become zero, therefore the energy of the electron cannot equal E_i, that is, the electron cannot absorb all of the energy of the incident ray. Obviously, the photoelectric effect is not a limiting case of the Compton effect; it must constitute a different type of interaction.

The energies of the Compton electrons are dissipated in the absorber by ionization and excitation, and some of the scattered rays may undergo γ-ray interactions with other absorber molecules depending on their points of origins and directions and the thickness of the absorber.

Compton scattering is a predominate mode of interaction for γ-rays with energies in the range of about 0.5–4.0 MeV.

Pair Production

In *pair production,* a γ-ray interacts with a nucleus (probably not a direct collision) and causes the nucleus to emit a negatron–positron pair (Figure 5-7). Since the rest mass of an electron is 0.511 MeV, the creation of a pair requires a minimum energy of 1.022 MeV. Therefore, this mode of interaction does not occur with γ-rays having energies below 1.022 MeV. The extent of pair production is proportional to the square of the absorber's atomic number and becomes a prevalent mode of interaction for γ-rays with energies above about 4 MeV.

Because of conservation of momentum and energy during the interaction, the nucleus receives a small amount of recoil energy, but the quantity is usually insignificant when the dissipation of energy is considered. Practically, the combined kinetic energies of the negatron and positron are equal to the energy of the γ-ray minus 1.022 MeV. Dissipation of the kinetic energies of the negatron and positron is by ionization, excitation, and perhaps

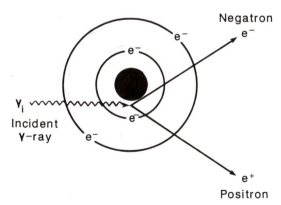

FIG. 5-7 Diagram of pair production.

bremsstrahlung to some extent. When the positron loses its kinetic energy, annihilation occurs. Thus, pair production causes 0.511 MeV photons to be produced in an absorber.

Absorption of γ- and X-Rays

Mathematically, the absorption of both γ- and X-rays is similar to that for other types of electromagnetic radiation in that Lambert's law is followed. Figure 5-8 depicts a beam of photons striking an absorber perpendicularly. The linear thickness of the absorber is X, which is divided into a series of infinitesimally small thicknesses dX. Let I be the intensity of the beam at any point and $-dI$ represent the absorption (reduction in beam intensity) when the beam passes through a thickness of dX. As shown by Equation 5-9, the rate of absorption with respect to distance $(-dI/dX)$ is directly proportional to the intensity of the beam. The proportionality constant (μ_ℓ) is called the linear absorption coefficient.

$$\frac{-dI}{dX} = \mu_\ell I \qquad (5\text{-}9)$$

Although different physical phenomena and different symbols are presented here, the form of Equation 5-9 is identical to the differential Equation, 4-2, presented for radioactive decay. Therefore, the following mathematical operations are similar to those that are given in more detail in Chapter 4.

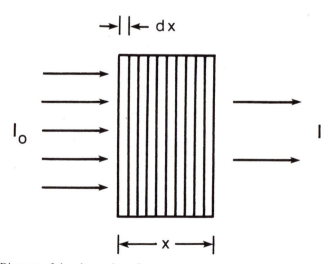

FIG. 5-8   Diagram of the absorption of γ-rays.

Equation 5-9 may be rearranged and integrated.

$$\int \frac{dI}{I} = \int - \mu_\ell dX$$

$$\ln I = - \mu_\ell X + C$$

To evaluate the integration constant C, let I_0 (a constant) be the intensity of the beam before striking the absorber, that is, when $X = 0$.

$$\ln I_0 = - \mu_\ell (0) + C$$

$$C = \ln I_0$$

Substitution of C in the equation yields a natural (base e) logarithmic form, Equation 5-10. Also, X_ℓ has been substituted for X to clearly show that thickness is expressed in linear units.

$$\ln I = \ln I_0 - \mu_\ell X_\ell \tag{5-10}$$

The common (base 10) logarithmic and exponential forms of Equation 5-10 are presented as Equations 5-11 and 5-12, respectively.

$$\log I = \log I_0 - 0.4342 \, \mu_\ell X_\ell \tag{5-11}$$

$$I = I_0 e^{-\mu_\ell X_\ell} \tag{5-12}$$

From Equations 5-10 and 5-11, it is apparent that a plot of $\ln I$ (or $\log I$) versus absorber thickness (X_ℓ) yields a straight line with a slope of $-\mu_\ell$ (or $-0.4342 \, \mu_\ell$) and an intercept of $\ln I_0$ (or $\log I_0$). Experimentally, data suitable for such plots may be obtained by counting a γ-source repeatedly with different thicknesses of absorbers between the source and detector.

Half-Thickness and Absorption Coefficients. *Half thickness*, symbolized as $X_{1/2}$, is defined as the thickness of an absorber of a given composition that reduces the intensity of a beam of radiation to one-half of its incident intensity. Although linear absorption coefficients for absorbers are readily determined, absorption characteristics are frequently reported in terms of linear half-thickness ($X_{\ell 1/2}$). The relationship between linear absorption coefficients and linear half-thicknesses is indentical to the relationship between decay constants and half-lives.

$$\ln \left(\frac{1}{2} I_0 \right) = \ln I_0 - \mu_\ell X_{\ell 1/2}$$

$$\ln 2 = \mu_\ell X_{\ell 1/2}$$

$$\mu_\ell = \frac{\ln 2}{X_{\ell 1/2}} = \frac{0.693}{X_{\ell 1/2}} \tag{5-13}$$

$$X_{\ell 1/2} = \frac{\ln 2}{\mu_\ell} = \frac{0.693}{\mu_\ell} \tag{5-14}$$

TABLE 5-2   Linear Half-Thicknesses and Absorption Coefficients for Different Absorbers and Different Energy γ-Rays

γ-ray energy (MeV)	Absorber			
	Water	Aluminum	Iron	Lead
		$X_{\ell_{1/2}}$ (cm)		
1.0	9.76	4.13	1.57	0.88
1.5	12.16	5.10	1.73	1.17
2.0	13.86	5.92	2.10	1.38
		μ_ℓ(cm^{-1})		
1.0	0.071	0.168	0.441	0.790
1.5	0.057	0.136	0.400	0.592
2.0	0.050	0.117	0.330	0.504

As shown by Equation 5-13, the units for linear absorption coefficients are reciprocals of the half-thickness units.

An absorption coefficient or its corresponding half-thickness is a characteristic of (1) the composition of the absorber and (2) the energy or wavelength of the radiation. Table 5-2 gives linear half-thicknesses and absorption coefficients for some absorbers with different energy γ-rays. From these data it may be concluded that the absorption of γ-rays decreases with increasing γ-ray energies and increases with absorbers of increasing density.

Example Problem 5-1.   For a source of 1.5 MeV γ-rays, what percentage of the rays would be absorbed by a lead shield 3.0 cm thick?

Using Equation 5-12,

$$I = I_0\, e^{-\mu_\ell X_\ell} \qquad\qquad \mu_\ell = 0.592 \text{ cm}^{-1}$$

$$\frac{I}{I_0} = e^{-\mu_\ell X_\ell} \qquad\qquad X_\ell = 3.0 \text{ cm}$$

$$I/I_0 = \text{fraction transmitted}$$

$$\frac{I}{I_0} = e^{-0.592 \text{ cm}^{-1}(3.0 \text{ cm})} \quad = 0.169$$

Since 16.9% of the rays are transmitted, the amount absorbed by the 3.0 cm thickness of lead would be 83.1%.

As discussed earlier in this chapter, thicknesses may be expressed in units other than linear ones. This also applies to half-thicknesses, which in turn affects values and units of absorption coefficients. According to Equation 5-3, a linear thickness (X_ℓ) is converted to a mass thickness (X_m) by multiplying the linear thickness by the density of the absorber (ρ), and half-thicknesses are treated likewise.

$$X_{m\frac{1}{2}} = X_{\ell\frac{1}{2}} (\rho)$$

Let μ_m be the mass absorption coefficient. It is related to the mass half-thickness in the same way linear values are.

$$\mu_m = \frac{\ln 2}{X_{m\frac{1}{2}}}$$

The units of μ_m are reciprocals of X_m units, (cm^2/g) or (cm^2/mg), and linear absorption coefficients can be converted to mass coefficients by dividing the linear coefficient by the absorber's density.

$$\mu_m = \frac{\mu_\ell}{\rho}$$

The mass coefficient, used with mass thickness, can replace the linear coefficient and linear thickness in Equations 5-10, 5-11, or 5-12 without changes because

$$\mu_m X_m = \frac{\mu_\ell}{\rho} \cdot (X_\ell)(\rho)$$

As may be seen in Table 5-3, the mass absorption coefficients for the various absorbers do not vary nearly as much as the linear absorption coefficients. However, for γ-rays of a given energy, the mass absorption coefficients are not identical, which indicates that the absorption of γ-rays is dependent on factors other than the quantity of mass they pass through.

Since two of the three major interactions of γ-rays involve interactions with electrons, it seems reasonable to imagine that the absorption of γ-rays may be more closely related to the density of electrons in the absorber than its mass density. The moles of electrons per cm^3 can be calculated readily from the density of the absorber.

TABLE 5-3 Mass and Electronic Absorption Coefficients for Different Absorbers and Different Energy γ-Rays

γ-ray energy (MeV)	Absorber			
	Water	Aluminum	Iron	Lead
		μ_m (cm^2/g)		
1.0	0.071	0.062	0.062	0.070
1.5	0.057	0.050	0.056	0.052
2.0	0.050	0.043	0.046	0.046
		μ_e (cm^2/mol electrons)		
1.0	0.095	0.129	0.133	0.177
1.5	0.076	0.104	0.120	0.131
2.0	0.067	0.089	0.099	0.116

$$\text{mol } e^-/\text{cm}^3 = \frac{\rho Z}{\text{At. wt.}}$$

Multiplication of the linear thickness by the electron density yield an *electronic thickness* (X_e).

$$X_e = \frac{X_\ell \rho Z}{\text{At. Wt.}}$$

Division of μ_ℓ by the electron density yields the *electronic absorption coefficient* (μ_e).

$$\mu_e = \frac{\mu_\ell(\text{At. Wt.})}{\rho Z}$$

As shown for mass coefficients, electronic coefficients and thickness may replace linear ones in Equations 5-10, 5-11, and 5-12 because units cancel.

Table 5-3 shows some electronic absorption coefficients for several absorbers. For a given energy of γ-ray, the electronic absorption coefficients may be somewhat closer in value than the mass coefficients among the different types of absorbers, but they certainly are not constant. Because of the lack of constancy for different absorbers, electronic absorption coefficients do not offer a significant advantage over mass absorption coefficients. In calculations, mass absorption coefficients are used more frequently than linear coefficients, but both are encountered commonly.

Example Problem 5-2.   What thickness of lead would be required to reduce the intensity of a source of 2.0 MeV γ-rays to 0.1%? Using Equation 5-10,

$$\ln I = \ln I_0 - \mu_m X_m \qquad \mu_m = 0.046 \text{ cm}^2/\text{g}$$

$$\ln \frac{I}{I_0} = -\mu_m X_m \qquad \frac{I}{I_0} = 0.001$$

$$X_m = \frac{\ln I/I_0}{-\mu_m} \qquad X_m = ?$$

$$X_m = \frac{\ln(0.001)}{-0.046 \text{ cm}^2/\text{g}}$$

$$X_m = 150.17 \text{ g/cm}^2$$

Since the density of lead is 11.3437 g/cm³, the linear thickness of the lead is

$$X_\ell = \frac{X_m}{\rho} = \frac{150.17 \text{ g/cm}^2}{11.3437 \text{ g/cm}^3} = 13.24 \text{ cm}$$

Partial Absorption Coefficients.   As indicated earlier, the relative proportion of γ-rays that interacts by each of the three processes is dependent on photon energies and the nature of the absorber. Some generalities were stated, but a better understanding of the relative contributions of the three processes to the absorption of γ-rays may be gained by considering data shown in Figure 5-9. If μ is the linear absorption coefficient for a given absorber and given energy of γ-rays, then the fraction of the rays that interacts by a given process may be multiplied by μ to yield a partial absorption coefficient. Let f_p be the fraction of photons that interact by the photoelectric effect, the partial absorption coefficient for the photoelectric process (μ_p) is

$$\mu_p = f_p \mu$$

Similar partial absorption coefficients could be calculated for the Compton effect (μ_c) and for pair production (μ_{pp}). The sum of the partial coefficients would equal the total.

$$\mu = \mu_p + \mu_c + \mu_{pp}$$

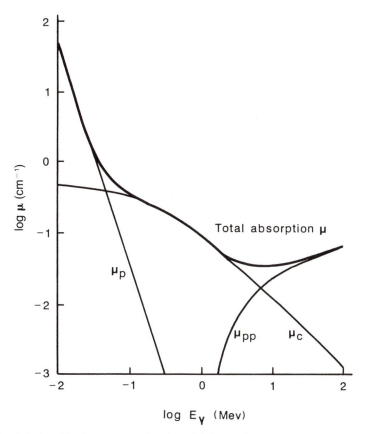

FIG. 5-9 Relationships between total and partial absorption coefficients and energy of γ-rays in an aluminum absorber.

Figure 5-9 shows how the absorption coefficients change with respect to energy when aluminum is the absorber. Although different absorbers show quantitative differences, the general shapes of the curves are representative. As may be seen, pair production does not occur until γ-ray energies exceed 1.022 MeV and predominates only at fairly high energies. The Compton effect is most prevalent for γ-rays with intermediate energies and the photoelectric effect occurs mostly with low-energy γ-rays.

PROBLEMS

1. What are the approximate ranges of α-particles in air when their initial kinetic energies are (a) 1.5 MeV; (b) 6.0 MeV?

2. In units of mg/cm^2, calculate the approximate ranges of the β-particles from ^{32}P, ^{14}C, and ^3H.

3. Using the answers to problem 2, calculate the β-particles' ranges in air for those three radionuclides. Assume the density of air is 1.2929×10^{-6} kg/cm^3.

4. Assuming the density of glass to be about 2.2 g/cm^3, determine if a test tube wall with a thickness of 1.0 mm would prevent the passage of β-particles from the three radionuclides specified in problem 2.

5. Determine the fraction of 2.0 MeV γ-rays that would be transmitted by a 30.0 cm thickness of water.

6. What thickness of iron would be required to attenuate a beam of 1.0 MeV γ-rays so that only 5% were transmitted?

7. If a 5 mm thickness of aluminum (density = 2.6989 g/cm^3) transmits 29.69% of the incident X-rays that have a quantum energy of 30 keV, what is the mass absorption coefficient for 30 keV X-rays with aluminum as an absorber?

8. NaI crystals are used in detectors for γ-ray spectrometers, and the counting efficiency of such instruments is dependent on the fraction of the source's rays that is absorbed by its crystal. If the mass absorption coefficient for 200 keV γ-rays is 0.3 cm^2/g in NaI and the density of NaI is 3.667 g/cm^3, what fraction of 200 keV rays would be absorbed by a 1.0 in. thickness of NaI?

9. For 2.3 MeV γ-rays and NaI, the mass absorption coefficient is about 0.04 cm^2/g. Calculate the fraction of such rays absorbed by a 1.0 in. thickness of NaI.

10. On the basis of appropriate absorption equations and problems 8 and 9, predict how the counting efficiency of a γ-ray counter with an NaI crystal would be affected by (a) increasing the quantum energies of the γ-rays; (b) increasing the size of the NaI crystal.

REFERENCES

Frielander, G., Kennedy, J. W., Macias, E. S., and Miller, J. M. (1981). *Nuclear and Radiochemistry,* 3rd ed. Wiley, New York.

Heckman, H. H., and Starring, P. W. (1963). *Nuclear Physics and the Fundamental Particles.* Holt, Rinehart and Winston, New York.

PART II

Detection and Measurement of Radioactivity

The measurement of radioactivity is dependent on the interaction of radiation with a device that responds by producing a signal that can be recorded in some manner. The responsive device is called a *detector,* and in most instruments, it creates an electronic signal, a voltage pulse, that may be amplified, evaluated with respect to pulse height, and finally recorded as a count in an electronic device, usually a *scaler* or computer. In autoradiographic methods, the detector is a photographic emulsion, and interactions are recorded by the film. In this part, the theory of operation and the practical uses of various types of detectors will be discussed.

CHAPTER 6

Detection of Radioactivity by Gas Ionization Methods

All radioactive decay processes yield ionizing radiation; α-particles, negatrons, and positrons create ions by direct interactions, whereas γ- and X-rays interact by processes that yield electrons and positrons that cause ionization. Instruments to be discussed in this chapter detect the creation of ions in gases.

In air, about 34 eV is required to produce an ion pair (M^+ and e^-); however, the amount of energy required varies for different gases. On the average, 26.2 eV per ion pair is required for argon, 29.2 eV for methane, and 44.4 eV for helium, regardless of the type of ionizing radiation. Specific ionization does vary with the type of radiation, type of gas, and gas pressure. In air at atmospheric pressure, the specific ionization values of α-, β-, and γ-radiation are in the order of 10^4–10^5, 10^2–10^3, and 1–10 ion pairs/cm, respectively. These factors have an important bearing on gas ionization detector design and operating characteristics.

IONIZATION CHAMBER DETECTORS

Ionization chambers, or more simply *ion chambers,* are not used extensively in biological and medical research for quantitative measurements of radioactivity, but are used for health physics applications.

Lauritsen Electroscope

The *Lauritsen electroscope* is a refined version of one of the first radioactivity measuring devices, the gold leaf electroscope. Although neither instrument is used today to assay radioactive samples, some types of pocket

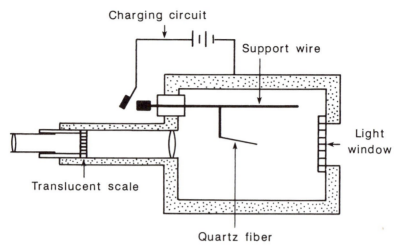

FIG. 6-1 Diagram of the Lauritsen electroscope.

dosimeters in current usage utilize a basic design nearly identical to the Lauritsen electroscope (Figure 6-1).

The chamber of the electroscope is a metal cylinder, originally made of brass, that has a support wire assembly (insulated from the cylinder) that extends into the chamber. Attached to the support wire, with electrical continuity between, is a flexible, gold-plated quartz fiber. The electroscope is charged by closing the switch to the battery momentarily, thus creating a potential difference between the chamber body and the wire assembly. Because of like charges, the quartz fiber is repelled from the support wire, and the extent of deflection is dependent on the charge placed on the wire assembly. When ionization occurs inside the chamber, the electrons migrate to the wire assembly thereby reducing its charge, and hence decreasing the fiber's deflection. Concomitantly, positive ions are attracted to the negative electrode, the chamber body, where they are neutralized by electrons.

The optical portion of the electroscope causes the fiber's image to be projected on a translucent disc that has a scale imprinted on it, which permits the fiber's deflection to be measured. Since a decrease in the fiber's repulsion is related to the number of ions created inside the chamber, calibration permits approximate disintegration rates to be determined when measurements are made with respect to time. Since radiation dose (see Chapter 15) is dependent on the amount of energy dissipated in an absorber, which is proportional to the extent of ionization, devices used for health physics applications, such as the pocket dosimeter, may be calibrated to read dose units (rad, rem, Gy) directly.

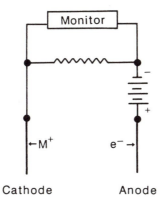

FIG. 6-2 Simplified circuit for an ionization chamber.

Ion Chambers with Electronic Monitors

To control gas composition and pressure, ion chambers are generally gas-tight cylinders or spheres. Frequently, the chamber's body constitutes one electrode and the second is a centrally located wire or rod.

A simplified circuit diagram for an ion chamber is shown in Figure 6-2. The two electrodes, insulated from one another, are charged by DC current. As ions are created, they migrate to their respective electrodes where charges are neutralized. Current will flow from the DC power supply to restore the charge on the electcrodes, but it must pass through a high ohm resistor. The IR drop (voltage) across the resistor is measured, which reflects the amount of current carried by the ions in the chamber, hence the number of ions formed.

The voltage applied to the electrodes has an important bearing on the collection of ions as shown by either curve in Figure 6-3. With no voltage,

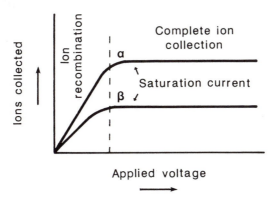

FIG. 6-3 The effects of applied voltage on the collection of ions in an ion chamber.

the electrons and positive ions will not be separated so recombination occurs after the electrons have lost their kinetic energies. At a low voltage, the charged particles are accelerated to a low velocity, but at low velocities recombination may still occur as the particles slowly migrate to their respective electrodes. Higher voltages induce higher velocities, hence less recombination occurs, and at some point the voltage is sufficient for complete ion collection.

The current due to complete collection is called the *saturation current,* and it will remain constant as the voltage is increased further, perhaps for another 100–200 V. The geometry of the chamber's electrodes and the type and pressure of the gas will affect the exact voltage required to reach the saturation current, but it is usually in the range of 50–200 V.

The heights of the plateaus in Figure 6-3 reflect the magnitude of the saturation currents for particles of different energies. The lower curve might represent the response due to 0.5 MeV β^--particles while the upper curve may be due to 6.0 MeV α-particles.

If ion chambers are used for assaying radioactive samples, the samples are generally in, or converted to, a gaseous form. For example, ^{14}C may be assayed as $^{14}CO_2$ or $^{14}CH_4$. This may be acceptable for certain specialized studies, but producing, transferring, and controlling compositions of gaseous samples require vacuum line techniques and considerable time.

Gaseous samples in ion chambers do offer 4π *detection geometry.* The fronts of radiation emanating from a source describe a sphere, and if all portions of the sphere lie in the responsive region of a detector, the geometry is said to be 4π. If only one-half of the sphere interacts with a detector 2π detection is afforded.

Of course, radition from sources outside the chamber will be detected if the rays can penetrate the chamber's wall and happen to interact with the internal gas molecules. Thus, chambers with thin windows or walls are capable of detecting external sources of hard β-, or γ- and X-rays, but the efficiency of detection may be low because of geometric factors, and in the case of β^--particles, absorption by the window. Also, since the responsive absorber (a gas) has a very low density, the probability of γ- or X-rays interacting inside the chamber is quite low.

Although α-particles create a sufficient number of ions to produce electrically detectable pulses, other forms of radiation may not. Therefore, individual events (pulses) generally are not counted; instead, the amount of ions collected over a period of time is measured. Thus, the sensitivity of an ion chamber is dependent on the instrument's capability of measuring very small currents or voltages. In general, ion chambers are not as sensitive as other detectors, but do have more capacity, meaning that they remain responsive to higher levels of radiation. Many counters will be "blocked" (stop counting) when the pulse rate exceeds certain levels, and a detector that remains responsive is important for high radiation area monitoring. Although less sensitive, the electronics of ion chamber instruments can be quite simple and rugged, which is desirable for portable devices such as "cutie pies," which are hand-held battery-operated high-capacity ion chambers.

In summary, the characteristics of ion chamber are less than ideal for most sample assays, but are quite useful for measuring radiation fields and doses. Therefore, such instruments are better suited for certain health physics applications.

PROPORTIONAL DETECTORS

Our discussion of the effect of applied voltage on ion collection in the ion chamber section was incomplete because the consequences of raising the voltage above that needed to obtain a saturation current was not considered. The curve given in Figure 6-3 is extended in Figure 6-4, which shows the effect of higher applied voltages. To understand the theoretical basis for such a curve, the electrical phenomena that occur within a proportional tube must be considered.

Gas Amplification

Figure 6-5 shows the general construction details of a tube that might be used for proportional counting. The tube body may be metal or glass with a metallic coating on the inside that constitutes the cathode. A very fine wire, located centrally, serves as the anode, and a window that is thin enough to permit rays to enter, yet impervious to gases, is at one end of the tube. When a particle or ray enters the tube, it may interact with gas molecules creating ion pairs ($M^+ + e^-$).

The accleration force an electron experiences in such a tube is determined by the field strength (V/cm) of its surroundings. The field strength depends on the applied voltage as well as distance from the electrodes, and it increases rapidly as an electrode is approached. Therefore, an electron in the middle of the gas space will start to move slowly, but its velocity will increase greatly as it nears the anode. When an electron achieves a kinetic energy above the ionization potential of the gas (in the order of 20–50 eV) it may create another ion pair if it collides with a gas molecule. Thus, the primary electrons formed in the initial ionizing event may cause secondary ionizataion as they are accelerated toward the anode. This is referred to as *gas amplification*, which simply means that the number of ions collected exceeds the number formed by the primary ionization event.

As indicated in Figure 6-4, the applied voltage range where saturation currents are obtained is called the *ion chamber region*, and the voltage that is just sufficient to cause gas amplification is the beginning of the *propor-*

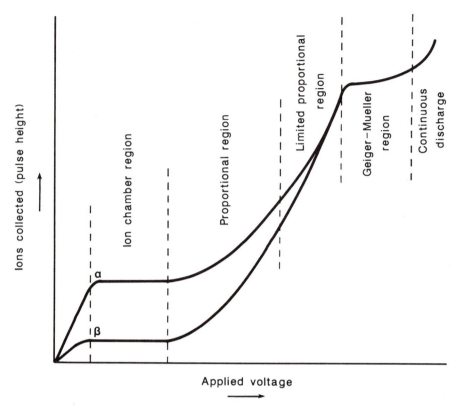

FIG. 6-4 The effects of applied voltage on the collection of ions (pulse height) in gas ionization detectors. (From C. H. Wang, D. L. Willis, and W. D. Loveland, *Radiotracer Methodology in the Biological, Environmental, and Physical Sciences,* © 1975, p. 107. Reprinted by permission of Prentice-Hall, Inc., Englewood Cliffs, NJ)

tional region. The proportional region includes a range of applied voltages, usually in the order of several hundred volts. Because of greater acceleration forces on the electrons, greater gas amplification occurs at higher applied voltages throughout the proportional region.

Pulse Generation

After a primary ionizing event, the newly freed electrons are accelerated toward the anode, and as they approach the anode, their kinetic energies become great enough so that collisions with gas molecules create secondary ions. The secondary electrons, being close to the wire, are accelerated rapidly and cause more ionization and so on. This produces an *avalanche* of electrons that are collected by the anode wire, reducing its charge momentarily.

Because of momentum, the velocities of the heavy positively charged gas ions are much lower than those of the electrons, and since most of the ionization occurs close to the wire anode, the gas ions form a cloud around the anode that migrates slowly to the cathode. At the cathode the ions pick up electrons and reduce the cathodic charge momentarily. The charges on both electrodes are restored quickly by a high voltage supply, but the temporary drop in voltage (neutralization of charge) constitutes an electrical pulse that can be amplified in another circuit and eventually be counted.

The *pulse height,* which is the magnitude of the voltage drop, is proportional to the number of ions collected. Therefore, for a given number of ions formed by the initial ionizing event, pulse heights increase with higher applied voltages because of greater gas amplification (Figure 6-4). For a given applied voltage, hence constant gas amplification, the pulse height will be *proportional* to the number of ions formed by the primary ionizing event. Thus, the pulse height is dependent on the energy of the ionizing particle. This explains why the α- and β-curves in Figure 6-4 remain separate throughout the proportional region. Because particles of different energies can be distinguished on the basis of their pulse heights, proportional counters can operate as *spectrometers,* instruments that measure the intensities of rays (cpm) at specified energy levels.

Dead Time and Resolving Time

After the initial ionizing event, an avalanche of electrons is created and collected at the anode fairly quickly, but the cloud of positive gas ions moves relatively slowly. Since the applied potential on the electrodes will not be restored completely until the positive ions are neutralized, there is a small but finite period when the tube is unresponsive to incoming radiation. This time period is called *dead time.* The term dead time implies that there is a period of time when a detector is totally unresponsive to incoming radiation. However, there are conditions in which the detector will respond, but produce a pulse too weak to be counted. For example, during the time the applied voltage rises following a pulse, the detector will be somewhat responsive to an ionizing event, but the pulse will be weak because gas amplification will be less due to the lack of a fully restored voltage. This period while the applied voltage is being restored is called the recovery time. *Resolving time* includes both the dead time and recovery time and is the time period following a pulse when the detector will not produce a recordable count if a second ionizing event occurs. More simply, resolving time is the period during which two ionizing events cannot be resolved into two counts.

For correcting counting data, resolving time rather than dead time is used; however, in many cases these terms are used synonymously. The consequences of resolving time on practical counting will be discussed in the section on G-M detectors.

Resolving times for proportional detectors vary depending upon tube geometry and other instrument parameters, but they may be as low as 0.2 μs (per count) and as high as 100 μs. One general advantage of proportional counters over G-M counters (to be discussed later) is a shorter resolving time. Usually counting rate corrections do not need to be applied to proportional counting data.

Proportional Detector Construction

There is much variation in construction design among proportional detectors, but end window tubes similar to the one shown in Figure 6-5 are common. The window might be split mica or Mylar plastic with a thickness as low as 150 μg/cm². Many α- and β-particles would not be absorbed by such a thickness.

Some tubes are designed for counting gaseous samples, which means the tube usually is removed from the instrument for gas filling and has no window.

A popular type of detector for proportional counters is a windowless *gas-flow* detector, an example of which is shown in Figure 6-6. The sample

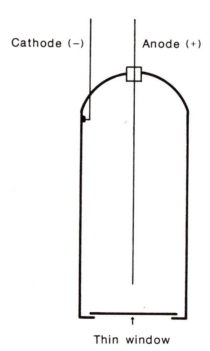

Cathode (−)   Anode (+)

Thin window

FIG. 6-5   Diagram of a gas ionization detector tube, end-window type.

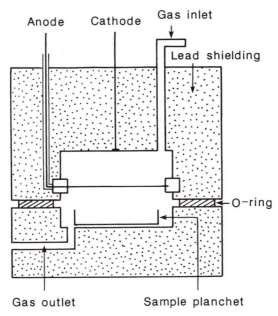

Anode   Cathode   Gas inlet

Lead shielding

O-ring

Gas outlet   Sample planchet

FIG. 6-6   Diagram of a gas-flow ionization detector.

is placed inside the chamber by raising the top part slightly then, usually by a slide mechanism, the lower part is exposed to the operator and the sample planchet is placed in the lower cavity. The lower part is returned to its position directly below the upper part, which is then lowered to give a gas-tight seal between parts. The chamber is flushed with a special gas, frequently referred to as *proportional gas*. Different gasses may be used, but a popular one is P-10 gas, 10% methane in argon. After the atmospheric air is flushed out, the gas flow is reduced to a very low rate and counting can begin. Sample changing and counting are performed automatically by most instruments used routinely. This type of detector offers a significant advantage over end-window detectors because absorption of weak α- or β-particles by the window is eliminated and 2π geometry is achieved.

Instrument Components

A block diagram for a proportional counter is given in Figure 6-7. The high voltage power supply provides the voltage applied to the electrodes of the proportional detector, which may be a sealed tube such as that shown in Figure 6-5. Operator controls permit the voltage to be adjusted over a fairly wide range (1500 V). The stability of the supply is very important for proportional counters because the applied voltage affects gas amplification and hence pulse heights.

The *preamplifier* amplifies the pulses from the detector by a constant

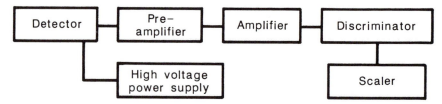

FIG. 6-7 Block diagram of a proportional counter.

factor and sends them on to the amplifer. Usually there are no operator controls for preamplifiers.

The *amplifier* amplifies the pulses further and shapes them as well. Its controls permit an operator to vary the degree of amplification within certain limits.

The *discriminator* may be viewed as an electronic selection device that allows pulses of specified heights to pass through to the scaler while quenching all others. Similar devices may be called pulse height analyzers or, more simply, analyzers. Most discriminators have controls that permit an operator to select the heights of pulses to be transmitted. The lower level control determines the minimum acceptable pulse height, and the upper level determines the maximum. The two levels of control provide a *discriminator window,* which is the range of pulse heights that will pass to the scaler and be counted. Pulses weaker than that specified by the lower level control, and those stronger than that specified by the upper level control will be rejected. Many units will have a means of deactivating the upper level of control, and, when deactivated, all pulses with heights that exceed the lower value will be counted. This condition is referred to as *integral counting.* Some simple instruments may have a discriminator with a lower level control only, and others may rely on the scaler for the lower level of discrimination. However, some degree of lower level discrimination is needed to reduce background counting rates by screening out weak *noise pulses* (spurious pulses that may arise from various electronic components of the instrument).

The *scaler* is simply a counting unit that records one count each time it receives a pulse from the discriminator. An accurate timing device is an integral component of a scaler. In the absence of a discriminator, a scaler does discriminate against very weak pulses because a minimum pulse height is required to trip the electronic mechanism that causes the scaler to advance one count.

Operating Characteristics

In the proportional counter, the heights of pulses are governed by (1) the number of ions formed in the initial ionizing event (energy of the ray), (2) the voltage applied to the tube (degree of gas amplification), and (3) the overall gain provided by the preamplifier and amplifier. Since the discriminator settings determine the heights of pulses to be counted, the operator can count different radionuclides selectively. However, optimum instrumental settings needed for a particular radionuclide usually must be determined experimentally. For this purpose, a standard of the desired radionuclide would be used, and although its exact disintegration rate is not important, it should be high enough so that sufficient counts for statistical accuracy (10,000 counts) can be obtained in a short counting period (30 s). Since there are three instrumental parameters that determine whether a pulse will be recorded as a count, curves are usually prepared by varying one parameter at a time.

Initially, one may set the lower level discriminator to a low value and invoke the integral counting mode. The amplifier control might be set at a low value also, then starting at the lowest voltage setting, the sample is counted repeatedly and the voltage is increased by small increments between counting periods. A plot of counting rate versus voltage should produce a curve with a readily discernible plateau. Such a plot is shown in Figure 6-8a, which represents data from an α-emitter. At the lowest voltages no counts are recorded, but as the voltage is increased the pulses are amplified (via gas amplification) more, and when the pulse heights exceed the value of the lower discriminator, counting begins. The voltage where counting begins is called the *starting voltage,* whereas the voltage at the edge of the plateau is called the *threshold.*

Since α-particles are monoenergetic, the pulse heights for all should be nearly the same, causing a rapid rise in the counting rate as the voltage is increased beyond the starting voltage, but then the rate levels off to form a plateau. For counting a particular α-emitter, a voltage setting in the middle of the plateau should be selected because small fluctuations in the applied voltage would have the least effect on the counting rate (hence counting efficiency).

Suppose the lower level of the discriminator had been set an increment higher, how would have the curve been affected? Greater amplification is needed for the pulse heights to exceed the value of the lower level setting, and this may be accomplished by increasing the applied voltage. As a result, the curve is shifted to the right and expanded slightly.

What if the amplifier had been set one increment higher initially? All pulses, regardless of the voltage, will be amplified more, therefore less gas amplification is required. This shifts the curve to the left and compresses it slightly.

Figure 6-8b shows a similar plot for a β⁻-emitter. Note that the starting voltage appears at a higher voltage than it did in the case of the α-emitter. The difference is due to the difference in energies between the two types of particles. Although the counting rate rises rapidly after the starting voltage is passed, the slope for the β-curve is not as steep as the α-curve. But

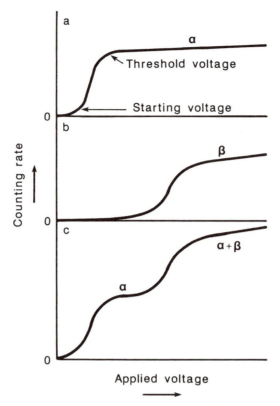

FIG. 6-8 Counting rates of a proportional counter as affected by applied voltage and type of sample.

remember the energy spectra of β^--emitters is continuous, not discrete. When the voltage is high enough that the pulses from all the β^--particles are amplified sufficiently, the counting rate levels off creating a plateau.

Counting plateaus are not perfectly flat, primarily because the energy spectrum of background radiation is continuous. Also some noise pulses may contribute to the slope. Shielding around the detector would decrease background and reduce the slope of the plateau somewhat.

If a mixture of the α- and β^--emitters of Figure 6-8a and b had been used for these operations, a curve similar to the one in Figure 6-8c would have been obtained. This curve shows that the α-radiation could be counted in the presence of β^--radiation by setting the voltage in the middle of the α-plateau. The counting rate due to β^--particles could be determined by raising the voltage to the β^--plateau and subtracting the counting rate at the α-plateau. However, the same could be accomplished more easily by invoking the upper level of the discriminator and adjusting it so the stronger α-pulses were rejected.

Applications

Proportional counters, particularly those with a gas-flow detector, are quite efficient for assaying α- and β-radiation. The energy of an α- or β-particle will be absorbed completely by the detector's gas; therefore, the greatest limitation in counting efficiency is sample geometry. With 2π geometry, counting efficiencies approaching 50% can be achieved. In contrast, γ-rays are not detected efficiently because only a small fraction will interact with the detector's gas. Increasing the volume of a detector improves counting efficiency somewhat, but the low density of the gas cannot be circumvented.

Preparation of samples for counting is relatively simple and inexpensive unless gaseous samples or samples with an appreciable mass are involved. Proportional counters are well suited for assaying environmental samples that may contain α- and β-emitters, and with high grade electronics, the energies of the particles can be determined, which permits the identification of radionuclides. Consequently, the proportional counter can be used for qualitative as well as quantitative purposes.

In the past decade, multiwire proportional detectors have become commercially available, and one of their principal uses is to obtain visual images from distributions of radioactivity in planer specimens, such as electrophoresis gels and thin-layer chromatograms. The so-called gel scanner is an example. A computer is an important component of instruments employing multiwire detectors, and images comparable to those produced by autoradiography are attainable. The principles of operation of such instruments are discussed in Chapter 17.

The Limited Proportional Region

The limited proportion region, shown in Figure 6-4, is a range of voltages where pulse heights rise steeply and converge. This region is not used for counting operations normally because at these voltages, the detector is less stable and offers no compensating advantage.

G-M DETECTORS

The Geiger–Müller (G-M) detector bears the names of the two German physicists who developed it, and it has been used extensively in a variety of detection applications. Although G-M detector construction will be discussed later, tube forms are similar to those described for proportional counters (Figure 6-5).

Pulse Generation

G-M detectors operate with applied voltages that are higher than those of proportional detectors. At the upper voltages of the limited proportional region (Figure 6-4), the curves for α- and β-particles rise steeply and converge. Convergence of the curves indicates that the number of ions collected (pulse height) is independent of the number of ions formed by the initial ionizing event. The G-M region extends over a fairly narrow range of applied voltages (100–300 V) and ends in a region where tubes undergo continuous discharge. Immediately after convergence, the curve in Figure 6-4 has its steepest slope then forms a plateau, called the *G-M plateau,* where the pulse heights remain essentially constant over a short range of applied voltages.

The production of a pulse starts the same way as that described for proportional detectors, but with a higher applied voltage, gas amplification is increased greatly. Electronically induced secondary ionization starts further from the anode and is propagated along the wire's complete length. The intensity of ionization is so great that un-ionized molecules are depleted in the cylindrical volume surrounding the anode, and this stops the ionization process. Thus, the extent of ionization reaches a maximum that depends on the tube's "ion space," the volume of completely ionized gas. For a weakly ionizing β-particle, more gas amplification occurs than for a strongly ionizing α-particle, but in both cases gas amplification continues until the ion space has been used. Consequently, the magnitude of the electron avalanche is dependent on the ion space of a tube rather than the number of ions created initially. Obviously, if gas amplification is extensive enough to use up the ion space, higher applied voltages will not increase the final number of ions formed, and this explains the generation of the G-M plateau in Figure 6-4.

The applied voltages above the G-M region are so high that the tube will undergo continuous discharge. After a pulse, the potential builds up high enough to cause gas amplifications before all ions are neutralized, and one stray ion is sufficient to create another avalanche.

Quenching

The extensive ionization in G-M detectors creates an additional problem. As an ion approaches the cathode it picks up an electron, but the electron usually enters an outer higher energy orbital. That is, after picking up an electron the neutral molecule exists in an electronically excited state $(M^+ + e^- \rightarrow M^*)$.

Several deexcitation processes are possible but nobel gases tend to emit a photon when the electron falls to a lower energy level (ground state). The wavelength of the photon depends on the energy levels involved, but it may be in the ultraviolet or low-energy X-ray portion of the electromagnetic spectrum. Such photons are capable of interacting with other gas molecules by the photoelectric effect creating an ion pair that can precipitate another electron avalanche. Consequently, one pulse can lead to a series of echo pulses unless the process is *quenched;* that is, excited gas molecules are prevented from causing delayed ionization.

Quenching can be done electronically by reducing the electrode potential until the gas molecules are restored to their ground states, but an internal quenching agent is used more commonly.

The gas in internally quenched G-M detectors consists of a nobel gas (He, Ne, Ar) plus a fractional percentage of a quenching agent, which may be either a halogen (Cl_2 or Br_2) or an organic compound, such as ethyl alcohol or butane. Since the ionization potential of a nobel gas is higher than that of the quenching gas, charge transfers occur so that most of the positive ions neutralized at the cathode are those of the quencher. In the case of an organic quencher, molecular deexcitation is likely to occur by disruption of the quencher's bonds, which causes the quenching agent to be used up, so the lifetime of an organically quenched tube is dependent on the number of discharges it undergoes. Obviously, if the applied potential is allowed to become high enough for continuous discharge, an organically quenched tube can be "burned out" quickly.

Diatomic halogen molecules undergo disproportionation (bond breakage) also, but the resulting single atoms lose their energies by thermal processes and recombine readily to regenerate the diatomic form. Because the quenching gas is not irreversibly destroyed, halogen quenched tubes have much longer useful lifetimes than organically quenched tubes. However, the quenching efficiency of halogen tubes is more variable than the organic type. For those reasons, organic quenching agents are preferred when highest counting precision is needed. Halogen quenched tubes are more commonly used in applications that require rugged portable detectors, such as those used in survey meters.

Resolving Time

As in the case of proportional detectors, the avalanche of electrons is collected by the anode quickly, but since ionization is much more extensive in the G-M region, a denser cloud of positive ions is formed around the anode. Migration and subsequent neutralization of a greater mass of ions require more time. Of course, the tube will not yield a count from incoming radiation until the ions are neutralized and the electrode potential is restored. Therefore, the resolving times for G-M detectors are considerably longer than those for proportional detectors and generally are in the range of 100–300 μs, but may be as long as 1 ms. Such periods of detector insensitivity can introduce significant errors in counting data, however, they are easily corrected as will be shown later in this section (Resolving Time Correction).

G-M Detector Construction

The basic construction of G-M detectors is similar to those described for proportional detectors. However, for end-window G-M tubes, the anode wire generally is a bit thicker and has a small glass bead at the end of it for insulation purposes. The field strength around the point of an uninsulated wire would be much greater than around the remainder of the anode. Therefore, the bead helps create a uniform electric field in the center of a cylinder that yields more even gas amplification throughout the tube.

Gas-flow detectors that operate in the G-M region are also used commonly. Although most detectors are designed for use in either the proportional or G-M region some can be used for both applications; however, different gasses are employed. *Geiger gas* consists of 99% helium and about 1% of butane or isobutane and is used extensively with gas flow detectors operating in the G-M region.

Radiochromatogram scanners frequently employ a gas-flow G-M detector. Detector design varies among manufacturers, but one that provides a large window area is shown in Figure 6-9. The body, which serves as the cathode, has a cylindrical cavity in which the anode wire is located. Thus, the entire face of the half-cylinder may serve as the window opening. In normal applications, the detector is mounted so the window is down rather than up as shown in Figure 6-9. Because the amount of gas needed to maintain a constant atmosphere inside the detector would be excessive, a thin plastic film attached to a loosely fitting face plate covers the detector opening. The film is thin enough so it will not absorb β⁻-rays excessively, but such thinness negates the possibility of using it to seal the detector. Nevertheless, with a very low gas pressure and slow flow rate a constant atmosphere inside the detector can be maintained. Gas leakage around the face

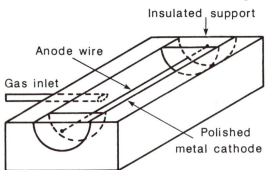

FIG. 6-9   A perspective view of the body of a wide window, gas-flow detector.

plate prevents rupture of the film and provides a means for gas exhaust. Several face plates may be provided to meet sensitivity and resolution requirements. Normally, the openings in face plates are slits of varying widths. A wider slit increases sensitivity (more counts), but decreases the detectors ability to resolve radioactive areas on chromatograms.

Instrument Components

The electronic components of G-M counters are generally simpler and less expensive than those of proportional counters, but most of the same electronic functions are performed by both types of counters.

A high-voltage supply provides the potential for the electrodes, but compared to proportional counters, an exceptionally stable supply is not needed because when operating in the middle of the G-M plateau, small fluctuations in voltage will not affect the number or heights of pulses. Although the pulses from G-M detectors are larger than those from proportional detectors, amplification is still needed. However, the extent of amplification, or gain factor, is usually lower and is held constant. Therefore, the amplifiers of G-M counters do not have operator controls normally. The only discrimination used is provided by the scaler.

The simplicity of the electronics needed for a G-M detector allows their use as portable battery-operated survey meters. However, such devices may have a *count rate meter* rather than a scaler as an output device. A count rate meter acts as though it records the number of counts over a short interval of time, perhaps 0.1 to 10 s, and electronically multiplies the number by an appropriate factor to yield cpm or cps. Sometimes a recorder is used for a graphic display, and radiochromatogram scanners frequently use this form of output. An auditory output may be employed for which the intensity of the signal is correlated with the frequency of a clicking noise or the pitch and volume of a "howler." By using different meter scales, rate meters may read in exposure or dose rate units (Chapter 15) such as R/hr or Gy/hr.

The time constant of a count rate meter refers to the length of time counts are recorded before the signal for the meter is generated. A short time constant yields variable readings when counting rates are low because of the random nature of radioactivity, but a short time constant is desirable when surveying areas where high counting rates may be encountered because the meter responds more quickly. For these reasons some survey meters have a control that provides several time constant options and almost all have multiple ranges of rates that usually differ by factors of 10. The accuracy of count rate meters is not comparable to a scaler-timer for measuring radioactivity, primarily because of the different lengths of "counting" periods.

Operating Characteristics

In our discussion on the applied voltages corresponding to the various gas ionization regions depicted in Figure 6-4, precise values were not given, because gas amplification depends on the field strength, which is affected by the geometry of the electrodes as well as the applied voltage. Also the composition of the gas in a detector will affect the voltage needed for it to operate in a particular region. All of these factors do not need to be controlled precisely if one is willing to accept the fact that each detector will have its own unique characteristics. A particular lot of G-M tubes from a manufacturer will have very similar operating characteristics, but there will be slight variation among the lot.

When first employing a G-M detector its counting plateau should be determined. This is accomplished readily by counting a typical sample repeatedly and adjusting the high voltage between counting periods. A plot of cpm versus voltage should appear as shown in Figure 6-10. Before beginning the counting process, the high voltage should be reduced to its lowest value and then raised until counting begins. This is called the *starting voltage* or starting potential. For G-M tubes, starting voltages vary between about 700 and 1200 V; generally, halogen quenched tubes have lower starting voltages than organically quenched ones. As the voltage is increased by small increments (10–20 V), the counting rate will rise rapidly then level off to form a plateau, which is analogous to the *G-M plateau* shown in Figure 6-4. Although Figure 6-10 shows the curve extending into the region of continuous discharge, an operator should avoid such voltages because of potential detector damage. The length of the plateau is in the order of 200–300 V, and the voltage to be selected as the *operating voltage* should be a value about 100 V beyond the leading edge of the plateau, which is called the *threshold voltage*.

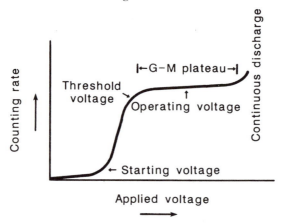

FIG. 6-10   Operating characteristics of a G-M detector.

In subsequent assays the operating voltage would be set at the value determined from the counting plateau. The counting plateau should be re-determined periodically (weekly or monthly depending on use) and whenever maintenance is performed on the instrument.

Simple, portable instruments employing G-M detectors, such as survey meters, normally do not have operator controls for the applied voltage, and the detector's operating characteristics are not easily determined by the operator. Consequently, it is important to have qualified electronic technicians "calibrate" such instruments on a semiannual basis at least.

Resolving Time Correction

For counting rates above 10,000 cpm, significant counting errors can occur due to the period of time during the counting process when the detector is insensitive to incoming radiation. Suppose the resolving time (τ) is 1.7×10^{-7} min/count (ct) (about 10 μs/ct), and the observed counting rate (R) is 54,000 cpm. The time during 1 min when the detector is insensitive is $R\tau$:

$$R\tau = (5.4 \times 10^4 \text{ ct})(1.7 \times 10^{-7} \text{ min/ct}) = 9.18 \times 10^{-3} \text{ min}$$

The result represents a loss of more than a half of a second counting time and a timing error of 0.918%. However, if the resolving time is known, a corrected counting rate (R_c) is easily calculated.

$$R_c = \frac{R}{1 - R\tau} \tag{6-1}$$

Using the data presented above, the corrected counting rate is

$$R_c = \frac{54,000 \text{ cpm}}{1 - (54,000)(1.7 \times 10^{-7})} = 54,500 \text{ cpm}$$

Determination of Resolving Time

Resolving times are usually determined by using "split sources," which consist of two sources that can be counted separately, then together while maintaining constant positions relative to the detector. Blank sources are used to determine background. Let R_1, R_2, $R_{1,2}$, and R_b be the observed counting rates (not net counting rates) of sources 1, 2, 1 plus 2, and the blank or background, respectively. Their respective corrected gross counting rates are R_{c_1}, R_{c_2}, $R_{c_{1,2}}$, and R_{c_b}.

$$R_{c_1} + R_{c_2} = R_{c_{1,2}} + R_{c_b}$$

$$\frac{R_1}{1 - R_1\tau} + \frac{R_2}{1 - R_2\tau} = \frac{R_{1,2}}{1 - R_{1,2}\tau} + \frac{R_b}{1 - R_b\tau}$$

Derivation of a common denominator yields a cubic equation with respect to τ, but since τ is small, all terms containing τ^2 or τ^3 can be disregarded. This leads to Equation 6-2.

$$\tau = \frac{R_1 + R_2 - R_{1,2} - R_b}{2(R_1 R_2 - R_{1,2} R_b)} \tag{6-2}$$

Another equation may be derived from the bionomial expansion of $(1 - R\tau)^{-1}$, which is the series $1 + R\tau + R^2\tau^2 + R^3\tau^3 + \cdots$. Elimination of terms with powers higher than one for τ yields

$$(1 - R\tau)^{-1} = 1 + R\tau$$

$$R_1 (1 + R_1\tau) + R_2 (1 - R_2\tau) = R_{1,2}(1 - R_{1,2}\tau) + R_b (1 - R_b\tau)$$

and

$$\tau = \frac{R_1 + R_2 - R_{1,2} - R_b}{R_b^2 + R_{1,2}^2 - R_1^2 - R_2^2}$$

Since the terms are squared and R_b is very small compared to the other counting rates, R_b^2 may be eliminated in the denominator yielding Equation 6-3.

$$\tau = \frac{R_1 + R_2 - R_{1,2} - R_b}{R_{1,2}^2 - R_1^2 - R_2^2} \tag{6-3}$$

Example Problem 6-1. Calculate the resolving time for a G-M counter that yields the following gross counting rates:

$$
\begin{array}{lr}
\text{Source 1} & 8{,}321 \text{ cpm} \\
\text{Source 2} & 31{,}168 \text{ cpm} \\
\text{Source 1 with 2} & 39{,}409 \text{ cpm} \\
\text{Blank source} & 52 \text{ cpm}
\end{array}
$$

Using Equation 6-2,

$$\tau = \frac{[8{,}321 + 31{,}168 - 39{,}409 - 52] \text{ cpm}}{(2)[(8{,}321)(31{,}168) - (39{,}409)(52)](\text{cpm})^2}$$

$$\tau = 5.44 \times 10^{-8} \text{ (cpm)}^{-1} \text{ or } 5.44 \times 10^{-8} \text{ min/ct}$$

This is equivalent to about 3.3 μs/ct.

Example Problem 6-2. Calculate the corrected net counting rate for a sample that yielded a gross counting rate of 10,723 cpm with the same instrument and same sample geometry as described in example problem 6-1. Starting with Equation 6-1,

$$R_c = \frac{R}{1 - R\tau}$$

$$R_c = \frac{10{,}723 \text{ cpm}}{1 - (10{,}723 \text{ cpm})(5.44 \times 10^{-8} \text{ cpm}^{-1})}$$

$$R_c = 10{,}729 \text{ cpm}$$

$$R_{cb} \approx R_b \approx 52 \text{ cpm}$$

$$\text{net } R_c = (10{,}729 - 52) \text{ cpm} = 10{,}677 \text{ cpm}$$

Applications for G-M Counters

For precise measurements of radioactive samples, G-M counters are not used extensively because higher counting efficiencies can be obtained by other types of instruments generally. The low density of the filling gas yields low counting efficiencies with γ-emitters, and with end-window tubes, the counting geometry is somewhat less than 2π, which means that the counting efficiency for a single emissions is below 50% at best. Also, for α- and β-radiation, some absorption by the window material is likely, which may have profound effects. The efficiencies for some common β^--emitters may be in the range of ^{32}P ($E_{max} = 1.71$ MeV) $\approx 35\%$, ^{14}C ($E_{max} = 0.156$ MeV) $\approx 3\%$, ^{3}H ($E_{max} = 0.018$ MeV) $\approx 0\%$. By the elimination of the window, efficiencies for weak emitters can be improved dramatically; for example, in a gas-flow detector ^{14}C may be counted with an efficiency approaching 40%, and ^{3}H, which is not detected with an end-window tube, may have a counting efficiency of 5–15%.

Counting efficiency is an important consideration when selecting the type of detector for assaying a particular radionuclide, but, as will be shown in Chapter 13, the statistical accuracy of counting data depends on the number of counts collected. Therefore, longer counting periods can compensate for a lower counting efficiency. The cost of an instrument is also a factor worthy of consideration. For example, if a person wished to assay only ^{32}P, a G-M counter might be the best choice. It would be much less expensive than a liquid scintillation counter, sample preparation costs are likely to be lower, and the difference in counting efficiencies (33% compared to 95%) could be overcome by counting samples for periods three times longer.

Because of versatility, simplicity, and relative cost, G-M detectors are used widely in specialized instruments. Their use in radiochromatogram scanners and survey meters has already been discussed. G-M detectors are also employed in radiation area surveillance monitors (see Chapter 16) and devices to measure approximate quantities of radioactive materials during transfer operations; for example, the dose of a radioactive drug in a loaded syringe. Almost everyone working with radioisotopes will use a G-M detector in some form during the course of their work.

PROBLEMS

1. For a single 5.0-MeV α-particle, approximately how many electrons should be collected at the anode of an ion chamber that is filled with helium and operated with an applied potential that yields saturating currents? Assume the average energy necessary to produce an ion pair is about 44.4 eV in helium.

2. Using the same conditions described in problem 1, calculate the number of electron that should be collected from a single negatron with a kinetic energy equal to the average energy of β$^-$-particles from ^{14}C ($\bar{E} = 0.045$ MeV).

3. Calculate the resolving time for a G-M counter that yields the following data when split sources are counted. Use Equation 6-2.

Source 1	6,271 cpm
Source 2	28,994 cpm
Source 1 plus 2	35,195 cpm
Blank source	48 cpm

4. Repeat the calculation specified in problem 3 using Equation 6-3.

5. Using the answer to problem 4, calculate the corrected counting rate for uncorrected rates of (a) 60,000 cpm and (b) 4000 cpm.

6. For each of the counting rates given in problem 5, calculate the relative error in the counting rate measurements. That is, the error in the measurement $(R_c - R)$ divided by the corrected rate (R_c).

CHAPTER 7

Detection of Radioactivity by Solid Scintillation Methods

INTRODUCTION

The general meaning of the verb "scintillate" is to give off sparks or flashes of light, to sparkle; but in radiation detection, scintillation refers to processes that yield pulses of visible light when ionizing radiation is absorbed. Apparently, Sir William Crooks was the first scintillation counter because he reported in 1903 that α-particles striking a ZnS screen produced flashes of light. For the next decade or so, visible counting of flashes became an established method of assaying strong α-emitters, but the method fell into disuse as gas ionization methods were developed and refined. In the 1940s, the *multiplying phototube* (MP-tube) was developed and its sensitivity to light easily surpassed that of the human eye. With MP-tubes, electronic counting of weak light pulses became feasible.

All major types of ionizing radiation, α-, β-, γ-, and X-rays, are capable of causing scintillation events when absorbed by certain kinds of materials, however scintillation detection has not become the method of choice for all applications.

Scintillation detection methods are classified according to the physical state of the absorber. Thus, solid scintillation instruments employ a solid absorber coupled to a photosensitive detector such as an MP-tube. For the detection of γ- and energetic X-rays, crystals of inorganic compounds are used. Their densities (3.7 g/cm^3 for NaI) make them effective absorbers of γ-rays, but prevent the penetration of α- and β-particles below surface layers. Solid organic compounds, such as anthracene, have lower densities and are more effective in detecting α- and β-rays, but less effective for γ-

rays. Plastic scintillators, organic compounds imbedded in plastic resins, have low densities compared to inorganic salts and are used for special applications, such as in flow-cell detectors for high-performance liquid chromatographs. Such detectors are discussed in Chapter 17.

By far the most commonly used solid scintillation counter is one that utilizes a sodium iodide crystal and is designed for the detection of γ-rays. Therefore, primary emphasis in this chapter will be on that type of instrument.

OVERVIEW OF THE SCINTILLATION PROCESS

The sodium iodide crystals used for γ-ray detection contain about 0.1% thallium as a crystalline impurity and are designated as NaI(Tl) crystals. The major portion of such crystals consist of regularly spaced Na^+ and I^- ions, but the Tl^+ ions, which replace some of the Na^+ ions, create slight deformations of the regular crystal lattice at the sites of substitution. The crystalline defect is sometimes called an *activation center* because energy dissipated in such a crystal tends to migrate to the defect sites and then be emitted as visible light. A multiplying phototube detects the resulting flashes of light.

Band Theory and the Scintillation Event

Band theory deals with the energy levels of electrons and their movement in solids. This topic is discussed more completely in Chapter 11, and the reader is encouraged to consult the appropriate section in that chapter to augment the simplified discussion presented below.

Although electrons in solids have quantized energy levels, the spacings between sublevels are so small that virtually continuous bands exist. However, differences between levels, corresponding to principal quantum numbers, may result in gaps between energy bands. In solids two energy bands are of most concern, the *valence* and *conduction* bands. The valence band is that band of energies possessed by the valence electrons, those electrons involved in bonding between atoms and located in the outermost orbitals. The electrons in the conduction band have sufficient energy to be disassociated from their parent atoms; effectively they are free to wander or move in a particular direction if an electric field is applied.

For materials such as NaI, an energy gap exists between the valence and conduction bands, and this is referred to as a *forbidden gap*. Such a gap represents energy levels an electron may not occupy as a result of the quantum nature of the atoms in the solid. However, electrons may acquire sufficient energy to jump the gap, to be promoted from the valence to conduction band. Ionization in a solid, the separation of an electron from its parent atom or molecule, occurs when valence electrons receive sufficient energy to move from the valence band to the conduction band. Recombination of an ion and electron involves the demotion of an electron from the conduction band to the valence band.

When an electron is promoted to the conduction band, a positively charged "*hole*" in the valence band is created, and although a specific ion cannot migrate, its hole can. A valence electron from a neighboring atom may be trapped by the hole, but this creates another hole one atom removed. Because of their charge, holes migrate in a direction opposite to electrons when a potential gradient exists, and electrical current is carried by both electrons and holes. Since no bias (voltage) is applied to NaI (Tl) crystals, conduction electrons and holes migrate in a random fashion and eventually recombine.

The crystalline impurity (Tl^+) has a smaller forbidden gap than Na^+ or I^- so recombination is more likely to occur at activation centers. A thallium ion with a hole corresponds to a Tl^{2+} ion, and when it gains an electron it reverts to Tl^+ with the electron entering one of the higher energy levels in the valence band. The result is an excited thallium ion, Tl^{+*}, which will subsequently lose its energy of excitation by one of several mechanisms. The mechanism of interest is *fluorescence,* the emission of a photon of visible light.

An alternate process can occur in NaI(Tl) crystals; *excitons* can be produced, and they too can migrate randomly in the crystal lattice. An exciton may be visualized as an excited iodide ion (I^{-*}) and its migration is due to energy transfers between neighboring ions. If a neighboring ion happens to be a thallium ion, the energy of excitation is readily transferred to it yielding an excited thallium ion, (Tl^{+*}). As described above, the Tl^{+*} then loses its excitation energy by the emission of a photon. Thus, ionization and excitation of crystalline ions due to the absorption of radiation yield excited thallium ions that fluoresce as they revert to their ground states. The dissipation of about 30 eV of radiation energy is required to produce one photon in NaI(Tl) crystals. The maximum of the emission spectrum is about 410 nm, which corresponds to about 3 eV per photon. Obviously, only about 10% of the energy dissipated in the crystal is emitted as light; therefore, the flow of energy to activation sites involves losses along the way, and deexcitation mechanisms other than fluorescence occur. *Scintillation efficiency* is defined as the yield of light energy divided by the energy absorbed by the crystal, and, in the case cited above, it is about 10%. Although energy losses may seem large, a scintillation efficiency of 10% is acceptable, but more importantly it is fairly constant over a wide range of energies, which means that the number of photons of light emitted is proportional to the energy dissipated in the crystal. This constant proportionality permits the energies of γ-rays absorbed by the crystal to be ascertained, and as will be discussed later, opens the prospect of γ-ray spectroscopy.

DETECTION OF LIGHT

Several different types of detectors may be used to detect the visible light emitted by a scintillation event. Semiconductor photodiodes and microchannel plate multiplying phototubes are continually being perfected, and, for certain applications, specific detectors offer desirable advantages. However, the most commonly used photodector today is the dynode type, multiplying phototube (MP-tube).

MP-Tube

Figure 7-1 is a diagram of an MP-tube with eight dynodes. It consists of a cylindrical tube, perhaps 3 in. in diameter, with a window, usually made of glass. Quartz windows are more expensive, but they transmit radiation over a wider region of the electromagnetic spectrum. Both glass and quartz are transparent to visible light (400–700 nm), but the transparency of quartz extends well up into the near-UV region. For glass, 75% transmission is encountered around 350 nm, while for quartz it occurs around 200 nm. Whether quartz windows are justified depends on the wavelengths of light encountered and whether the photocathode responds efficiently to near-UV light. For NaI(Tl) crystals, which yield light around 410 nm, glass windows are quite acceptable.

Photocathode.   The *photocathode* is a thin wafer located next to the window and a high negative electropotential is applied to it. When light strikes the photocathode, electrons are released from its opposite side. Such electrons are referred to as *photoelectrons*. The photosensitive material of photocathodes varies depending on manufacturers and needs for certain appli-

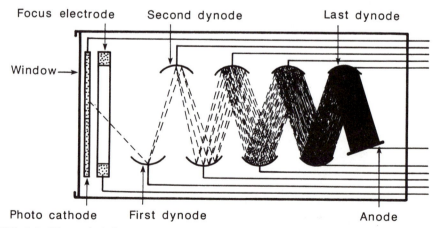

FIG. 7-1   Theoretical diagram of a multiplying phototube (MP-tube) with eight dynodes.

cation, but Sb–Cs and Ag–Mg alloys have been used. The so called bialkali ($Sb–K_2–Cs$) and trialkali (Sb–K–Na–Cs) tubes offer higher sensitivities in most applications. The spectral response of an MP-tube (its sensitivity to various wavelengths of light) depends on the photocathode, and the response of a tube with a ($Sb–K_2–Cs$) photocathode is shown in Figure 7-2. Note that maximum sensitivity is in the region of 400 nm, which matches the wavelengths of light from scintillation events quite well. The sensitivity may be expressed as the quantum efficiency of the photocathode, which is the percentage of photoelectrons released by a number of photons. Quantum efficiencies of 25–30% are possible, which means that when a photon strikes the photocathode there is a 25–30% chance that it will release a photoelectron. For an MP-tube operating at a constant voltage and with light of given wavelengths, the quantum efficiency is relatively constant.

Dynodes. A *dynode* may be visualized as a small plate with electrons on its surface. When an electron impinges on the plate its momentum is sufficient to dislodge several surface electrons. In an operating MP-tube, the first dynode (closest to the photocathode) has a negative potential, but it is positive relative to the photocathode so photoelectrons are accelerated toward the first dynode. The focus electrode aids in directing all photoelectrons toward the first dynode regardless of their points of release from the photocathode. A similar positive potential difference exists between the first and second dynodes so electrons escaping from the first dynode are accelerated toward the second. The curvature of the dynode aids in estab-

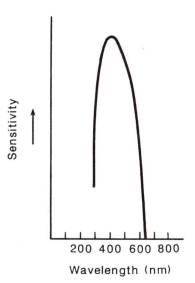

FIG. 7-2 Spectral response of a bialkali ($Sb–K_2–Cs$) multiplying phototube. (Redrawn with permission from *Applications of Liquid Scintillation Counting,* by D. L. Horrocks. Copyright © 1974 by Academic Press, Inc., San Diego, CA)

lishing electron trajectories that will cause the electrons to strike the next dynode, but focusing grids (not shown in Figure 7-1) may also be used. Multiplication occurs as electrons are attracted to successive dynodes and subsequently collected at the anode.

The arrangement of dynodes in an MP-tube varies among manufacturers, but a linear placement where the dynodes appear as slats in a venetian blind is popular.

Anode. The positively charged *anode* collects the cascade of electrons from the last dynode, and the cascade constitutes an electrical pulse. To visualize the multiplication of electrons that yields the cascade, assume that each electron that impinges on a dynode causes two electrons to escape. For each photoelectron that strikes the first dynode, two electrons escape and subsequently strike the second dynode; four electrons strike the third dynode, eight the fourth, and so on until 256 electrons are collected at the anode. Actually greater multiplication usually occurs because most MP-tubes have more dynodes and the gain factor is greater than two. At usual operating voltages the gain factor is between three to four, which means that the overall multiplication or gain for a 10 dynode tube would be $3^{10}(6 \times 10^4)$ to $4^{10} (1 \times 10^6)$. That is, for each photoelectron (assuming a gain factor of four) 10^6 electrons would be collected at the anode. Obviously, the magnitude of a pulse is proportional to the number of photoelectrons that initiated the multiplication process. Since constant proportionality factors exist for the multiplication process and the quantum efficiency of the crystal, the magnitude of the pulse (pulse height) is proportional to the energy of the γ-ray, assuming its energy is completely absorbed.

Effect of Operating Voltage on Gain. The potential difference between the photocathode and anode may be as low as 800 V or as high as about 2500 V depending on the characteristics of the MP-tube. This overall voltage is divided in steps with the difference between successive pairs of dynodes being in the order of 50–150 V. The potential difference between the photocathode and first dynode is higher, perhaps in the range of 300–500 V. Assuming a difference of 350 V for the photocathode-first dynode potential and 75 V for dynode pairs as well as for the difference between the last dynode and anode, the overall potential for a 10 dynode MP-tube would be 1100 V. Most MP-tubes will operate over a fairly wide range of applied voltages.

The overall gain of an MP-tube (number of electrons collected at the anode per photoelectron) is affected by the applied voltage. A higher MP-tube voltage causes the potential difference between dynodes to be higher, which increases the gain factor for dynode pairs. The velocity, and hence kinetic energy, that an electron gains when accelerated between two electrodes (dynodes in this case) is a function of the potential and distance between the electrodes. A higher voltage accelerates electrons more causing them to impinge on the successive dynode with greater momentum, which increases the number of surface electrons likely to receive sufficient energy to escape. Of course, there is a minimum operating voltage below which electrons will not be accelerated enough for multiplication to occur. At very high voltages, thermionic emission (to be discussed later) by the dynodes causes the tube to produce excessive noise pulses.

Noise Pulses.   Pulses not initiated by a scintillation event are referred to as *noise pulses*. Practically all noise produced by MP-tubes is due to a phenomenon called *thermionic emission* that occurs at the dynodes. An electron on the surface of an electrode at a high potential (a dynode) may gain sufficient energy to escape, and such an electron is said to be thermionically emitted. When an escape occurs, the electron is accelerated toward the next dynode and on through the remaining sequence of dynodes, thereby producing a cascade of electrons at the anode. Suppose an electron escaped from the third dynode and the gain factor between dynodes was four, the multiplication by succeeding dynodes would yield about 4^7 (1.6×10^4) electrons at the anode. Under the same conditions, a photoelectron would have produced about 4^{10} (1×10^6) electrons. This illustrates an important point. Noise pulses generally are much weaker than those arising from scintillation events and their heights depend on from where in the dynode sequence they originated.

The probability of thermionic emission increases with the potential applied to the dynode, therefore higher operating voltages increase the frequency of noise pulses. This causes a conflict of interest because higher applied voltages increase pulse heights from scintillation events that enhance their detection. On the other hand, high background counting rates (noise pulses) decrease our ability to measure the amount of radioactivity accurately from a statistical point of view (see Chapter 13).

Another factor that affects noise pulse production is the temperature of the dynodes. Higher temperatures increase the energy levels of electrons and the chances of an electron gaining sufficient energy to escape, hence cooling MP-tubes lowers their noise levels. The effects of temperature on thermionic emission were more pronounced in tubes manufactured before the late 1960s, but the current generation of MP-tubes produces much less noise than their predecessors at all operating temperatures. In some applications it is desirable to maintain a constant operating temperature, but current MP-tubes do not need to be cooled to be operated effectively.

As alluded to above, MP-tube noise will yield a high background counting rate. As will be seen in Chapter 13, a background of several thousands cpm is statistically unacceptable in most situations. Other electronic components in scintillation counters are quite effective in distinguishing noise pulses from those arising from scintillation events so the problem associated with high background counting rate can be circumvented.

INSTRUMENT DESIGNS

A block diagram for a simple single channel solid scintillation counter is shown in Figure 7-3.

Crystals

As indicated earlier, for the detection of γ- and X-rays, an NaI(Tl) crystal is used commonly. Generally, the crystals are produced in a cylindrical shape with a well drilled at the axis to the center of the cylinder. Crystal size may vary and is a significant factor in instrument cost. A 3 × 3-in. crystal has a diameter and height of 3 in. and is a desirable size, but smaller crystals (1.5 × 1 in.) are effective detectors, although their counting efficiencies will be lower than those of larger crystals. One end of the cylinder is coupled directly to the window of the MP-tube with "optical grease," which has an index of refraction such that little loss of light occurs when photons pass from the crystal to the MP-tube. The remainder of the crystal, including the well, is covered with a thin layer of aluminum that protects the crystal from moisture and mechanical shock. It also excludes external

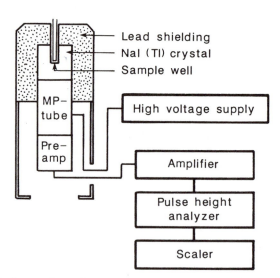

FIG. 7-3   Block diagram for a single channel solid scintillation counter with an NaI (Tl) crystal.

light and reflects light generated inside the crystal back toward the MP-tube. The junction between the tube and crystal is also protected to exclude moisture and external light.

Normally, samples are placed in bottoms of test tubes, so that when placed in the well, the sample is slightly below the geometric center of the crystal. This yields nearly 4π-counting geometry because, except for perfectly vertical emanation, all rays would enter the crystal. The aluminum covering of the crystal in the well area not only protects the crystal from abrasions when samples are changed, but it is sufficiently thick to absorb α- and β-rays, yet thin enough to not absorb an appreciable fraction of γ- and X-rays.

For the production of a scintillation event, the ray must interact with the crystal and the three predominate processes involved in γ-ray absorption are the photoelectric effect, Compton effect, and pair production. Equation 5-10 is appropriate for calculating fractional absorption of γ-rays by NaI.

A γ-ray emitted from the bottom of the well must traverse a minimum distance equal to the radius of the crystal minus the radius of the well to escape absorption. For 1- and 3-in. crystals, this represents absorber thicknesses of approximately 1 and 3.5 cm, respectively. Both thicknesses are nearly sufficient to completely absorb 150-keV γ-rays. However, for 1-MeV rays ($\mu_\ell = 0.21$ cm^{-1} in NaI) about 19% are absorbed by a 1 cm thickness and 52% by a 3.5 cm thickness. Because of differences in absorption, the counting efficiency of 1 MeV γ-rays with a 1-in. crystal is approximately half of that with a 3-in. crystal. Thus, except for low energy rays, counter efficiency is affected by crystal size. Unfortunately, the exponential nature of γ-ray absorption means that a 17-in. crystal would be required to reach an absorption percentage of 99% for 1-MeV rays, and even larger crystals are needed for 2- or 3-MeV rays. The cost of producing fracture-free crystals increases dramatically with size and represents a practical limit for crystal size and counter efficiency. Nevertheless, an efficiency of about 40% is not a severe limitation for accurate radioactivity measurements, and a 3-in. crystal will absorb about that percentage of 3-MeV γ-rays. Most γ-emitters used in biological research methods do not exceed 3 MeV so γ-counters with 2- or 3-in. crystals have quite acceptable counting efficiencies.

Shielding

To reduce background due to cosmic rays or γ-rays from natural sources, lead, usually several inches thick, is used to shield the crystal. Generally, the shielding and holder for the crystal, MP-tube, and preamplifier are constructed as a unit with electrical cables connecting it to the other electrical components of the instrument. Although not shown in Figure 7-3, automatic sample changers are commonly employed.

MP-Tube and Preamplifier

As mentioned above, the MP-tube is mounted so its window is coupled to the crystal with a thin film of optical grease, and the preamplifier is connected to the base of the MP-tube. This arrangement increases the signal strength of the pulses before being transferred by cable to other electronic components.

High-Voltage Supply

The high-voltage supply provides the voltages applied to the MP-tube. Stability of these voltages is important because of its effect on MP-tube gain and pulse heights. Since there is a range of voltages that permit tube operation, high-voltage supplies have controls to adjust the voltage applied. With early manufactured tubes that had high thermionic emission rates, the operating voltage was adjusted to as low a level as possible to reduce noise yet high enough to give adequate multiplication and detection efficiency.

In modern instruments, the MP-tube noise is much less of a problem, and high-voltage adjustments are usually done by electronics technicians instead of being controlled by the operator.

Amplifier

Pulses from the preamplifier are fed to the amplifier where further amplification and pulse shaping occurs. Formally, the *gain of an amplifier* is the ratio of the heights of the output pulse relative to the input pulse; however, in some instruments the term "gain" may represent an arbitrary unit of amplification. Generally, controls are present so an operator can adjust the gain within certain limits.

A *linear amplifier* is one in which the height of the output pulse is linearly proportional to that of the input pulse. Such amplifiers are used commonly with solid scintillation spectrometers. *Logarithmic amplifiers* are used for some types of instruments, and their output pulse heights are proportional to the logarithm of their input pulse heights. In some instruments the amplifier may be referred to as an *attenuator,* which means that the amplification scale has been reversed.

An amplifier that increases the magnitude of all pulses it receives will increase the range of pulse heights or produce a broader spectrum of pulse heights. This is particularly noticeable with a linear amplifier, but a logarithmic amplifier compresses the upper ranges of the spectrum relative to the lower ranges.

Pulse Height Analyzer

Pulse height analyzers and discriminators are similar electronic components in that they determine which pulses are sent on to the scaler or readout device, and pulse height is the basis for selection or rejection. Recall from Chapter 6 that the discriminator described had two pulse height limits. The lower limit determined the minimum height (voltage) of pulses that are sent to the scaler, while the upper specified the maximum acceptable height. For a pulse to be counted, its height must be in between the lower and upper limits, and this range of pulse heights is referred to as the *window*. The usual distinction between discriminators and pulse height analyzers is the way the window is controlled. In discriminators the upper and lower limits are adjusted independently from one another, whereas in analyzers the window moves with the lower limit, which is called the *baseline* or *base*. In general, narrower windows are used in analyzers, but the window still represents the range of pulse voltages that is sent on to the scaler. Adjustments of the baseline control permits the window to be moved to any area of the pulse spectrum. A second control is used to adjust the width of the window. Since the window width represents the narrow range of acceptable pulse voltages it may be abbreviated as ΔV.

Figure 7-4 shows three sets of analyzer settings, an initial setting, one with a higher base, and one with a wider window. Under initial conditions only pulse number 2 is sent to the scaler. When the window width is held constant, but the base is raised as shown in the middle panel, only pulse number 3 is counted. In the third panel the base is at the same position as initial conditions, but the window width has been increased. Under these conditions a larger portion of the pulse spectrum is being covered by the window, and both pulses number 1 and 2 are counted by the scaler. Most

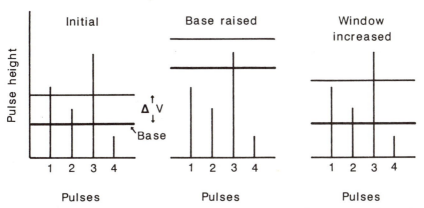

FIG. 7-4   The effects of pulse height analyzer settings on rejection of pulses with various heights.

pulse height analyzers have an additional control that effectively eliminates the upper limit of the window; therefore, all pulses that exceed the baseline height are counted and such conditions are referred to as integral counting.

An important function of discriminators and pulse height analyzers is to screen out pulses that are not due to scintillation events arising from the emitter being assayed. A very high percentage of the weak MP-tube noise pulses can be eliminated by raising the baseline slightly. Of course, a desirable position of the baseline depends on the gain of the amplifier, but using the amplifier and baseline controls while counting a blank sample, an operator can determine appropriate settings easily.

Lead shielding around the crystal and MP-tube reduces the number of cosmic rays that may interact with the crystal, but they cannot be eliminated entirely. Generally, pulses resulting from cosmic rays are much larger than those from most radioactive sources, and a pulse height analyzer provides a means of eliminating them.

Scaler

A scaler with an associated precision timer may be used to perform the counting function, but other readout devices, such as rate meters and computers, may be used also.

Pulse Spectra

In our discussion of the scintillation event and operation of MP-tubes, it was concluded that the heights (voltage) of an MP-tube's pulses were proportional to the amount of energy dissipated in the crystal. Since the amount of energy absorbed can vary, a range of pulse heights may be observed. The distribution of pulses over a range of pulse heights is referred to as a *pulse spectrum*. Graphically, a pulse spectrum represents a plot of pulse frequency (number of pulses, or counts, of a given height) versus pulse height.

A spectrum can be obtained with an instrument such as that shown in Figure 7-3. Assuming that the amplifier has been adjusted appropriately, the baseline of the pulse height analyzer is adjusted just high enough to exclude noise pulses, and a fairly narrow window, for example, $\Delta V = 0.1$ V, is selected. A source is placed in the sample well and counted for a period of time. Then the baseline is raised by an increment equal to the window width (0.1 V in our example), and the source is recounted. This process is repeated until the entire range of baseline settings has been covered. One-half the width of the window is added to each of the baseline settings and the data are plotted; counting rate versus base $+ \frac{1}{2}\Delta V$. Spectra will be produced by using a wider window and increasing the baseline by increments less than the window width. Resolution will not be as good and some smaller peaks may be broadened to such an extent that they will not be apparent in spectra. However, it may be a useful technique for locating the center of an intense, symmetrical peak.

Spectra may also be obtained by using a count rate meter connected to a graphic recorder or readout device. With the sample in the crystal well, the baseline (initially at zero) is driven slowly and continuously by a synchronous motor upward until the maximum baseline setting is reached.

The terms "counter" and "spectrometer" are frequently used interchangeably because all spectrometers can function as counters and most counters can yield spectral data. The instrument diagram shown in Figure 7-3 represents a single channel spectrometer. Today, γ-ray spectra are commonly obtained with instruments that employ a *multichannel analyzer.*

Multichannel Analyzer

The basic features of an instrument employing a multichannel analyzer are similar to those shown in Figure 7-3 except that the multichannel analyzer replaces the pulse height analyzer and scaler. The multichannel analyzer contains a dedicated computer that stores counting data for each of its channels. There may be 1000 to 16,000 channels, and each channel functions as if it was a separate pulse height analyzer with its own scaler. Analog to digital converters are used with a computer to store data for each channel, and similar devices are discussed with greater detail in Chapter 8. Suppose an instrument had 1000 channels and the range of the baseline settings was 2000 arbitrary units. The window widths of all channels would be set at 2 units. The base of the first channel might be set at zero, therefore it would select pulses that had height between 0 and 2 units. The base of the second channel would be set at 2 units so it covered the region of the spectrum corresponding to 2 to 4 units. Successive channels would have bases that increase by 2 units. In this way the entire range of the pulse spectra could be covered. A plot of counting rate versus channel base position (plus 1 unit) would yield a bar graph-type spectrum. Of course, with the availability of computers, such plots are rarely performed by hand. The counts in each channel may also be displayed with an oscilloscope.

γ-RAY SPECTROSCOPY

The pulse spectrum that results from the exposure of a NaI(Tl) crystal to a source of γ-rays is easily related to γ-ray interaction phenomenon. As described in Chapter 5 the primary modes of interactions are the photoelectric effect, Compton effect, and pair production. Figure 7-5 shows the

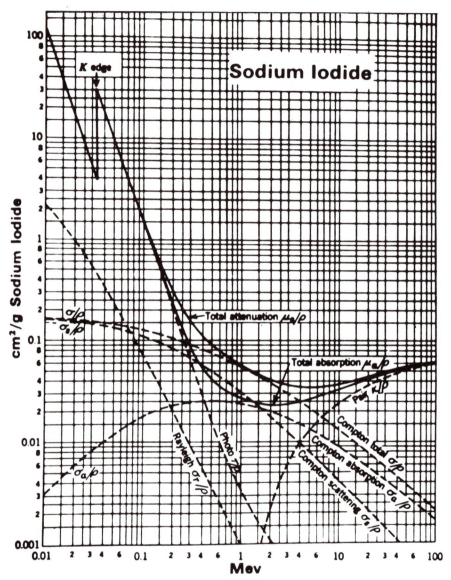

FIG. 7-5   Partial and total mass absorption coefficients for various energies of γ-rays with NaI as an absorber. (Reprinted with permission from *Handbook of Radioactive Nuclides*, by Y. Yang. Copyright © 1969 by CRC Press, Inc., Boca Raton, FL)

partial absorption coefficients for these three processes as a function of γ-ray energy with NaI as an absorber. From these curves several important generalities useful in the interpretation of spectra may be summarized.

1. The photoelectric effect is the predominate interaction mode for rays in the energy range of 0–100 keV.

2. The absorption of rays by a given thickness increases rapidly as energies progressively below 100 keV are encountered, and most of the increased absorption is via the photoelectric mode.

3. If energy is sufficient, the photoelectric effect occurs more frequently with inner shell electrons. The energy of K shell electrons in iodide ions is about 28 keV, therefore the probability of interactions occurring with such electrons increases sharply when γ-ray energies slightly exceed 28 keV. This is shown as the K edge in Figure 7-5.

4. Absorption by the photoelectric effect becomes a minor mode above about 1 MeV, nevertheless absorption by this mode is discernible at higher energies.

5. The percentage of rays penetrating NaI that interact by the Compton effect decreases rather slowly yet continuously as γ-ray energies increase.

6. The Compton effect is the predominate mode of interaction in energy range 0.4–4 MeV.

7. Pair production does not occur unless γ-ray energies exceed 1.022 MeV, and for energies typical of commonly used γ-emitters it is not a predominate mode of interaction.

8. Even for low energy γ-rays, a several-centimeter thickness of NaI will not absorb 100% of the rays. Similarly, secondary rays generated within a crystal have a reasonable probability of escaping a crystal without interacting.

Photoelectric Effect

The interaction of a γ-ray by the photoelectric effect yields a photoelectron with a kinetic energy equal to the energy of the ray minus the binding energy of the electron. The most likely electron to be affected is one in the K shell of an iodide ion, which has an energy of 28 keV.

Suppose the source emitted 600 keV γ-rays. The kinetic energy of the photoelectron would be 572 keV and that energy would be dissipated in the dense crystal. However, the iodide ion from which the photoelectron was removed has a vacancy in its K shell, and when an outer electron fills the vacancy, a 28-eV X-ray will be emitted. It is very likely that such a low-energy ray will be absorbed by the crystal, dissipating an additional 28 keV of energy. Within the time frame of a scintillation event, the dissipation of energies (572 and 28 keV) occurs simultaneously, therefore the total absorption is 600 keV. Earlier it was suggested that scintillation efficiency might be around 10%, which translates to one photon for each 30 eV absorbed. Therefore, about 20,000 photons would be produced by the interaction as described. Because such an interaction is highly probable, a spectrum for a 600-keV γ-emitter should show a significant peak at a pulse

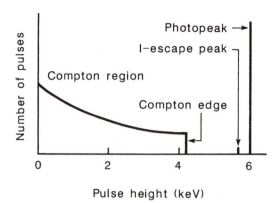

FIG. 7-6   A theoretical spectrum for a 600-keV γ-emitter with a NaI (Tl) crystal.

height corresponding to 600 keV. Such a peak is referred to as the *photo-peak.*

There is a small but significant probability that the 28-keV X-ray may escape the crystal, and, if this occurs, the total energy dissipated in the crystal is only 572 keV. In our example, a small peak, called the iodine escape peak, should appear at a pulse height corresponding to 572 keV in the spectrum.

Theoretically, the photopeak and iodine escape peak should be vertical lines in the spectrum as shown in Figure 7-6, but peak broadening occurs because of slight variation in the scintillation efficiency from interaction to interaction. Consequently, the iodine escape peak is usually obliterated by the photopeak except when γ-ray energies are low and excellent spectral resolution is achieved.

Since under constant operating conditions the heights of pulses are directly proportional to the energy absorbed by the crystal, pulse height analyzers and multichannel analyzers are easily calibrated in energy units. Calibration methods will be discussed in a later section of this chapter.

Compton Effect

A Compton interaction yields a Compton electron and a secondary γ-ray (discussed in Chapter 5). The kinetic energy of the electron depends on the emission angle of the secondary ray (θ is the angle between the directions of the incident and secondary rays). Maximum kinetic energy is realized when the secondary ray is scattered backwards or when θ = 180°. Equation 7-1 is easily derived from Equation 5-8 and it gives the maximum kinetic energy of the Compton electron (E_e) as a function of the energy of the incident γ-ray (E_γ) when all energy values are in units of MeV.

$$E_e = \frac{E_\gamma}{1 + 0.511/2\ E_\gamma} \tag{7-1}$$

For 600-keV γ-rays the maximum Compton electron energy is 421 keV, but Compton electrons will form a continuous energy spectrum from zero to the maximum. The total energy dissipated in the crystal will include the energy of the Compton electron plus any energy that might be absorbed from the secondary γ-ray. Three alternative fates for the secondary γ-ray should be considered.

1. The secondary ray escapes the crystal without further interaction. In this case the height of the pulse would correspond to the energy of the Compton electron. Spectra show a continuum from zero to the pulse height corresponding to the Compton electron maximum, and this is referred to as the *Compton region,* or perhaps less elegantly, as the Compton smear. Generally, the Compton region appears as a plateau that drops sharply at the Compton electron maximum. The plateau edge, which in our example is at 421 keV, is called the *Compton edge.* See Figure 7-6.

2. The secondary ray interacts in another area of the crystal by the photoelectric effect. Such an interaction would most likely result in the complete dissipation of the secondary ray's energy, which when added to the Compton electron's energy would be equal to the energy of the primary γ-ray. Consequently, pulses resulting from such a sequence of interactions would appear in the photopeak.

3. The secondary γ-ray interacts by the Compton effect in another region of the crystal. Assuming the second, secondary ray escapes, the energy absorbed by the crystal would be the sum of energies of the two Compton electrons, or the energy of the primary ray minus the energy of the second secondary ray. Pulses arising from this series of events could appear at heights near zero to somewhat less than the photopeak. The majority of the pulses would be in the Compton region, but a few may be beyond the Compton edge. This helps explain why the edge is not as sharp as might be expected and why actual spectra show a few pulses between the Compton edge and the photopeak.

Spectrum of ^{137}Cs

Actual spectra do not show the sharpness of the theoretical diagram presented in Figure 7-6 for several reasons: (1) small variations in the scintillation efficiency from pulse to pulse, (2) the overlay of secondary Compton γ-rays, (3) incomplete resolution by the spectrometer, and (4) the random nature of radioactivity. The later two reasons can be circumvented partially by collecting data over a longer observation period and by using a higher quality pulse height analyzer or multichannel analyzer. Other features being equal, resolution (capability to distinguish between pulse heights that differ slightly) is improved when more channels are used in a multichannel analyzer.

A spectrum for ^{137}Cs, which emits 662-keV γ-rays, is shown in Figure 7-7. The photopeak at 662 is easily recognized, but it covers the iodine escape peak at 633, and the Compton edge (477) is not prominent. The sharp peak at 32 keV is the photopeak of a K shell X-ray emitted by ^{137}Ba. To explain its origin, the decay of ^{137}Cs must be considered. Cesium-137 decays by the emission of a β⁻-particle (E_{max} = 0.514 MeV) yielding ^{137m}Ba, which emits a 0.662-MeV γ-ray. The half-life of ^{137}Cs is 30.17 yr while that for ^{137m}Ba is 2.55 min. This means that γ-rays are always associated with ^{137}Cs decay. Of course, the β⁻-particle is absorbed before reaching the crystal so it contributes nothing to the spectrum. However, about 4% of the ^{137m}Ba atoms undergo internal conversion that results in the ejection of a K shell electron. Again, the energy of the conversion electron would be dissipated outside the crystal, but when an L shell electron fills the K shell vacancy of ^{137}Ba, a 32-keV X-ray is emitted, and it can interact with the crystal, most likely by the photoelectric effect.

Pair Production

In pair production, the entire energy of the incident γ-ray ($E_γ$) is absorbed, and the sum of the positron and negatron's kinetic energies is equal to $E_γ$ − 1.022 MeV. That amount of energy would be dissipated in the crystal, but when the positron comes to rest it will undergo annihilation with an electron yielding two 0.511-MeV rays. If both of the annihilation rays are absorbed completely (via photoelectric effect or sequence of Compton effect and photoelectric effect) the total energy dissipated in the crystal

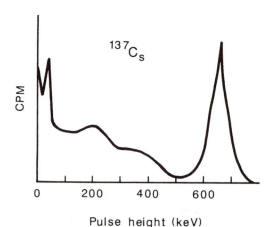

FIG. 7-7   A spectrum for ^{137}Cs with a 2 X 2-in NaI(Tl) crystal.

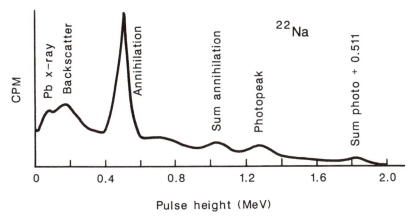

FIG. 7-8 A spectrum for ^{22}Na with a 2 X 2-in NaI(Tl).

would be equal to E_γ. Pulses resulting from such events would appear in the photopeak.

However, there is a probability that *one* or *both* of the 0.511-MeV rays may escape the crystal without an interaction. If one ray escapes, the resulting pulses would be consistent with an energy of $E_\gamma - 0.511$ MeV, and, if both escape, the energy deposited in the crystal would be $E_\gamma - 1.022$ MeV.

Small peaks, called *pair peaks* or *escape peaks,* are usually discernible in the Compton region, and they are located the equivalent of 0.511 and 1.022 MeV from the photopeak. Single and double escape peaks for 1.276-MeV γ-rays from ^{22}Na appear to contribute to the spectrum in Figure 7-8. There is a low "hump" near 765 keV, and the backscatter peak around 200 keV seems to have a slight shoulder near 253 keV.

Other Spectral Features

Peaks Around 200 keV. Small broad peaks in the region of 200 keV may be *backscatter peaks* due to interactions of γ-rays with the lead shield around the crystal. A significant fraction of energetic γ-rays will not be absorbed by the crystal and are likely to enter the lead shielding where they may interact by the Compton effect. If this occurs near the inner surface of the shield, those secondary rays that are directed backward toward the crystal may escape absorption by the lead and strike the crystal. The geometry is such that for a secondary ray to be directed toward the crystal, it must have an energy around 200 keV regardless of the primary ray's energy.

Backscattering can be reduced by increasing the inside diameter of the shield, which decreases the fraction of secondary rays reflected toward the crystal.

Peaks Around 73 keV.   Recall that the photoelectric effect is likely to yield an absorber ion with a vacancy in its K shell and that an outer electron, via several shell transitions, will fall to fill the vacancy. For lead, the most common L to K transition results in the emission of a 72.8-keV X-ray. Thus interactions of γ-rays with lead yield "lead X-rays."

A γ-ray passing completely through the crystal will strike the lead shield, and because of the shield's thickness, undoubtedly undergo an interaction. Some of the lead X-rays generated near the inside of the shield are likely to be directed toward the crystal and be absorbed. Consequently, the photopeaks of lead X-rays around 73 keV are sometimes apparent in spectra. Graded shields, such as one consisting of an inside layer of copper, then a layer of cadmium and finally the thick layer of lead, are used to reduce the intensity of lead X-rays. The Pb X-rays are absorbed readily by the Cd layer and the Cd X-rays by the Cu shield.

Peaks at 511 keV.   Although γ-rays having energies around 511 keV might be responsible for spectral peaks in that region, generally such peaks are the signatures of annihilation reactions between positrons and negatrons.

One source of annihilation radiation is from very energetic γ-rays that pass through the crystal then interact with the lead shield by pair production. The energies of the particles are dissipated in the shield, but after annihilation, some of the 511-keV rays will be directed back toward the crystal. Photopeaks of annihilation radiation are frequently prominent features in spectra of positron emitters as illustrated with ^{22}Na (Figure 7-8). Because of the ranges of positrons, annihilation reactions usually occur in the source itself, its container walls or the crystal casing, and the two 511-keV rays generated at such locations are quite likely to interact with the crystal.

Bremsstrahlung.   As discussed in Chapter 3, bremsstrahlung is a mode of interaction for very energetic β-particles. Consequently, hard β-emitters may yield spectra that show an unexpectedly high level of pulses in the Compton region. The X-rays produced as the β-particles are being stopped form a continuous energy spectrum from zero to E_{max}, which would overlay that region of the γ-ray spectrum. Bremsstrahlung may obliterate spectral features that would otherwise be apparent, but it usually can be reduced by using a low atomic weight absorber between the source and crystal. Recall that bremsstrahlung occurs in high atomic weight absorbers. Having the source dissolved in water or an organic solvent may be sufficient to eliminate the problem, but plastic shields between the source and crystal are effective as well.

Sum Peaks.   A radionuclide that emits more than one γ-ray, or perhaps a γ-ray plus annihilation radiation, may exhibit *summation effects* or *sum peaks*. When more than one ray is emitted there is a chance that the rays may interact with the crystal simultaneously or within a time period so short that the MP-tube cannot resolve the light pulses. Pulse heights from such events depend on the types of interactions that take place, but if the entire energies of both rays are absorbed, a pulse equivalent to the sum of the rays' energies would result. Sum peaks are usually very small and obviously located in the high energy portion of the spectrum. A sum peak is shown in the spectrum of ²²Na (Figure 7-8) and will be identified in the next section. Cobalt-60 emits two γ-rays with energies of 1.17 and 1.33 MeV. The photopeaks for these rays are prominent spectral features in Figure 7-9, and a small sum peak for the two rays is apparent around 2.5 MeV.

Spectrum of ²²Na

A decay diagram for ²²Na was presented in Chapter 3, and the major decay routes are (1) 90% by β⁺-emission (E_{max} = 0.545 MeV) with subsequent γ-emission (1.27 MeV), and (2) 10% by electron capture followed by γ-emission (also 1.27 MeV).

The dominate feature in the spectrum of ²²Na is the annihilation peak at 511 keV and it is much larger than the photopeak of the 1.27-MeV γ-ray. In the lower energy region, backscatter radiation (slightly below 200 keV) and lead X-rays, around 73 keV, are apparent. Two summation effects are discernible, one at 1.022 MeV, which represents two annihilation photons and a γ-ray plus 0.511 MeV near 1.78 MeV.

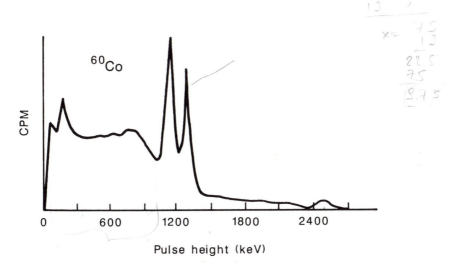

FIG. 7-9   A spectrum for ⁶⁰Co with a 3 X 3-in NaI(Tl) crystal.

Spectrometer Calibration

A useful feature of γ-ray spectrometers is the capability of providing information about γ- and X-ray energies. Because of the direct proportionality between energy deposition in the crystal and pulse height, calibration is accomplished easily and, in some instruments, automatically.

Graphic Calibration with a Single Channel Analyzer.   The instruments' overall gain is adjusted so that the desired energy range is covered by the pulse height units (baseline settings) of the analyzer. This may be accomplished by adjusting the high voltage on the MP-tube, but more likely by adjusting the amplifier. The pulse heights of the photopeaks of at least two standards are determined carefully from spectra generated under the same instrument operating conditions. The standards should have γ-energies that are well defined and produce photopeaks in different regions of the base setting range. In determining photopeak location, the maximum of the curve is sought and one-half the width of the window must be added to the base setting if the spectral plot does not have such corrections applied to the base settings already (see the section on pulse spectra). Finally, a plot of γ-energies (E_γ) versus photopeak locations in baseline units (U) is made, and it should yield a straight line with a slope of m and an intercept of b. Baseline units (pulse heights) and γ-ray energies may be interconverted by reference to the graph or calculated according to Equation 7-2.

$$E_\gamma = mU + b \tag{7-2}$$

A calibration graph for an instrument with a multichannel analyzer could be prepared in a similar manner except channel numbers would be used in place of baseline settings.

Instrument Calibration.   Instruments may have baseline units that are easily transposed into energy units, but such instruments may not have sufficient stability over a period of months for permanent factory calibration of the analyzer's scales. Calibration may be done by using an appropriate γ-ray standard and setting the baseline (or channel number in a multichannel analyzer) so it corresponds numerically to the γ-ray energy of the standard. Then the overall amplification (or gain) of the instrument is adjusted so that the photopeak is precisely situated in the center of the window.

Automatic Calibration.   Some instruments preform a calibration similar to the one described above automatically. Usually a separate analyzer channel is dedicated to calibration, and it has fixed baseline and window settings that are appropriate for a particular standard γ-source. With the standard source in the sample well the instrument's overall gain is adjusted automatically so that the source's photopeak precisely fits the dedicated channels' window. The working channel's scales (perhaps in units of keV or MeV) are such when the gain causes the photopeak to fit the dedicated channel's window the working channel'(s) scale is properly aligned.

Identification of γ-Emitters

Many unknown γ-emitters can be identified simply by determining the energies corresponding to their photopeaks and comparing them to published values for various radionuclides. However, ambiguities do arise, particularly when mixtures of γ-emitters are involved. Fortunately, other decay characteristics may be used to resolve such problems.

Emitters that have indistinguishable γ-ray energies may differ in half-lives. Measurement of photopeak areas (counting rate when the photopeak is centered in the window) as a function of time will allow half-lives to be determined. In neutron activation analysis where mixtures of short-lived radionuclides are usually produced, this technique is particularly useful. The energies of the rays are determined and photopeak areas are measured at various intervals (minutes or days).

A number of radionuclides emit more than one γ-ray. The ratio of their photopeak areas, for a given detector, is a characteristic of the radionuclide. Consider the spectrum shown in Figure 5-9 for ^{60}Co, which reveals two photopeaks, at 1.17 and 1.33 MeV. Although the emission rates of both rays are equal their photopeak areas differ somewhat, but, most importantly, their ratios should remain constant over time. Peak ratios that change with time obviously involve more than one radionuclide.

These three characteristics, peak energy, peak half-life, and peak ratios, are usually sufficient to identify radionuclides, even in complex mixtures. Of course, the process of comparing experimental data to known decay characteristics is easily accomplished by computers, and for such purposes the output from a multichannel analyzer may be converted to a form suitable for direct processing.

PRACTICAL ASPECTS OF QUANTITATIVE MEASUREMENTS

In many research studies the appropriate measurements require a counter to determine the quantity of a known γ-emitter. Practical equipment and procedures for such studies are discussed in this section.

General Instrument Features

A single-channel solid scintillation spectrometer with an NaI(Tl) crystal is perfectly adequate to measure the quantities of known γ-emitters. Instruments with two or three channels are useful for dual label studies, but the expense of a multichannel analyzer is usually warranted. Most modern in-

TABLE 7-1   Absorption of γ-Rays by NaI

Energy (MeV)	Approximate absorption (%) at a thickness of			
	1 cm	2 cm	3 cm	4 cm
0.1	>99	>99	>99	>99
0.2	70	91	97	>99
0.6	26	45	59	69
1.0	19	34	47	57
2.0	14	26	36	47
3.0	12	23	32	40

struments can be purchased with operator convenience features, such as automatic sample changing, data printing and storage, background subtraction, and capability for simple to complex calculations, for example, conversion of cpm to dpm.

The crystal is an important component affecting counting efficiency. Flat faced crystals yield about one-half the counting efficiency that the same size well-type crystal provides. Unless only low-energy rays are being counted, crystal size is an important determinant of counting efficiency. Table 7-1 shows the approximate absorption percentages for various thicknesses of NaI and different energy γ-rays. Although counting efficiency depends on instrumental conditions, the percentage absorbed represents the maximum counting efficiency attainable. Obviously, for energies between 0.6 and 3.0 MeV, a 3 × 3 in. crystal offers a significant advantage over a 1 × 1 in. crystal.

Spectral Regions for Counting

Integral counting conditions, where all pulses above the baseline are counted, yields the highest counting efficiency for a particular radionuclide. Unfortunately, it also yields the highest background. As will be developed in Chapter 13, the statistical validity of counting data is dependent on counting efficiency (CE) and background counting rate (R_b). Generally, one should strive to improve counting efficiency and reduce background. However, when the sample's counting rate is not more than about 10 times the background counting rate, one should maximize $(CE)^2/R_b$, which is called a figure of merit. Table 7-2 shows some experimental data for an instrument with a 2 × 2-in. crystal when various regions of a ^{137}Cs spectrum (see Figure 7-7) are counted. As indicated in Table 7-2, the highest figure of merit is usually obtained by counting the photopeak, primarily because the background is very low compared to a large region of the spectrum that includes the Compton region. Another advantage of using the photopeak region is that there is less likelihood that unsuspected contamination from other emitters would affect counting results.

TABLE 7-2 Experimental Counting Performance for Regions of a ^{137}Cs Spectrum

Spectral region (keV)	Counting efficiency (%)	Background (cpm)	Figure of merit $(CE)^2/R_b$
0–60	12.6	17	9.3
60–550	28.7	37	22.3
550–750	18.5	5	68.5
60–750	47.2	42	53.0
0–750	59.8	59	60.6
0–∞	59.9	65	55.2

Nevertheless, statistically valid results can be obtained using a larger region of the spectrum that includes the Compton pulses, for example 60–750 keV. For net counting rates above about 300 cpm, the statistical advantage of using the photopeak region as opposed to the larger region would be slight.

Amplifier and Analyzer Settings

To optimize counter performance the amplifier and analyzer must be adjusted properly. The amplifier affects the position of a photopeak, and the analyzer baseline and window must be set so the photopeak is centered in the window. The initial baseline setting is somewhat arbitrary unless one wishes to calibrate in energy units, but suppose the baseline may be adjusted between 0 and 1000 units.

For a medium energy emitter, such as ^{137}Cs, the base might be set at 500 and the window at 10% of the full base scale (100 units) initially. Amplifier settings are lowered to zero, and a relatively high activity sample is counted for a short period of time. Then the amplifier setting is increased by an increment and the sample recounted. This process is continued until the counting rate is observed to rise then fall, which indicates that amplification has caused the photopeak to enter, and with greater amplification, past beyond the window. Finer amplifier adjustments are used, either up or down, until the maximum counting rate is found. At this point the amplifier setting is established.

The optimum window width and base setting are determined by counting the sample and background with various window widths. Baseline settings need to be adjusted when window widths are changed to keep the photopeak centered in the window. For example, if the window is increased by 50 units, the base should be reduced by 25 units. From the counting data obtained, the figure of merit is calculated for each window width, and the one that yields the highest $(cpm)^2/R_b$ should be selected.

Preset switches for specific radionuclides are available for many instruments. These function switches involve amplifier gain, base and window settings appropriate for the particular radionuclide. Because an instrument's operating characteristics change over periods of time, such function switches may not provide absolute optimum settings for statistical purposes, but generally, they will be quite acceptable if the instrument is serviced and calibrated regularly. Sealed standards are available for many γ-emitters, and the practice of including a standard with each group of samples counted is a good practice since an unexpected change in the standard's counting rate provides a signal that instrument performance characteristics have changed.

Sample Preparation

Compared to liquid scintillation counting, the preparation of samples for solid scintillation counting is fairly simple. Sample self absorption is rarely a problem because of the penetrating nature of X- and γ-rays. Usually the mass of a sample is small, and little difference will be noted if it is counted as a solution, suspension, or in a dry form. However, it is important to prepare all samples to be compared in an identical manner so even slight variations in sample absorption are avoided. Disposable test tubes are frequently employed as sample containers for well-type crystals, and for a particular type of tube, the small variations in wall thickness have an insignificant effect on γ-ray absorption by the tube. In fact, for most γ-emitters absorption by the container walls is difficult to detect.

The placement of the sample in the tube can have an effect on counting efficiency because of geometric factors. Theoretically, the highest counting efficiency is realized when the sample exists as a point source at the geometric center of the crystal. Therefore, the ideal preparation technique is to have as small a mass as possible located on the bottom of the tube. However, confinement of the sample to the bottom 1 to 2 cm of a tube is quite acceptable in virtually all cases. Counting efficiency does decrease as the height of the sample in a tube increases, but the decrease is small. Therefore, small differences in sample heights will cause insignificant variations in counting efficiency.

Dual Label Counting

The assay of two radionuclides when present as varying mixtures can be accomplished with reasonable accuracy if respective photopeaks are separated sufficiently. Generally, the radionuclide with the higher photopeak energy will contribute pulses in the spectral region corresponding to the photopeak of the other γ-emitter. However, the lower energy emitter will contribute nothing in the photopeak region of the higher energy emitter. An example of such a situation is the assay of ^{125}I and ^{131}I mixtures.

The major photopeak of ^{131}I is at 364 keV, but a minor peak occurs at 32 keV, and pulses due to Compton interactions also occur in the lower energy portion of the spectrum. The major peak for ^{125}I is at 35 keV, and it cannot be separated from pulses that arise from ^{131}I, but the relative contributions of pulses in that region can be ascertained.

Operating conditions require two different channel settings for which a two channel instrument would be useful, but counting a sample twice with different settings accomplishes the same task. One channel (A) is set to count the photopeak of ^{131}I exclusively, and the other (channel B) is set to count the photopeak area of ^{125}I. Of course, ^{131}I, if present, would contribute counts in that channel also.

Using an ^{131}I standard, the gain, base and window settings are determined for photopeak counting of the major peak in channel A. The amplification should be high enough to separate the major and minor ^{131}I peaks. Using energy units, appropriate settings might be 314 keV for the base and 100 keV for the window width. The counting efficiency of ^{131}I in channel A (CE_A^{131}) would then be determined. Using an ^{125}I standard, channel B settings are established similarly, optimizing photopeak counting for the 35 keV ray of ^{125}I. Appropriate settings might be base, 20 keV and window width 50 keV. Then the counting efficiency of ^{125}I in channel B (CE_B^{125}) is determined. Without changing the settings of channel B, the ^{131}I standard is counted to ascertain the counting efficiency of ^{131}I in that channel (CE_B^{131}). For an experimental sample yielding net counting rates in channels A and B of cpm_A and cpm_B, respectively, the amounts of radioactivity of each isotope (dpm_{131} and dpm_{125}) may be calculated using Equations 7-3 and 7-4.

$$dpm_{131} = \frac{cpm_A}{CE_A^{131}} \tag{7-3}$$

$$dpm_{125} = \frac{cpm_B - (dpm_{131} \times CE_B^{131})}{CE_B^{125}} \tag{7-4}$$

A multichannel analyzer is a very effective device for counting samples that contain a dual label. Instead of adjusting analyzer and amplifier settings, the operator selects for read-out the slots that collect counts from the desired regions of the combined spectrum. Calculations similar to those described earlier can be programmed into the computer that is normally associated with a multichannel analyzer to yield data in dpm directly.

CHAPTER 8

Liquid Scintillation:
Theory of Detection
and Instrument Design

In biological research, basic and applied, liquid scintillation counting is used extensively and is particularly valuable for weak β^--emitters such as ^3H, ^{14}C, and ^{35}S because of excellent counting efficiencies, up to 60% for ^3H and 90% for ^{14}C and ^{35}S. Hard β-emitters, such as ^{32}P, yield efficiencies near 100%. Other types of emitters, α-, β^+-, γ-, and X-ray, can be detected also (see Chapter 10), but generally they are assayed by other methods.

OVERVIEW OF THE SCINTILLATION PROCESS

A radioactive sample, generally consisting of a particular radioisotope incorporated into an organic compound, is dissolved in a *liquid scintillation cocktail* or *liquid scintillation solution*. Cocktails are solutions of one or two species of fluorescent compounds, called *fluors* or *scintillators,* in an organic solvent. The solution is contained in a small vial (5–20 ml), which, when being counted, is placed between two multiplying phototubes (MP-tubes). Thus, a radioactive atom is surrounded by cocktail so when it disintegrates its emission(s) enter the cocktail. The kinds of interactions the particles or rays undergo are the same as those described in Chapter 5, but the production of electronically excited molecules is the critical process. Since β^--emitters are among the most common types of radionuclides encountered in liquid scintillation counting, the scintillation process initiated by a negatron will be considered.

After emission, all of the kinetic energy of a negatron will be dissipated in the surrounding solution creating many excited molecules. The excited solvent molecules transfer their energies to fluor molecules, which promotes them to excited states. The fluor molecules lose their energy of excitation by emitting one photon of light per molecule (the scintillation event), and when the photons strike the photocathode of the MP-tubes, the tubes produce an electrical pulse that, after electronic manipulation, is registered as a count.

THE SCINTILLATION PROCESS

The properties of solvents and fluors that are needed to prepare a suitable liquid scintillation cocktail will be discussed later, but most solvents are aromatic liquids and toluene is a common example. A fluorescent compound, abbreviated as PPO, is a common fluor. The concentration of the fluor is usually less than 1%, and the volume of the radioactive sample relative to the volume of cocktail (5–20 ml) generally is negligibly small. Consequently, solvent molecules are by far the most abundant type of molecule in the liquid scintillation cocktail.

When a negatron is emitted, its kinetic energy will be dissipated in the surrounding medium, primarily by causing ionization and excitation of the molecules of the medium. Because solvent molecules (toluene) are the most prevalent species, these interactions occur with the solvent predominately. No voltage is applied across the solution, so recombination of ions and free electrons occurs readily leading to more excited toluene molecules. Although excited molecules may lose their energies by several different processes, such as breaking internal bonds and emitting electromagnetic radiation (fluorescence), toluene is a useful solvent because a significant pathway for its deexcitation is the transfer of energy to a neighboring molecule. Because of relative concentrations, energy transfers between toluene molecules are the most prevalent. However, after numerous transfers it is likely that an excited toluene molecule may be produced next to a PPO molecule. The PPO molecule will readily accept the energy from an excited toluene molecule and become excited.

The energy transfer between toluene molecules is not a purely random process since fluor molecules seem to induce dipole moments in neighboring solvent molecules. An excited toluene molecule located 3 to 10 molecules from a PPO molecule will transfer its energy to the PPO by a highly direct route instead of a random route. That is, the fluor molecule acts as an energy sink when it is just a few molecules away from an excited solvent molecule.

After excitation, the PPO molecules give up their energies of excitation by emitting one photon each. Normally, fluorescent compounds are considered to be compounds that readily absorb light of one wavelength and reemit the energy as light with a slightly longer wavelength. The excited (singlet) state is an intermediate in the process that is completed in a few nanoseconds. A suitable fluor molecule, such as PPO, may be excited to the singlet state by energy transfer as well as by absorption of light, and, importantly, lose its energy of excitation by the emission of a photon of light rather than by other possible deexcitation processes.

The quantity of energy needed to produce the excited state and the wavelength of light emitted are important characteristics of fluors. Because the response of an MP-tube is dependent on the wavelength of light that strikes the photocathode, the wavelength of the emissions must overlap the response spectcrum for the MP-tube. That is, the light emitted must have a wavelength capable of causing an electrical pulse to be generated by the MP-tube.

Horrocks (1974) has summarized some of the data concerning the overall efficiency of the scintillation process. Measurements can be made by bombarding a cocktail with monoenergetic elecrons. Since the kinetic energies of the electrons are completely absorbed by the cocktail, from a measurement of the number of photons produced per electron and the wavelength of the emitted light, the overall scintillation efficiency (S_{eff}) can be calculated.

$$S_{eff} = \frac{\text{fluorescence energy}}{\text{electron energy}} \tag{8-1}$$

For the most efficient cocktails the scintillation efficiency is only about 4%, which means that on the average 96% of a β-particle's energy is dissipated by other processes. There are competing deexcitation processes for the solvent molecules, and the extent of competition is dependent on the composition of the solution. This is the basis for the phenomenon called chemical quenching, which will be discussed later.

The scintillation efficiency for a given cocktail is constant over a broad range of energies for electrons. Horrocks (1974) calculated that with a 4% scintillation efficiency, about 12.5 photons are produced per keV of energy dissipated. Thus, a 1.0-MeV negatron should yield about 12,500 photons, a 0.01-MeV negatron 125 photons, and on the average about 1 photon by a 100-eV electron. An important concept to remember is that although the scintillation efficiency is not high, the magnitude of the light pulse (number of photons) is directly proportional to the amount of energy dissipated in the solution.

The range of a 1.0-MeV negatron in a scintillation solution is less than 1.0 mm, so the scintillation process occurs within a small volume surrounding the location where the negatron's emission occurred. Although the photons from the fluor molecules are not emitted simultaneously, the emissions occur in a sufficiently short time period to be indistinguishable to an MP-tube. Also, photons may be emitted in all directions since the fluor molecules are not aligned in any way. A simplified picture of the scintillation event is as follows: The dissipation of the negatron's kinetic energy causes a flash of light, emitted in all directions, from a pinhead-sized volume somewhere in the vial, wherever the primary interactions took place. Further, the amount of light (number of photons) produced is directly proportional to the initial kinetic energy of the negatron that caused the event.

The flash of light is so small it cannot be seen by the human eye; in fact, the flash might be only one photon. Obviously, when only one or a few photons are produced, emission directions have an important bearing on whether the light will reach the MP-tubes. The simplified picture is a fair approximation for reasonably energetic negatrons, but somewhat lacking when very weak negatrons are considered.

The Scintillation Process with a Waveshifter

Many liquid scintillation cocktails contain two different fluorescent compounds. One present in the higher concentration is called the *primary fluor* or *primary scintillator;* the other is called a *waveshifter, secondary fluor,* or *secondary scintillator.* The purpose of the waveshifter is to change the wavelength of light emitted by the scintillation process. Although waveshifter molecules may absorb energy from solvent molecules, the primary fluor molecules are much more likely to receive the energy from the solvent because primary fluor molecules are more numerous than waveshifter molecules.

A suitable waveshifter must be capable of absorbing energy from the primary fluor, and the principal route for such a transfer is by nonradiative transfer. Because the emission spectrum of the primary fluor matches the excitation spectrum of an effective waveshifter, it appears that a primary fluor molecule emits a photon that is subsequently absorbed by a waveshifter. However, the transfer of excitation energy between the two types of molecules occurs in a time period much shorter than that required for radiative transfer (photon emission and absorption). Therefore, a more direct energy transfer process must be involved.

The inclusion of a waveshifter extends the scintillation process described in the previous section only slightly. The energy of the particle is dissipated in the solution by exciting solvent molecules that transfer their energies to other solvent molecules and eventually to the primary fluor molecules, some of which may emit photons. Most of the excitation energy of the primary fluor is absorbed by the waveshifter and the waveshifter emits the light at a longer wavelength. The proportionality of the scintillation process remains intact because, for a given cocktail, the number of photons produced is directly proportional to the amount of energy dissipated in the solution.

Waveshifters may provide an advantage if the shorter wavelengths of light produced by the primary fluor are not detected as efficiently by MP-tubes as are the longer wavelengths of light emitted by the waveshifter. PPO, a common primary fluor, has its fluorescence maximum at 365 nm, while that for dimethyl POPOP, a typical waveshifter, is 429 nm.

INSTRUMENT DESIGN AND FUNCTION

MP-Tubes for Liquid Scintillation

In Chapter 7 the general operating principles of an MP-tube were described, but some properties desirable for liquid scintillation counting will be discussed here.

The diameters of MP-tubes are about as large as the height of the normal 20-ml scintillation vials, therefore the tubes "view" nearly the entire vertical cross section of a vial.

Photocathodes of MP-tubes used in liquid scintillation counters generally consist of an alloy of alkali metals (Na, K, Cs) and antimony (Sb). Tubes referred to as bialkali tubes have two different alkali metals incorporated into their photocathode coating. A composition of K_2CsSb is one example.

The quantum efficiency of a photocathode is the fraction of photons that yields a photoelectron. Quantum efficiencies vary among MP-tubes and the wavelengths of the photons, but values as high as around 25–33% are common. Thus, on the average, three or four photons are required to produce a photoelectron, and hence produce an MP-tube pulse. Of course, one photon is capable of producing a photoelectron so another way of viewing the probability is that if single photons strike the photocathode, one of three or four times a pulse should be produced.

Quantum efficiency and, hence, sensitivity of an MP-tube are dependent on the wavelength of light. Consequently, MP-tubes show a spectral response and an example was shown in Figure 7-2 for a tube with a K_2CsSb photocathode. The maximum sensitivity (highest quantum efficiency) is near 400 nm for the MP-tube shown. Although sensitivity appears to drop sharply, it is quite acceptable in the region between 350 and 475 nm. Since commonly used fluors emit light with wavelengths in this region, an MP-tube with such a spectral response is effective in detecting scintillation events.

The magnitude of the voltage pulse produced by an MP-tube is dependent on the number of electrons released by the photocathode and the MP-tube's overall gain, which are functions of the number of dynodes and the operating voltage of the tube. However, the number of photoelectrons produced is dependent on the amount of light that strikes the photocathode and the wavelengths of those photons. For a given cocktail, the wavelengths of light will be determined by the fluors used, and the amount of light produced will be determined by the initial kinetic energies of the particles. Therefore, if an MP-tube is operated with a constant high voltage, the magnitude of its electrical pulses will be proportional to the energies of the particles that produce the scintillations. After linear amplification, the pulse spectrum produced by a particular β^--emitter will show a distribution that is similar to that of the emitter's energy spectrum.

Electronic Components in Liquid Scintillation Counters

Refinements in instrumental design have surpassed those shown in Figures 8-1 and 8-3, but they are presented to aid in describing basic functions of liquid scintillation counters. Figure 8-1 is a block diagram of a simple single MP-tube counter that shows high similarity to the single channel solid scintillation spectrometer presented in Chapter 7 (Figure 7-3). The solid scintillation crystal is replaced by a vial containing the radioactive sample and scintillation cocktail. Also, the pulse height analyzer of Figure 7-3 is labeled as a discriminator in Figure 8-1, although their basic functions are the same.

Via an elevator mechanism, the sample vial is placed close to the MP-tube's face in a light tight chamber that is shielded with lead to reduce background radiation, primarily cosmic rays. The MP-tube is mounted horizontally and is attached directly to the preamplifier that extends out from the counting chamber. In older instruments the counting chamber and sample changing mechanisms were located inside freezer chests to reduce thermionic emission (hence background pulses) by the MP-tube, and cables connected the preamplifier and MP-tube to the other instrumental components outside the freezer. With the improvement of MP-tubes, refrigeration became unnecessary, but it is still used sometimes to maintain a constant temperature. The scintillation process and MP-tube performance are affected by temperature, but, in most laboratories, ambient temperature fluctuations are not large enough to cause significant changes in counting efficiency or background counting rates.

The interior of the counting chamber, where the vial is placed, is painted with a reflective material or polished so that light that does not strike the MP-tube directly is likely to do so after one or two reflections. Of course, the shower of photons, or the light pulse, is converted to an electrical pulse by the MP-tube. The pulse is amplified initially by the preamplifier, then amplified and shaped by the amplifier, which may be called an attenuator in some instruments. The discriminator performs a pulse selection task similar to those described for discriminators in Chapter 6 and pulse height analyzers in Chapter 7. Therefore, depending on the magnitude of an in-

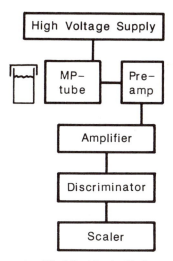

FIG. 8-1   Block diagram for a simplified liquid scintillation counter.

coming pulse, it is either rejected or sent on to the scaler to be registered as a count.

Most liquid scintillation counters have discriminators with separately adjustable lower and upper limits, and only pulses with heights between those limits are permitted to pass to the scaler. Recall that the range of pulse heights between the lower and upper limits is called the discriminator window, and its function is illustrated in Figure 8-2. Pulses numbered 2, 3, 5, and 6 are between the lower and upper limits (10–70) and would be allowed to pass to the scaler. Pulses numbered 1 and 7 would be screened out because they do not exceed the lower limit, while pulses 4 and 8 exceed the upper limit.

Because of MP-tube noise (weak, spurious pulses produced in total darkness, primarily by thermionic emission in the MP-tube) an instrument such as that shown in Figure 8-1 would have a high background counting rate. However, by proper adjustment of the lower limit of the discriminator most of the noise pulses can be eliminated without eliminating pulses arising from scintillation events.

A β-ray spectrum is continuous from zero to E_{max}. Therefore, negatrons with energies close to zero, say 0.1 keV or less, probably will not provide sufficient energy to yield a scintillation event, that is, cause the emission of at least one photon. Such low energy negatrons will not be detected regardless of the lower level discriminator setting.

Having an upper level limit in the discriminator also permits the reduction of background counts. Cosmic rays are very energetic and if one interacts with the cocktail, a very strong pulse is produced. For a particular radionuclide the upper level limit of the discriminator can be set slightly

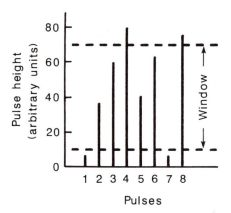

FIG. 8-2 Diagram of discrimination of pulses based on their heights.

above the pulse height that would result from a negatron being emitted with an energy of E_{max}. All pulses that exceed that setting are eliminated, and this reduces the instrument's background.

Since the E_{max} values for different negatron emitters vary, a discriminator window that would be appropriate for a weak emitter such as ^{3}H would not be satisfactory for a strong emitter such as ^{32}P. In fact, the energy differences between the negatrons from ^{32}P and ^{3}H are great enough that a sample containing a mixture of these radionuclides can be assayed for each type of activity. More will be said about dual-label counting in Chapter 9.

An important consideration in determining appropriate discriminator settings is the amount of amplification the electrical pulses receive before discrimination. Pulse amplification can occur at three points according to the block diagram of Figure 8-1.

1. **MP-tube voltage.** The high voltage applied to the MP-tube, and hence the voltage across each dynode, affects the overall gain of the MP-tube. For a given light pulse, higher voltages yield larger electron avalanches at the MP-tube's collector anode. For earlier instruments, operator controls were provided to adjust the high voltage of the MP-tube. Current instruments still have a means of adjusting the high voltage, but such settings are done at the factory or by service technicians. In most cases, the MP-tube gain is held constant.

2. **Preamplifier gain.** The preamplifier normally has no operator controls, and its function is to amplify the pulses before they leave the counting chamber. The amount of amplification by this component is constant for all counting conditions.

3. **Amplifier.** Depending on the brand of liquid scintillation counter, the amplifier may be a *linear* or *logarithmic* amplifier (see the discussion on amplifiers in Chapter 9). Amplification is normally under operator con-

trol when linear amplifiers, which may be called *attenuators,* are employed. For a particular radionuclide, correct adjustment of the amplifier will depend on the discriminator settings and vice versa. Optimum counting conditions require particular amplifier and discriminator settings, which will be different for different radionuclides. Optimum settings are determined experimentally by counting standard samples and will be discussed in Chapter 9.

Coincidence Counting

Coincidence counting has always been an important feature in liquid scintillation counters because it markedly reduces background due to MP-tube noise. Two MP-tubes are needed for coincidence counting, and a simple arrangement is illustrated in Figure 8-3. The components to the right of the sample vial (analyzer circuit) are identical to the ones shown in Figure 8-1 except for the addition of a coincidence gate located between the discriminator and scaler. On the left side, a second MP-tube has been added along with a preamplifier, amplifier, and discriminator. These components are called the monitor circuit.

If a scintillation event occurs, light will be radiated in all directions, and

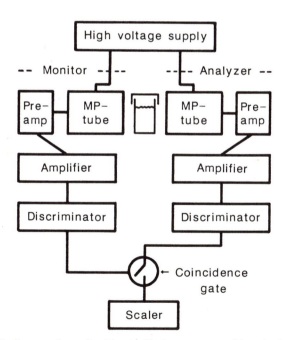

FIG. 8-3 Block diagram for a liquid scintillation counter with coincidence counting. Pulses from the analyzer circuit are sent to the scaler if the coincidence gate receives a pulse from the analyzer and monitor circuits simultaneously.

both MP-tubes should "see" the radiation simultaneously; therefore the pulses from both tubes should be *in coincidence*. Noise pulses are random events, so the probability of both tubes producing a noise pulse simultaneously is small, particularly if discriminators are in the circuit to screen out the very weak pulses. Determining whether pulses from the MP-tubes are in coincidence provides an excellent means of distinguishing pulses that arise from a scintillation event from those due to MP-tube noise. This is accomplished by the use of a device called a coincidence gate.

The gate can be visualized as a switch that when open prevents the flow of a pulse from the analyzer's discriminator to the scaler. When closed, a pulse in the analyzer circuit would flow to the scaler and be registered as a count. Normally, the switch is open, but a pulse from the monitor channel closes the switch momentarily and allows a pulse from the analyzer circuit to pass to the scaler. Obviously, noncoincident pulses in either channel will not produce a count.

The use of discriminators and coincidence counting circuits almost completely eliminates the counting of weak noise pulses. In modern instruments, MP-tube noise pulses probably contribute less than 1 cpm to the background counting rate.

Pulse Summation

Older instruments with designs comparable to that shown in Figure 8-3 yielded counting efficiencies for ^{3}H and ^{14}C in the order of 35 and 60%, respectively. The incorporation of an electronic function called *pulse summation* causes a significant improvement in counting efficiencies. MP-tubes have been improved and have contributed to higher counting efficiencies also. The concepts of pulse summation, coincidence counting, and pulse discrimination are illustrated in the design shown in Figure 8-4. The function of components labeled as channel 2 should be ignored for the time being.

The logic in this design is that the pulses from both MP-tubes are sent to the summing amplifier where they are added together. The coincidence circuit simply monitors the pulses from both MP-tubes, and if the pulses are in coincidence, the coincidence gate allows the summed pulse to pass on to the amplifier. If a pulse from only one tube is produced, pulse summation will not increase its magnitude and the coincidence circuit will not allow the pulse to pass. Consequently, the coincidence circuit performs a function similar to that described previously, but it is located ahead of the amplifier and discriminator.

The pulse summation function is particularly useful in distinguishing between very weak scintillation events and MP-tube noise pulses, and it serves to amplify all pulses sent to the amplifier and discriminator. To see

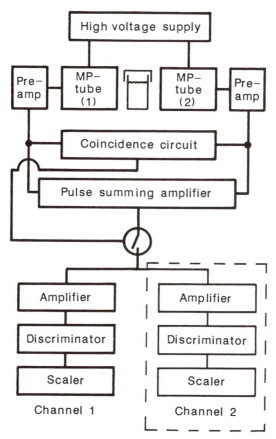

FIG. 8-4 Block diagram for a liquid scintillation counter with pulse summation and dual channels.

the advantage of pulse summation, compare the designs in Figures 8-3 and 8-4.

Although it takes several photons to produce one photoelectron at the photocathode, the process is proportional and is more easily visualized if a 1 to 1 ratio is assumed; thus pulse heights may be related in photons units. Suppose five photons are emitted in a scintillation event. Because of the small number, the distribution of photons between the MP-tubes could be very unbalanced as indicated in Table 8-1. For the instrument in Figure 8-3, four of the six possible distributions would produce pulses capable of being counted, but the pulse height varies from 1 to 4. For the instrument in Figure 8-4, the same four distributions would produce countable pulses, but all have an equal value that are higher than any of those produced without pulse summation. For a weak scintillation event, pulse summation increases the pulse height and thereby increases the probability of distin-

TABLE 8-1 The Effect of Pulse Summation on the Magnitude of Pulses Produced by Weak Negatrons in Liquid Scintillation Counters

Number of photons collected		Pulse height in photon units	
Monitor or MP-tube 1	Analyzer or MP-tube 2	Analyzer only (Figure 8-3)	Summed (Figure 8-4)
5	0	0^a	5^a
4	1	1	5
3	2	2	5
2	3	3	5
1	4	4	5
0	5	5^a	5^a

[a]No pulse would be counted because of a lack of coincidence.

guishing it from rare coincident noise pulses when it reaches the discriminator.

Figure 8-4 indicates that the coincident pulses may be analyzed, rejected, or counted by an additional channel. Effectively, the output from the coincidence gate may be sent to each of two or more channels simultaneously. Each channel consists of identical sets of components including an amplifier, discriminator, and scaler or device for recording counts. All channels, including the amplifiers and discriminators, can be adjusted independently. Therefore, a two-channel instrument allows a sample to be counted under two different conditions simultaneously. Similar data could be obtained from a one-channel instrument by counting the sample under one set of conditions, changing the settings, and recounting, but it would be more time consuming.

Most modern instruments have more than one channel. Such instruments are quite useful for counting samples that contain two different radionuclides (dual label counting) and for determining quenching. Quenching refers to processes that result in lower counting efficiencies for samples. This topic will be discussed in a later section.

Pulse Spectra

A pulse spectrum shows the distribution of electronic pulse heights for a sample, and graphically it represents a plot of the number of pulses versus pulse height. Since the amplifier changes the magnitude of pulses it receives from the preamplifier, the general features of spectra differ depending on the type (linear or logarithmic) of amplifier employed.

Linear amplification will be discussed first, and a typical pulse spectrum for ^{14}C is shown in Figure 8-5. This spectrum was obtained by repeatedly counting a ^{14}C-sample with different discriminator settings. The range of settings for both the lower and upper level controls was 0 to 1000 pulse height units (arbitrary), and the amplifier had been adjusted so that a majority of the pulses had heights between 50 and 1000 units.

To initiate the process, the lower discriminator was set at 50 and upper at 75, then the sample was counted for a minute. The result represented the number of pulses having a pulse height between 50 and 75 units, and these data were used to form the first bar in the graph (Figure 8-5). Then the upper discriminator limit was raised to 100 units, and the sample was recounted for the same period of time yielding the number of pulses having heights between 50 and 100 units. Subtraction of the previous counts gave the number of pulses having heights between 75 and 100 units, which was plotted as the second bar. The process was continued by repeatedly raising the upper level discriminator by 25 units, recounting the sample, and subtracting the previous count until the entire discriminator range had been covered.

An alternative method would have been to start as before, but raise both discriminators by 25 units between counting periods. This method eliminates the need for subtraction of counts and should yield similar results; however, its success depends on perfect alignment of the lower and upper discriminator limits and high precision in reproducing settings. By either method, considerable variation in counts for narrow discriminator windows is to be expected because of the lack of precision in setting the discriminators and the difficulty in overcoming statistical counting errors (see Chapter 13).

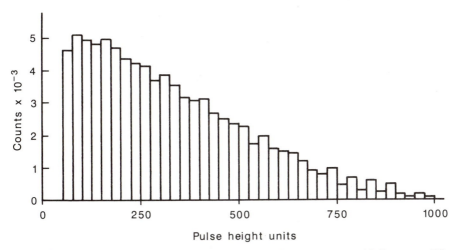

FIG. 8-5 A pulse spectrum of ^{14}C from a liquid scintillation counter with linear amplification.

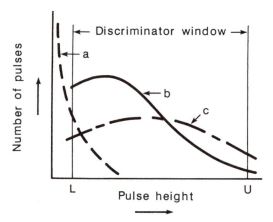

FIG. 8-6   Pulse spectra for ^{14}C with different levels of linear amplification.

Theoretically, if the discriminator controls are set at 50 (lower) and 1000 (upper), the number of counts expected (within statistical probabilities) would be the sum of the values for the series of bars covering that range of pulse heights (50–1000). Thus, for curves such as those shown in Figure 8-6, the number of counts for a given discriminator window is proportional to the area under the corresponding portion of the curve.

For pulses with a given distribution of pulse height values, greater amplification causes all pulse heights to be increased and increases the width of their distribution. Thus, with greater amplification pulse spectral curves are shifted to the right (higher pulse heights) and become flatter. Curves a, b, and c in Figure 8-6 represent increasing degrees of amplification. The pulses of curve a have a fairly narrow range of heights and the majority of them have heights below the lower limit of the discriminator window. Counting efficiency would be very low under such instrument settings. Increasing the degree of amplification, curve b, shifts the curve to the right (higher pulse heights) and lowers the maximum of the curve. The area under the curves between the discriminator limits is considerably larger for curve b than curve a, which means that a much higher counting efficiency would be obtained with curve b settings. Excessive amplification, curve c, causes many of the pulses to have heights that exceed the upper limit of the discriminator window, and a very wide range of pulse heights is obtained.

For a given discriminator window, the amplification should be adjusted so that the area under the pulse spectrum curve between the discriminator's limits is at a maximum, because that yields the maximum counting rate for the sample, hence, maximum counting efficiency. The amplifier setting that optimally situates the curve between the discriminator limits is called the *balance point*.

Balance Point. The balance point for a particular discriminator window can be determined experimentally by alternately adjusting the amplifier and recounting a standard sample. Plotting counting rate versus amplifier setting yields a curve that exhibits a maximum called the *balance point*. A family of curves may be produced by using different discriminator windows as shown in Figure 8-7.

Consider the curve for the 50–1000 window. In the region between 0 to 5 amplifier units, the counting rate increases rapidly, which indicates that large portions of the pulse spectrum are being shifted above the lower discriminator limit as the amplifier settings are advanced. The maximum of the curve is at 8.5 units, and this represents the balance point for the 50–1000 window. As the balance point is approached, the pulse spectrum should extend across the discriminator window, and at the balance point, a few pulses should be above the upper discriminator limit and an equal

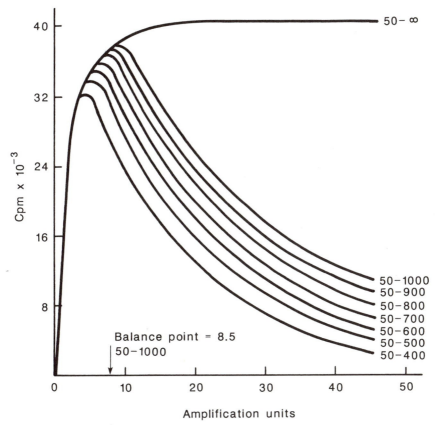

FIG. 8-7 Balance point curves for various discriminator windows from an instrument with linear amplification.

number below the lower limit. Amplification beyond the balance point causes more pulses to be lost above the upper limit than are gained from below the lower limit, therefore the counting rate falls as the amplification is increased further.

Counting samples at their balance point is desirable because that yields the maximum counting efficiency for those particular samples and window combinations. The slope of the curve is a minimum at the balance point, so if small changes in the instruments' amplification were to occur it would have the least effect on counting efficiency at the balance point setting.

Comparison of the curves in Figure 8-7 shows that narrower windows yield lower counting rates and, hence, lower counting efficiencies. If the lower discriminator limit is held constant while the upper limit is reduced, the balance points shift to lower values, and all curves are coincident at very low amplification because the amplification is not sufficient to cause any of the pulse heights to exceed the upper limit.

Most counters have discriminators that permit the upper limits to be deactivated so that all pulses that exceed the lower limit are counted. Since a pulse of infinite magnitude would be recorded as a count, such a discriminator window is referred to as L–∞, and counting under these conditions is called integral counting. For the curve in Figure 8-7 labeled 50–∞, no balance point exists because pulses cannot be amplified too much. Obviously, slightly higher counting efficiencies are possible by integral counting, but integral counting also yields a significantly higher background counting rate due to very energetic background radiation.

For statistical reasons, it is desirable to select counting conditions that maximize $(CE)^2/R_b$, where CE is the counting efficiency and R_b the background counting rate (see Chapter 13). This ratio is called a "figure of merit." Because the background counting rate for integral counting is much greater than that for a given window such as 50–1000 and the difference in counting efficiencies is small, the figure of merit is usually lower. For that reason integral counting is not used extensively.

The same statistical considerations are applied in the selection of discriminator windows. Wider windows yield higher counting efficiencies and higher backgrounds, but the question becomes a matter of relative differences in background and efficiency for different windows. Usually, a wide window (50–1000 or 50–900) yields the highest figure of merit, but the difference between the figure of merit for similar windows, e.g., 50–1000 and 50–900, may be almost trivial for many radionuclides.

Logarithmic Amplification

Instruments that employ logarithmic amplifications do not have separate amplifiers for each channel as shown in Figure 8-8. The output pulse from the summing amplifier is converted to a pulse proportional to the logarithm

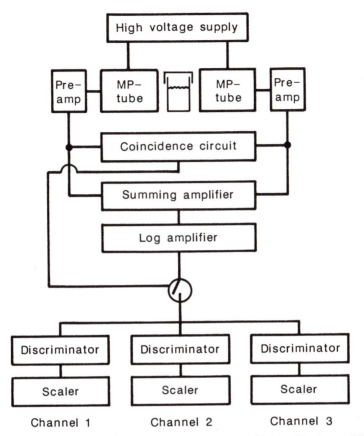

FIG. 8-8 Block diagram for a liquid scintillation counter with logarithmic amplification.

of the summed pulse. All pulses are subject to the same conversion and are amplified sufficiently for adequate discrimination or pulse height analysis. A wider range of pulse heights results, although the logarithmic scale makes pulse spectra appear more compressed. In terms of β-particle energies, the range of pulse heights sent to the discriminator may be between zero and about 2 MeV, which is appropriate for weak emitters such as ^{3}H (E_{max} = 0.018 MeV) and strong emitters such as ^{32}P (E_{max} = 1.71 MeV). Superimposed, logarithmic pulse spectra for ^{3}H, ^{14}C, and ^{32}P are shown in Figure 8-9.

In instruments with logarithmic amplification pulse selection (the pulses sent to the scaler) is determined by the discriminator exclusively, whereas instruments with linear amplification use the amplifier and discriminator in conjunction with one another. As far as practical performance merits, such as counting efficiencies, background counting rates, and dual label counting sensitivities, neither type of amplifier is significantly superior to the other.

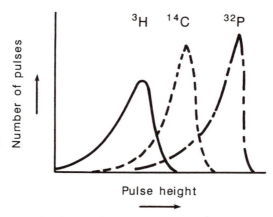

FIG. 8-9 Pulse spectra for three radionuclides from an instrument with logarithmic amplification.

Preset Controls for Instruments

Many modern instruments do not have adjustable controls for the discriminators or amplifiers; instead they have push button switches or plugs for various radionuclides. Such switches simply activate certain discriminator windows and amplifier settings. Using appropriate standards, electronics technicians may adjust the preset counting conditions (discriminator and amplifier settings) to optimize instrument performance, but normally an operator cannot alter those adjustments.

Since the balance point for a particular discriminator window may be affected by the composition of the liquid scintillation solution, counting may not occur at the balance point, nor with the highest figure of merit under preset conditions. However, except for counting rates close to background, counting conditions slightly less than optimum generally are quite acceptable from a statistical point of view.

Instruments with Microprocessors

With the development of microprocessor technology, data acquisition, storage, and analysis have become integral functions in many types of modern instruments, including liquid scintillation counters. Figure 8-10 shows a block diagram of such a counter. Although components are somewhat different, coincidence, pulse summation, and amplification units operate in the way described for instruments, such as the one diagrammed in Figure 8-4. The difference is that the pulses from the amplifier go to an analog to digital converter (ADC) instead of to a discriminator and scaler, and the ADC converts heights of pulses to a digital value. Although energy units, such as keV, are not correct for pulse height measurements, the digital values can be expressed in keV because of proportionality factors. Since there is no discriminator in the system before the ADC, all summed coincident pulses are received by the ADC, and the coincidence circuit provides a signal that indicates whether a particular pulse should be rejected or processed. Consequently, the few weak noise pulses that happen to be in coincidence, other background pulses as well as pulses from sample emissions are received and processed by the ADC.

The ADC feeds data to the analyzer, which consists of a bank of memory slots arranged in order of increasing pulse height values. A count is registered in the slot that corresponds to its digital height value (energy). In effect, the analyzer is like a series of several thousand discriminators, with tiny windows and each with its own scaler. The upper level setting for a particular discriminator matches the lower level setting for the next higher discriminator and its lower level setting matches the upper level setting for the discriminator immediately below it. The analyzer may contain 4000 slots or more. If the width of each slot corresponds to 0.5 keV and the analyzer has 4000 slots, a range of energies between zero and 2000 keV can be accommodated. Such conditions would be appropriate for ^{32}P

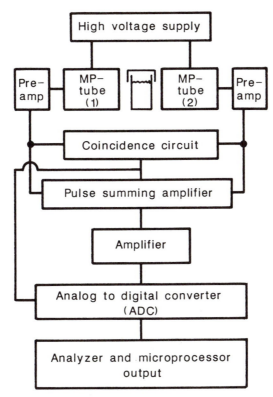

FIG. 8-10 Block diagram for a liquid scintillation counter with an analog-to-digital converter and microprocessor.

which has an E_{max} of 1710 keV. After sufficient counting time, the number of pulses or counts registered in each slot are determined. If the number of counts for each slot are plotted against its energy position, a pulse spectrum similar to Figure 8-5 is obtained. Of course, the pulse height increments would be smaller and the shape of the spectrum would be different since ^{32}P has an energy spectrum different from ^{14}C. In effect, what has been described is a multichannel analyzer dedicated to liquid scintillation counting.

For assays, it is desirable to exclude as many of the background counts and include as many of the sample pulses as possible. For a ^{32}P-sample, one might select the readout to include the sum of pulses between 1.0 and 1710 keV. The slots below 1 keV would have mostly noise counts, whereas any count above 1710 keV would not be due to a ^{32}P-emission, so those counts should be rejected. In instruments with discriminators, the operator controls which pulses will be rejected by adjusting discriminator settings. Rejection of pulses with inappropriate energies occurs during the counting process and are never recorded. With the ADC and analyzer, all pulses are recorded and rejection can be performed after counting is complete. The major advantage is that, by selecting different readout parameters for the same sample, the operator can determine what he or she might be rejecting and, hence, detect possible counting problems, such as unusual quenching (reduced counting efficiency due to sample compositions) or contamination with other radionuclides. Of course, with a discriminator-type instrument, recounting a sample with different discriminator settings would yield similar diagnostic information, but that would be more time consuming. Predetermined readout parameters for different radionuclides may be provided so simple operator controls are readily available on instruments with microprocessors.

QUENCHING

The composition of the liquid scintillation sample, including the solvent, fluor(s), radionuclide, and other components that might be present, affects its scintillation efficiency. Quenching refers to conditions that lower the scintillation efficiency and, hence, the counting efficiency as a result of changes in sample compositions. At first glance it would appear that this problem could be circumvented easily by keeping compositions constant. However, the radionuclide sample to be assayed must have a finite size and be mixed with the scintillation solution. Therefore, if samples to be assayed differ at all in size or composition, quenching is a potential problem. Oxygen is a quenching agent, and counting solutions saturated with air are quenched relative to the same solution that has been flushed with an inert gas such as N_2 or Ar. Fortunately, there are good methods for detecting quenching and applying counting data corrections, so the problem is not as severe as it might appear. There are two types of quenching: color and chemical quenching.

Color Quenching

The samples to be assayed may contain chemicals that absorb light of the wavelengths emitted by the scintillation fluors. If a significant fraction of the light emitted by the fluor molecules is absorbed and dissipated in the solution instead of being absorbed by the photocathode of an MP-tube, the magnitude of the pulse from the MP-tube will be lower than it would have been if light absorption had not occurred. Absorption of some fraction of the scintillation light by the sample itself causes a fairly uniform decrease in pulse heights. If an instrument is adjusted to its balance point for an unquenched (noncolored) sample, (i.e., the pulse spectrum is optimally situated in the discriminator window), then the inclusion of a light absorber (colored compound) has an effect similar to lowering the amplifier gain. This reduces the counting efficiency because a slight reduction of all pulse heights causes some of the weaker pulses to fall below the lower discriminator limit.

Chemical Quenching

The inclusion of certain chemicals, such as water, alcohols, and chlorinated hydrocarbons, may cause quenching if those compounds are capable of absorbing some of the energy from excited solvent molecules and do not transfer the energy to fluor molecules. Thus, chemical quenching agents compete with the fluor molecules in the absorption of energy from the excited solvent molecules, and quenchers use that energy to undergo chemical reactions rather than emit light. In this kind of quenching it is the efficiency of the energy transfer process that is affected. Quenching causes a decrease in the amount of energy that is absorbed by the fluor molecule relative to the amount of energy dissipated in the solution. Consequently, the amount of light emitted is reduced, which reduces the magnitude of the pulses and the sample's counting efficiency. The concentration of the quenching agent and its chemical properties are important factors in determining the degree of quenching. Compounds containing halogens (chloroform) or oxygen (water and acetone) generally are strong chemical quenching agents. As mentioned previously, the oxygen in air is a chemical quenching agent.

Quench correction methods generally involve determining the counting efficiency for each sample, then converting the counting rate (cpm) to a disintegration rate (dpm). Methods for determining the counting efficiencies of a sample during the time it is being assayed are available; so quench corrections are readily made on a routine basis. Those methods will be discussed in Chapter 9.

REVIEW OF THE SCINTILLATION PROCESS

Figure 8-11 is a diagram of the events and flow of energy in the scintillation process. All of the types of radioactive rays (α, β, γ) are capable of interacting with liquid scintillation cocktails and producing ions and excited

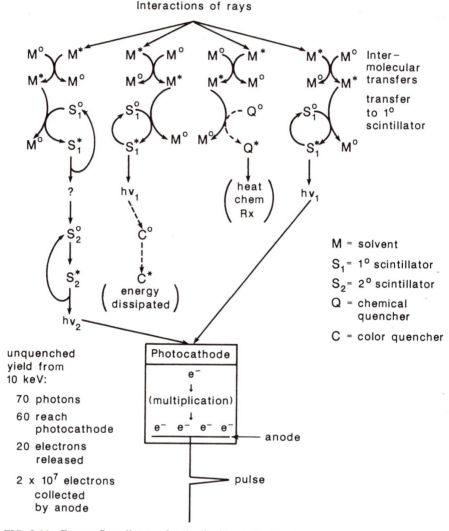

FIG. 8-11   Energy flow diagram for the liquid scintillation process.

molecules. Recombinations of ions and electrons yield more excited molecules. Since the major molecular species present is solvent, most of these interactions occur with the solvent molecules. The excited solvent molecules transfer their energies to surrounding molecules that, most likely, are other solvent molecules, and they in turn will transfer their energies of excitation to surrounding molecules. Several intramolecular transfers increase the probability that an excited solvent molecule will be located next to a molecule that is something other than a solvent molecule. If the neighboring molecule is a chemical quenching agent and it receives excitation energy, that portion of energy may be dissipated as heat or by a chemical reaction. If the neighboring molecule is a primary fluor molecule it may become excited and subsequently loose its energy of excitation by emitting a photon of light or by transferring energy to a secondary fluor molecule that subsequently emits a photon of light. The photons may be absorbed by a color quenching agent, or the photocathode of one of the MP-tubes. A large fraction of the photons emitted by the waveshifter molecules will reach the photocathode of the two MP-tubes if color quenching agents are absent. Color quenchers may absorb energy from either the primary fluor or waveshifter molecules, and hence attenuate the amount of light emitted by the scintillation event.

The photocathodes absorb light and release electrons that are multiplied as they pass through the series of dynodes. An avalanche of electrons collected at the anode of the MP-tube creates an electrical pulse that may then be amplified, discriminated against, and perhaps recorded as a count by the instrument.

REFERENCES

Birks, J. B. (1964). *The Theory and Practice of Scintillation Counting,* pp. 241–244. Pergamon Press, Oxford.

Hasing, J. W., and Weber, G. (1963). *J. Opt. Soc. Am.* 53:1410.

Horrocks, D. L. (1974). *Applications of Liquid Scintillation Counting,* pp. 26–31. Academic Press, New York.

Skarstad, P., Ma, R., and Lipsky, S. (1968). *Mol. Crystallogr.* 4:3.

CHAPTER 9

Liquid Scintillation:
Practical Aspects

LIQUID SCINTILLATION SAMPLES

For our discussions, a *liquid scintillation sample* refers to a vial containing a *counting solution* that is placed in a counter for assay. The counting solution may be a true solution or a homogeneous dispersion of a *radioactive sample* in a *cocktail*. Generally, a radioactive sample contains more than just a radionuclide, and its composition, which frequently includes quenching agents, contributes significantly to the performance characteristics of the counting solution. To a large extent the properties of the counting solution are determined by the characteristics of the cocktail that consists of a *solvent* and a *fluor* or *fluors*. The solvent constitutes the bulk of the counting solution, and consequently its molecules are the principal ones involved in the absorption of energy and its transfer to the fluor molecules that convert it to light.

A desirable counting solution has a high scintillation efficiency, which means it readily absorbs energy from β-particles and efficiently converts it to light of a wavelength suitable for detection by the MP-tubes of a counter. Counting efficiency is affected by scintillation efficiency, but the difference in terms deserves further clarification.

Consider the production and fates of pulses from 10-keV β-particles in two different counting solutions. In one solution each particle yields a 125-photon scintillation event, while in the second solution each particle yields a 100-photon event. Obviously, the second solution has a lower scintillation efficiency, but in both cases there is a sufficient number of photons produced so that some light strikes both of the MP-tubes, yielding coincident pulses that subsequently are summed, amplified, processed by the discriminator, and counted. The pulse heights from 10-keV particles would be different for the two counting solutions. For example, suppose the instrument's gain (amplification) factor was 1.0 arbitrary unit, each particle in the high efficiency solution should produce a pulse with a height of about 125 units, while in the low efficiency solution the height of pulses from 10-keV particles would be about 100 units.

The pulse selection parameters may be (1) amplifier and discriminator settings for a linearly amplified counter, (2) discriminator settings for a logarithmically amplified counter, or (3) channels or slots selected for readout with a multichannel analyzer. If pulse selection parameters were set at 110 to 1000 units, the counting efficiency of 10-keV particles in the high efficiency solution would be 100%, while it would be 0% for the low efficiency solution. The counting efficiency for the low efficiency solution could be rectified by using either of two methods. (1) Increase the amplification so the gain factor is at least 1.1. This condition would yield pulses with heights of 137.5 and 110 units in the high and low efficiency solutions, respectively. (2) Lower the lower level of the pulse height limit to a value slightly below 100 units. Either of these changes in the pulse selection function would have caused all of the pulses in the low efficiency solution to be sent on to the scaler or otherwise be registered as counts. The point is, if coincident pulses are produced by the MP-tubes, the summed pulse will be "countable," and whether it is or not depends on the pulse selection function, i.e., the gain of the instrument and position of the discriminator window or analyzer pulse selection parameters.

The situation is different when very low energy particles are involved. Theoretically, the minimum light pulse that yields a countable pulse is 2 photons, when one strikes each of the two MP-tubes, and each causes the release of one photoelectron from the photocathode. Keep in mind that the proportionality factors used above are arbitrary average values (80 and 100 eV per photon); also among individual scintillation events the photon yield varies somewhat. The energy of a photon is discrete, and when particle energies are low the variation in photon yield per particle becomes more noticeable. Obviously, a 2.5-photon event cannot happen; it is either 3 or 2 photons.

In the high efficiency solution, 160-eV particles should yield 2 photons on the average, but some individual particles might yield 5 photons each and some none. In the low efficiency solution the average yield would be 1.6 photons per 160-eV particle. Some particles may yield sufficient photons for a countable pulse while others would not. However, a greater percentage of the 160-eV particles would produce countable pulses in the high efficiency solution than in the low one. Therefore, the high efficiency solution would yield a higher counting efficiency for weak particles if the pulse selection function were set properly.

Obviously, β-particles must have some minimum amount of energy to produce a countable pulse in a given counting solution. However, the minimum is better represented as a "borderline range," and particles with energies within this range would yield countable pulses in some instances, but would not do so in 100% of the cases. Above the borderline range, the probability of producing a countable pulse is 100% while it is 0% below the borderline. The scintillation efficiency of a counting solution determines the borderline range, therefore a higher efficiency solution would have a lower borderline range.

Beta ray spectra are continuous from zero to E_{max}; so every radionuclide will emit some β-particles with energies below the borderline range. For a given counting solution, the maximum counting efficiency is limited by the fraction of particles emitted with energies below the borderline range. Tritium ($E_{max} = 0.018$ MeV) emits a larger fraction of its β-particles with low energies that does ^{32}P ($E_{max} = 1.71$ MeV). Consequently, regardless of the borderline energy value of the counting solution, tritium should have a larger fraction of its particles with energies below the borderline range than would ^{32}P, hence higher maximum counting efficiencies should be expected for ^{32}P than ^{3}H. In the most efficient counting solutions available today, the maximum counting efficiencies for ^{3}H and ^{32}P are around 60 and 95%, respectively. If the borderline range is raised (a counting solution with a lower scintillation efficiency is used) the counting efficiency of ^{3}H will be reduced more than that for ^{32}P; for example, suppose ^{32}P counting efficiency is reduced by 5%, ^{3}H counting efficiency might be reduced by 20 or 30%.

The conclusion to be drawn is that the scintillation efficiency of a counting solution has a greater effect on the counting efficiency of the solution when weak emitters are involved. For that reason selection of cocktails and other facets of sample preparation are more important when a low energy emitter such as ^{3}H is to be assayed by liquid scintillation counting.

Cocktails

A desirable cocktail has a high scintillation efficiency and is capable of dissolving sufficient quantities of various radioactive samples. However, the chemical and physical properties of radioactive samples vary from substances readily soluble in organic solvents to aqueous solutions of salts and biological compounds. This precludes the formulation of an ideal universal cocktail. Nevertheless, efficient cocktails can be formulated for most sample applications, and a number of ready-to-use cocktails are available commercially. Since cocktails are laboratory expendables and frequently are used in large quantities, factors such as cost, handling hazards, and disposal problems deserve consideration in cocktail selection.

Solvents.   Desirable solvent characteristics include a capacity to dissolve suitable quantities of the fluors, which are organic compounds, and sufficient quantities of the samples to be assayed. However, solubility properties mean little unless the solvent is efficient in absorbing radiation energy and transferring it to fluor molecules.

In general, compounds with numerous functional groups containing heteroatoms are poor solvents. Alcohols, ketones, and esters are not used as primary solvents, although they may be used as secondary solvents to improve the solubility characteristics of the cocktail. Saturated aliphatics such as hexane and cyclohexane are acceptable solvents, but their scintillation efficiencies are not nearly as good as those for aromatic hydrocarbons. In aromatic compounds there are an abundance of π-bonds and the electrons involved in such bonds readily absorb energy to produce molecular excited states. In addition, the energy of excitation is readily transferred to surrounding molecules. This explains why aromatic compounds are among the most efficient solvents for liquid scintillation purposes.

Toluene has been the standard against which most solvents are judged for many years. It is available with adequate purity at a comparatively low cost and is an efficient solvent. However, it does not dissolve water appreciably and has a low flash point, 4°C (temperature at which its rate of vaporization is sufficient to form an ignitable mixture in air).

The xylene isomers have been used extensively and are also efficient solvents, in fact, *p*-xylene is somewhat more efficient than toluene while the *o*- and *m*-isomers are about the same as toluene. Because of the cost in separating isomers, "xylenes," an isomeric mixture, is frequently employed. The flash point of "xylenes" is about 29°C, which is above normal laboratory temperatures.

Among the solvents discussed in this chapter, pseudocumene (1,2,4-trimethylbenzene) has the highest scintillation efficiency, and it has a high flash point, 48°C. Although more expensive than toluene and xylene, it is becoming a popular solvent that is used in many commercial cocktails.

In the past, *p*-dioxane was used extensively to increase the water-holding capacity of cocktails. Its scintillation efficiency is about 65% of that for toluene when it is in a highly pure form. Unfortunately, *p*-dioxane decomposes readily to form strong chemical quenching agents that reduce scintillation efficiency; furthermore it is on the cancer suspect list and has a low flash point, 12°C. Generally, *p*-dioxane is used in combination with other solvents such as naphthalene and xylene, but good emulsifier systems have largely replaced dioxane-based cocktails.

Naphthalene, although a solid, behaves as a solvent when incorporated in high concentrations (10–50%) with other solvents. Frequently, it may improve solubility characteristics, but its primary use is to improve the scintillation efficiency of the primary solvent, for example, *p*-dioxane. The odor of moth balls is not uncommon among cocktails.

Detergents, primarily Triton X-100 (octylphenoxypolyethoxy ethanol), are used with aromatic solvents to form cocktails that hold large quantities of aqueous samples. In fact, Triton X-100 may be used as the only solvent. Although its scintillation efficiency is not high, it has excellent emulsification properties, is nontoxic, has low flammability, and is biodegradable, which simplifies disposal of spent cocktails. Counting solutions prepared with Triton X-100 and aqueous samples are not true solutions; instead under appropriate conditions they form stable emulsions.

Fluors or Scintillators. Some primary and secondary scintillators, along with their properties, are given in Table 9-1. Among those listed, PPO, butyl-PBD, and BBOT are considered primary fluors while bis-MSB, PO-POP, and dimethyl-POPOP are secondary scintillators. The secondary scintillators would be suitable primary fluors if their solubilities in aromatic solvents were larger. Note that their fluorescence wavelength maxima are considerably greater than those of PPO and butyl-PBD. Before the advent of the bialkali tubes, the spectral response of MP-tubes was not nearly as good in the 360–370 nm region compared to the 410–430 nm region, therefore the inclusion of a secondary fluor improved scintillation efficiency significantly.

BBOT gained popularity because it is quite soluble in aromatic solvents and has a longer fluorescence wavelength, negating the need for a secondary fluor. Usually, secondary fluors are not necessary if bialkali MP-tubes are used. For ^{14}C and ^{32}P samples, a simple cocktail consisting of 4 g PPO/liter of toluene yields excellent counting efficiencies that are not improved significantly by the inclusion of a waveshifter. Secondary fluors may improve counting efficiencies when tritium is assayed or when high levels of quenching, particularly color quenching, are encountered.

The optimum concentrations for the fluors vary somewhat, but in general counting efficiency rises with concentration up to about 3 g/liter, then becomes fairly constant over a considerable concentration range. In the case of PPO, counting efficiency decreases as concentration increases beyond about 7 g/liter. This is due to self-quenching. Fortunately, most fluors show little concentration dependence when used near their optimum concentration; therefore slight dilutions due to sample additions do not cause appreciable changes in counting efficiencies. The fluors mentioned in this section are available commercially and some are offered as blends, ready for dissolution in selected solvents.

TABLE 9-1 Some Fluors Used in Liquid Scintillation Cocktails

Abbreviated name	Chemical name and structure	Fluorescence maximum (nm)	Concentrations in cocktails (g/liter)
PPO	2,5-Diphenyloxazole	365	4–7
Butyl-PBD	2-(4-*tert*-Butylphenyl)-5-(4-biphenyl)-1,3,4-oxadiazole	367	7–20
BBOT	2,5-Bis-2-(5-tertbutyl benzoxazolyl)-thiophene	438	7–10
Bis-MSB	*p*-Bis-(*O*-methylstyryl)-benzene	412	0.2–1.5
POPOP	1,4-Bis-2-(5-phenyloxazoyl)-benzene	418	0.1–0.2
Dimethyl-POPOP	1,4-Bis-2-(4-methyl-5-phenyloxazoyl)-benzene	429	0.2–0.5

Cocktail Formulations

For organic soluble samples, inexpensive cocktails can be prepared, such as 4 g PPO/liter in toluene. Between 8 and 12 g butyl-PBD per liter of toluene or *p*-xylene has a slightly higher scintillation efficiency. A cocktail that will dissolve about 1 ml H_2O/15 ml cocktail consists of 1 liter *p*dioxane, 800 ml xylenes, 240 g naphthalene, 15 g PPO, and 150 mg dimethyl-POPOP. "Bray's Solution" has a greater utility for aqueous samples, and it contains 200 mg dimethyl-POPOP, 4 g PPO, 60 g naphthalene, 200 ml ethylene glycol, 100 ml methanol, and *p*dioxane to make 1 liter.

Some formulations with Triton X-100 as a solvent follow: (1) 5 g PPO, 0.1 g POPOP in 1 liter of a 2:1 (v/v) mixture of toluene and Triton X-100 (Reed, 1984); (2) 8 g PPO, 50 mg bis-MSB/liter of Triton X-100; (3) 10 g PPO, 100 mg POPOP/liter of Triton X-100.

As mentioned previously, complete cocktails may be purchased and a number of them have excellent performance characteristics as well as capacities to dissolve or disperse various sample types. Companies that offer such products include Amersham, Beckman, New England Nuclear (DuPont), Packard, and Research Products International.

For aqueous samples, excellent products containing surfactants are available from the companies cited above, and most undergo phase changes when water is added to them. For smaller quantities of water (10–15%), an apparent homogeneous liquid is produced. Intermediate amounts of water cause phases to separate, but the addition of more water will yield a stable gel. Such gels are suitable for counting purposes.

Vials and Sample Volumes

The 20-ml vial with an outside diameter of approximately 28 mm is the most common size, but smaller vials (5–12 ml) are also used. Some counters are designed for "small vials" only, but most "large vial" counters can handle small vials if sample changing conveyors are fitted with holders. Under certain conditions microfuge tubes may be used, but special reusable holders are required. Large vials have the advantage of accommodating more counting solution and this permits more radioactive sample to be dissolved. Smaller cocktail volumes and slightly cheaper vials are a cost advantage for small vials.

Glass vials manufactured from low potassium borosilicate glass have a slightly lower background than vials of ordinary borosilicate glass because most of the naturally abundant ^{40}K has been removed.

Plastic vials, primarily polyethylene, are available at a lower cost, and they yield slightly higher counting efficiencies and lower backgrounds than the common glass scintillation vial. However, polyethylene is somewhat permeable to aromatic solvents and this may cause problems. When cocktail permeates the vial wall it creates a scintillator that behaves separately from the solution inside the vial, and this can yield anomalous results when the vial is subjected to external radiation sources such as those used in some quench correction methods to be discussed later. Although no data are published concerning problems with sample changing mechanisms, some claim that plastic vials are more prone to jamming problems.

Normally, the radioactive sample to be assayed is added to the vial and an appropriate volume of cocktail is added; then the closed vial is shaken to effect solution or dispersion. The volume of cocktail is not critical in most situations. For example, samples that are completely soluble in the cocktail and do not contain quenching agents do not exhibit any differences in their counting efficiencies when 10 or 15 ml of cocktail is used. In fact, counting efficiency is virtually independent of cocktail volume (but not sample composition) between about 5 and 19 ml when large vials are used. Although the counting efficiency decreases by 2–3% for ^{14}C when cocktail volumes below 5 ml are counted, quench correction methods such as the sample channels ratio method are capable of yielding statistically valid counting results with volumes as low as 2 ml (Knoche et al., 1979). Under rigorously controlled conditions, scintillation solution volumes in the microliter range can yield meaningful data.

Appropriate radioactive sample volumes may depend on the capacity of the cocktail to dissolve the sample. If solubilities do not limit sample size, dilution of the fluors and the introduction of additional quenching agent become the primary concern. Most cocktail formulations have sufficient fluor concentrations to permit the addition of at least 1 ml of a sample to 15 ml of cocktail. Although such a dilution may not affect counting efficiency appreciably, most samples contain quenching agents and the concentration of quenching agents does affect counting efficiency. A good rule to follow is that if the composition of the final counting solution (sample plus cocktail) varies between samples that are to be compared, *quench correction methods* must be applied. Net counting rate, cpm, is an acceptable measure of a radioactive sample if counting solution compositions are held constant, but with the ease of handling data today quench correction calculations are not a burden. Consequently, quench correction methods are routinely used because with little additional effort assay errors from unforeseen sample composition changes are prevented.

PULSE SELECTION

As discussed previously, the electronic pulse spectra of liquid scintillation counters are affected by the particular radionuclide being assayed and, to a lesser extent, on the composition of the counting solution. Therefore, quite different pulse selection settings are required for counting ^{32}P, ^{14}C, and ^{3}H, which have E_{max} values of 1.71, 0.156, and 0.018 MeV, respectively, and if one wishes to count at the highest figure of merit (see Chapter 13) possible, the settings must be adjusted for the particular cocktail employed.

Predetermined Settings

Most instruments have preset function controls that provide appropriate pulse selection for commonly employed isotopes. In counters with linear amplifiers such controls invoke a certain amplifier gain and discriminator window, while in instruments with logarithmic amplification only the discriminator window is changed. Such preset controls rarely will provide the highest figure of merit, because the settings must be appropriate for a variety of counting solutions that may differ markedly in their scintillation efficiencies. Therefore, a fairly wide discriminator window is necessary, and this increases the background over that of a narrow window that precisely brackets the pulse spectrum in question. Nevertheless, predetermined pulse selection is convenient, particularly when there are multiple users for a single counter. For commonly employed cocktails with little to moderate sample quenching, the difference in figures of merit of predetermined settings as opposed to optimal settings is not substantial, and, furthermore, counting samples for a longer period of time can compensate for a lower figure of merit if statistical considerations become a concern.

Preset pulse selection is adequate for a majority of counting applications, but having operator-controlled pulse selection is desirable in certain situations such as (1) counting severely quenched samples, particularly if a low energy emitter is involved; (2) counting an isotope for which preset pulse selection is not provided; and (3) counting samples that have counting rates of only a few cpm above background.

Optimization of Pulse Selection Settings

A standard liquid scintillation sample containing the desired radionuclide should be prepared using a cocktail that is to be used for experimental samples. The amount of activity is not critical, but because many repeated countings will be performed, a high activity is desirable. A blank sample, prepared identically to the standard but without the radionuclide, is needed for background measurements.

For simplicity in describing operations, amplifier settings will be referred to as "% gain," which represents the percentage of maximum amplification. Discriminator window settings will be given in arbitrary units of 0–1000, where 1000 units represents 100% of the maximum pulse height detected. Normally, both lower and upper limit controls have the same range of settings.

Instruments with Linear Amplification.   For optimizing pulse selection settings, the first task is to determine an appropriate lower limit for the discriminator window. A setting of 50 (5% of maximum) is used frequently, but to be certain thermionic emission pulses are not contributing to the background counting rate, the following test can be made. Load an empty vial and set the sample changer for repeated counts, adjust the amplifier or attenuator for maximum gain, and set both the upper and lower discriminator windows to 500. No counting should be observed if the discriminator is aligned properly. Then alternately reduce the lower discriminator limit and count for several minutes. In the window range of 0–50, none to a few cpm may be expected depending on the age and make of the counter, but the lower level limit of the discriminator should be high enough so no counts are observed.

Assume an appropriate setting for the lower limit is 50 units. Since both amplifier and discriminator settings affect pulse selection, one components' settings must be held constant while the other's are varied. Thus, balance points (see Chapter 8) are determined for various windows such as 50–1000, 50–900, 50–800 and so on. For the first window, e.g., 50–1000, the upper limit would be set at 1000, the amplifier gain adjusted to 0%, then the standard is counted repeatedly with incremental increases of the gain (amplification) between counting periods. The data may be plotted as shown in Figure 8-7 using cpm versus % gain, and the balance point determined (the gain corresponding to the maximum of the curve). The counting rate at the balance point is the maximum counting rate and is directly proportional to the maximum counting efficiency for that sample with that window. The blank sample is loaded and the background counting rate is determined for each window with the amplifier set at the balance point for that window.

From the data collected, $(cpm)^2/R_b$ may be calculated for each window, where cpm is the maximum counting rate and R_b is the background under the same conditions. Since the amount of activity in the standard is constant for all samples, and $(cpm)^2/R_b$ is directly proportional to the statistically defined figure of merit, $(CE)^2/R_b$, window and amplifier settings that yield the highest $(cpm)^2/R_b$ will also yield the highest figure of merit, which of course is the most desirable setting for the standard sample.

Instruments with Logarithmic Amplification.   The process of determining pulse selection settings for instruments with logarithmic amplification is similar to that for instruments with linear amplifiers. The standard and blank are counted with different discriminator windows that are varied systematically. Then $(cpm)^2/R_b$ is calculated for each window and the window that provides the highest value, hence highest figure of merit, is selected.

Instruments with Multichannel Analyzers

Generally, the amplification in instruments with multichannel analyzers is not under operator control and the analyzer is set to have a range of channels that corresponds to β-particle energies of at least 0–2.0 MeV. Pulse selection involves finding the range of channels that yields appreciable counts when the standard is counted. Suppose the standard contained ^{14}C, and a 4000 channel analyzer is employed that has 0.5 keV per slot (channel). With an E_{max} of 156 keV, ^{14}C pulses would be expected in the slots from 0 to 312. The sum of the counts in those slots would be comparable to a cpm for a discriminator window covering the same range of pulse heights. Because of background, the counts in some of the lower slots should be excluded, and the same argument applies for the slots at the upper end of the channel range. Let cpm be the sum of cpm in a given range of slots when the standard is counted and R_b be the analogous value when the blank is counted. Then the simplified figure of merit $(cpm)^2/R_b$ can be compared for various slot ranges and the one that gives the highest figure of merit should be selected.

QUENCH CORRECTION METHODS

Recall that quenching is a process that lowers the scintillation efficiency of a counting solution either by reducing the amount of energy reaching the fluor molecules (chemical quenching) or by absorption of some of the light emitted by the fluor molecules (color quenching). The result in either case is that quenching causes the pulse spectrum to be shifted to lower pulse height values. If the pulse selection settings were derived using an unquenched standard (the spectrum optimally situated in the discriminator window), then quenching causes some of the weak pulses to be lost below the lower limit of the window yielding fewer counts and a lower counting efficiency. The problem could be corrected to a large extent by increasing the gain of the instrument so the pulse spectrum of the quenched sample is optimally situated in the window. This principle is used in some instruments for quench correction purposes.

Another effect of quenching occurs in the lower region of the pulse spectrum. Chemical quenching affects the scintillation efficiency of a solution (number of photons per eV of particle energy), and lowering the scintillation efficiency causes a greater percentage of the weaker particles to not produce countable pulses (see the Liquid Scintillation Samples section of this chapter). This kind of loss in counting efficiency cannot be recovered by gain adjustments.

Quench correction methods involve counting a sample, then, by an additional procedure, determining the counting efficiency of the sample and calculating its activity in dpm or dps. With modern instrumentation including data processors and multichannel analyzers, many of the quench correction methods can be performed automatically so that corrected data are provided. Nevertheless, the principles involved in quench correction methods are similar whether the methods involve operator assistance or not. There is one empirical method, internal standardization, but all other methods are based on one of two principles: (1) Quenching changes the pulse spectra of samples and parameters that are dependent on such changes may be correlated to counting efficiency. (2) An external source of γ-rays when placed close to a sample will produce a pulse spectrum. Quenching causes changes in the external source's spectrum and parameters that reflect such changes may be correlated with a sample's counting efficiency.

Internal Standardization (IS)

In the internal standardization method a sample is counted, then "spiked" with a *known* amount of activity of the same isotope (the standard) and recounted. The addition of the standard must not change the scintillation efficiency of the sample. For most cocktails that is not a problem because suitable standards such as [^{3}H]toluene and [^{14}C]toluene are available. Assuming small quantities are used, the only effect is an insignificant amount of dilution. After mixing, the spiked sample is counted, the counting efficiency (CE) is determined, and the activity in dpm or dps is calculated as indicated by Equations 9-1 and 9-2. R_s, R_b, and R_i represent the counting rates, in cpm, of the sample, background, and spiked sample, respectively, and A is the activity of the standard added in dpm.

$$CE = \frac{R_i - R_s}{A} \tag{9-1}$$

$$dpm = \frac{R_s - R_b}{CE} \tag{9-2}$$

Internal standardization is an appropriate method when only a few samples are involved, but the addition of a standard to a large number of samples is time consuming and expensive. For that reason this method is used rarely for routine quench corrections. The accuracy of the method is highly dependent on the accuracy of the standard's stated specific activity and the technique used to add the standard (see example problem 13-5). Nevertheless, the method compares favorably with other quench correction methods under common circumstances and should be the best method if due care and a high quality standard are used.

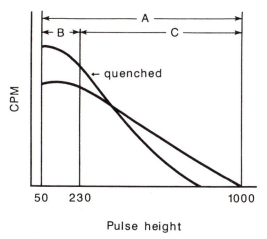

Pulse height

FIG. 9-1   Pulse spectra for two ^{14}C samples with different degrees of chemical quenching.

Sample Channels Ratio (SCR)

The *Sample Channels Ratio* (SCR) method of quench correction requires the sample to be counted with at least two different pulse selection settings. Although this can be accomplished by counting a sample twice with a single channel instrument, a two- or three-channel counter is preferred. The different pulse selection settings are used to measure the relative positions of pulse spectra of samples that subsequently are correlated with counting efficiency. Figure 9-1 shows spectra for two samples of ^{14}C with 20,000 dpm each. As shown, quenching causes the spectrum to be shifted toward lower pulse height values and compresses it somewhat. The curves in Figure 9-1 are typical for instruments with linear amplification, but the same general effect occurs when logarithmic amplification is used.

Between specified limits, the area under a curve is proportional to the cpm obtained when a discriminator window corresponding to those limits is used, and Table 9-2 shows counting data for each sample of Figure 9-1 when three channels (three different windows) were employed. Channel A represents the "counting channel," and it has the widest window (50–1000) in this example. The counting efficiency of the unquenched sample is 86.0% in that channel and 82.2% for the quenched sample. Channels B and C count a low and high range of pulse heights, 50–230 and 230–1000, respectively. As shown in Figure 9-1 and Table 9-2, quenching causes counts to be lost from the high range (channel C) and gained by the low range (channel B). Therefore the ratio of cpm in channel B divided by that in channel C (B/C) changes as the pulse spectrum is shifted, and the counting efficiency in channel A is related to the extent of such shifts. In more general terms, quenching causes a change in the distribution of pulse heights

TABLE 9-2 Counting Data for the Two Samples Whose Pulse Spectra Are Shown in Figure 9.1

Curve	Channel	Discriminator window	cpm	Counting efficiency %	Channels ratio
Unquenched	A	50–1000	17,202	86.0	B/A = 0.335
	B	50–230	5,757		B/C = 0.503
	C	230–1000	11,443		C/A = 0.665
Quenched	A	50–1000	16,489	82.2	B/A = 0.453
	B	50–230	7,463		B/C = 0.827
	C	230–1000	9,026		C/A = 0.547

(lowers them), and this change is reflected by a loss of counts in the counting channel. As will be discussed later, a channels ratio is not the only method for detecting a change in the pulse distribution, but it is an effective one. Note that different cpm ratios can be used to measure a change in pulse distribution, B/C, B/A, and C/A, or their reciprocals all change with a change in pulse distribution. Furthermore, windows different than those cited in this example could be used to divide a spectrum into low and high energy ranges; actually a myriad of ratios is possible, although many would not be reasonable choices. A method of determining appropriate settings for the channels follows.

First, the wide channel settings are determined as described in the Optimization of Pulse Selection Settings section. The standard used in this procedure should have the least amount of quenching one expects to encounter. Frequently, such a sample is referred to as an "unquenched standard," however, quenching is a relative term and one cocktail may be "quenched" with respect to another; also air-saturated cocktails exhibit quenching relative to ones prepared under a nitrogen or argon atmosphere. The term "least quenched standard" more accurately reflects the nature of an appropriate standard. Suppose channel A is optimized and the settings are gain = 6.00%, lower discriminator limit 50, and upper limit 1000. The gain for channel B is set at the same value as channel A, 6.00%, and the same lower discriminator limit, 50, is used for channel B. With the least quenched standard in the counting chamber, the sample is counted between adjustments to the upper limit of the discriminator in channel B. Adjustments by trial and error are continued until the counting rate of channel B is about one-third that in channel A. This yields an SCR of about 0.33 for the ratio of B/A. Note that the unquenched sample in Table 9-2 has a B/A ratio close to that value. The method just described is applicable to instruments with linear amplification, but for those with logarithmic amplification gains are not adjusted, only the discriminator limits.

Most instruments have predetermined settings for the narrow channel that are invoked when preset pulse selection switches are activated. That is, by appropriate communication with the instrument via switches or through a computer, all the predetermined channel settings will be invoked and the appropriate ratio may be printed along with other counting data. As mentioned previously, predetermined settings may not be ideal, but for the vast majority of applications they will be adequate, and they are very convenient.

Determining pulse selection conditions for an appropriate SCR would be a simple matter with instruments that have a multichannel analyzer, but the power of the analyzer can be utilized more effectively by using the SIS method of quench correction, which will be discussed later.

Quench Correction Curves. In practice, a series of standards (perhaps 10–20) with various degrees of quenching are prepared or purchased. The standards are counted, and the counting efficiency in the wide channel is plotted gainst the sample channels ratio. Generally, a smooth curve is obtained that is called a *quench correction curve*. An example of such a curve is shown in Figure 9-2. In this example the gain for both channels was 6.00%, and the discriminator windows were 50–1000 and 50–230 in channels A and B. The cocktail consisted of 4.80 g PPO/liter in reagent grade toluene plus 4700 dpm/ml of [^{14}C]toluene as the standard. Exactly 5 ml of this mixture was placed in each of 17 vials and various proportions of chloroform, the quenching agent, and toluene were added so that the final volume of each sample was 6.00 ml. These dilutions resulted in a final PPO concentration of 4.00 g/liter, and the concentration of chloroform varied between 0 and 13.3%.

The accuracy of quench correction curves depends on the accuracy of the standard, the quantitative methods used to prepare the samples, and the number of counts collected during the counting process (see example problem 13-3). Errors in the sample preparation steps are considerably greater than counting errors if samples are counted long enough to yield better than 10^5 counts in the wide (counting) channel.

After the preparation of a suitable quench correction curve an unknown and background may be counted, yielding net cpm in channel A and its SCR value (net cpm B/net cpm A). From the quench correction curve the counting efficiency corresponding to the SCR is determined, and counting efficiency is divided into the net cpm of channel A to yield the activity of the sample in units of dpm. Inspection of the quench correction curve in Figure 9-2 reveals that there is a portion of the curve where a small error in the SCR would cause a large error in reading the counting efficiency. The curve is particularly steep in the region between SCR values of about 0.8 to 1.0, consequently that portion of the curve should not be used. The samples with counting efficiencies below about 47% are so severely

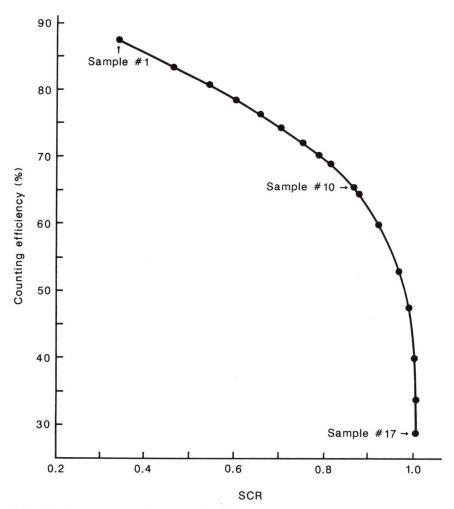

FIG. 9-2   A quench correction curve for ^{14}C samples using the samples channels ratio (SCR) method.

quenched that all of those samples yield an SCR close to 1.0. This means that virtually none of the pulses from these samples exceeds a height of 230 units, which is the value of the upper limit in the narrow channel. Obviously, after an SCR value of 1.0 is reached, greater quenching reduces counting efficiency, but does not change the SCR.

   In most counting applications such a large range of quenching is not likely to be encountered, but if so, more than one curve should be prepared. This can be done using an additional channel, or by recounting certain samples with different channel settings. For samples with an SCR between 0.3 and about 0.8 the appropriate portion of the curve shown in

Figure 9-2 should be used, but if the SCR is above 0.8, then the sample should be recounted with different channel settings and corrected with a different quench correction curve.

To prepare a second quench correction curve, sample 10 was used as the least quenched standard and instrument settings were redetermined as described above, that is, the counting channel was reoptimized and the settings for the narrow channel redetermined. Samples 10 through 17 were counted under the newly established settings, and a second quench correction curve developed. In this example the settings established for sample 10 were gain 18% for both channels, and discriminator windows as before, 50–1000 (channel A) and 50–230 (channel B). The quench correction curve resulting from the counting of samples 10 through 17 is shown in Figure 9-3. Note that under the previous conditions, sample 10 yielded a counting efficiency of about 65%, while under the second set of conditions it is slightly greater than 80%. Also the SCR values have been reduced significantly. With the two quench correction curves available, an unknown may be assayed with an instrument gain of 6%, and if the sample's SCR is below 0.8, the calculations described before are performed using the curve in Figure 9-2. If the SCR value is above 0.8 the unknown is recounted with an instrument gain of 18% and the appropriate calculations performed using the curve in Figure 9-3.

Note that having the gain at 18% allowed sample 10 to be counted with an efficiency approaching that of the least quenched sample (1) when it was counted with a gain of 6%. This illustrates two important points: (1)

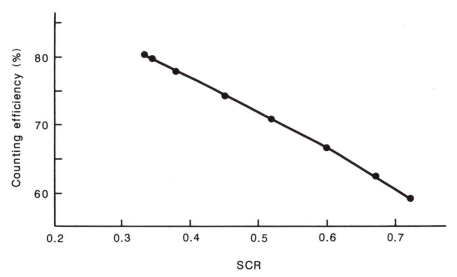

FIG. 9-3 A quench correction curve for severely quenched ^{14}C samples using the sample channels ratio (SCR) method.

Increasing the gain of an instrument can partially compensate for quenching in a sample, and this is the basis for an instrumental feature called *automatic gain restoration* to be discussed later. (2) One set of instrumental conditions that is optimum for one sample (e.g., sample 1) will be far from optimum for another (e.g., sample 10) if the samples have significantly different degrees of quenching. Obviously, preset function switches cannot provide near optimum conditions for counting samples that differ widely in their degree of quenching. For that reason, instruments may provide more than one preset function switch for a particular isotope. For example, there may be one function that provides channel settings for ^{14}C when quenching is low and another for ^{14}C when quenching is severe. The predetermined instrumental settings may be analogous to the two sets of conditions used to obtain Figures 9-2 and 9-3.

Instruments that possess significant data processing capabilities may be programmed to handle some or all of the quench correction calculations. Using preset or operator established channel settings, a series of quenched standards are assayed. The instrument stores the data and produces a mathematically described quench correction curve. When an unknown is counted its SCR is compared to the curve to give a counting efficiency for the sample that permits the activity of the unknown to be calculated and printed out.

When only chemical quenching is involved, all quench correction curves for a particular isotope that are produced with the same instrument settings are almost indistinguishable regardless of the quenching agent. Even different cocktail formulations yield interchangeable curves. Therefore, one complete set of quenched standards may be used for different applications if chemical quenching is the only type of quenching encountered.

Color quenching affects pulse height distributions differently than does chemical quenching, consequently separate quench correction curves should be developed for samples that exhibit color quenching. The procedures for producing and using such curves are analogous to those described for chemical quenching except a colored compound is used as the quenching agent.

The samples channels ratio method of quench correction is probably the most widely used method for routine assays. Its accuracy is good as long as a suitable number of counts is collected. A statistical analysis of the method is beyond the scope of this chapter, but the accuracy of the method depends on (1) the slopes of the correction curve at various SCR values, (2) the accuracy of the data used to produce and plot the curve, and (3) the accuracy of the SCR value for an unknown. The later factor is dependent on the number of counts collected in both channels used to compute the ratio. Assuming that the ratio is calculated as indicated in the example presented and the SCR value is about 0.33, the relative standard deviation of the computed ratio is about 1% when a sample is counted long enough to yield about 40,000 counts in the wide channel. Because the counts in the narrow channel increase relative to the wide channel as quenching increases, the total number of counts in the wide channel needed for a relative standard deviation of 1% decreases. Of course, the accuracy of the ratio is not directly proportional to the accuracy of the counting efficiency read from the curve, because, as indicated earlier, a steep slope can yield a large reading error, and for that reason more than one curve is needed if a large range of quenching is encountered.

Example problems 13-3 and 13-4 deal with standard deviations of SCR values, and they show that the SCR method is not good for samples having activities that are only a few cpm above background because excessively long counting periods are required for a reasonable degree of accuracy. Fortunately, other methods that involve *external standards* or internal standards do yield good estimates of counting efficiency under such conditions.

External Standard Counts (ESC)

If a γ-ray source is placed next to a liquid scintillation sample, the γ-rays can readily penetrate the glass or plastic wall of the vial and interact with the liquid inside the vial. Since the medium is not very dense, a large fraction of the γ-rays will not be absorbed, but of course, the energy of the γ-rays is an important parameter in determining the extent of absorption. A small but significant fraction of moderate energy γ-rays (200–700 keV) will interact with a thickness of scintillation solution comparable to that in a scintillation vial, and the primary type of interaction is the Compton process. Thus, the external source generates Compton electrons in the solution, and these have a continuous range of kinetic energies up to some well-defined maximum (see Chapters 5 and 7). Compton electrons will interact with the scintillation solution in the same way β^--particles do because their physical properties are identical. Each Compton electron will produce a scintillation event, and hence a pulse, provided it has at least the appropriate minimum energy. Therefore, an external γ-ray source will produce a continuous pulse spectrum corresponding to an energy range of zero to some maximum that is related to the maximum energy of the Compton electrons. The shapes of Compton pulse spectra are quite different from those of β^--emitters, but quenching will cause the same degree of attenuation of light pulses from Compton electrons as it does from β^--particles. The point is, quenching causes a shift in the β^--ray spectrum toward lower pulse height values, and the same is true for a pulse spectrum arising from Compton electrons. Consequently, the extent that a Compton spectrum is shifted is related to the extent a β^--spectrum is shifted, which, in turn, is related to counting efficiency. This is the primary basis for quench correction methods that utilize an external standard.

One physical arrangement for employing an external standard is shown diagrammatically in Figure 9-4. With the sample in the counting chamber, a small pellet containing the external standard is mechanically moved from its lead shielded storage compartment to a point just below the center of the bottom of the scintillation vial. When in place, short counting measurements are conducted, then the pellet is moved back to the storage compartment, and the sample is counted according to the operator's specifications. Several radionuclides may be used as external standards including ^{226}Ra, ^{133}Ba, and ^{137}Cs, and activities are high enough (usually in the mCi range) so that thousands of cpm are generated when the standard is in place. This yields a suitable number of counts for statistical accuracy in a minute or less of counting time.

In earlier instruments, one channel with a fairly wide window was dedicated to external standard counting measurements. The gain of such an instrument can be adjusted (or perhaps only the discriminator window) with a least-quenched blank so that an optimal counting rate for the external source is obtained. A quenched solution causes the Compton pulse spectrum to be shifted toward lower pulse heights and yields a lower counting rate for the external source. As a result, the cpm of the external standard (external standard counts) is related to the shift in its pulse spectrum, which is related to the shift that would occur if a β-emitter were in the same solution. Hence, cpm of the external standard is related to the counting efficiency of a β-emitter in the same solution.

The problem of distinguishing between pulses arising from Compton electrons and those from a β-emitter needs to be addressed. Pulses generated by the sample's activity are subject to being counted in the external standard's channel, and since different samples have various amounts of activity, unless those pulses can be excluded from the counting rate of the external standard, the method fails. One method of handling this problem is to treat the sample as background for the external standard measurement, that is, with the external standard in place, the total counting rate in the external standard's channel is due to the pulses generated by the standard plus those due to the sample. With the source stored, the cpm of the sample in the external standard's channel is determined and subsequently subtracted from the previous total cpm. Such calculations are not always needed because the energies of the Compton electrons and β-particles may be sufficiently different so that even a high activity sample contributes few if any counts to the external standard's channel. Also the counting rate due to the external standard is very high and a few cpm from the sample may represent a statistically insignificant difference in the total counting rate. In the discussion that follows the counting rate of the external standard's channel will refer to cpm due to the external standard (cpm_{ex}), irrespective of how it may have been obtained.

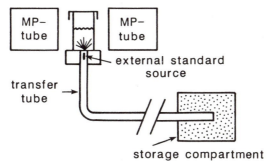

FIG. 9-4 Diagram of a physical arrangement to convey an external standard (γ-ray source) between a sample vial and its storage compartment.

Quench correction curves may be generated by preparing a series of quenched standards of the β-emitter to be assayed as described in the sample channels ratio method. Then cpm of the external standard is plotted against counting efficiency of the β-emitter. An unknown may be assayed under the same instrumental settings, and its counting efficiency determined from the curve by reading the efficiency corresponding to the cpm_{ex} value generated by that sample.

One serious deficiency with the ESC method is that the counting rate of the external standard depends on sample volume. Since the fraction of γ-rays absorbed is dependent on the thickness of the absorber (Equation 5-9), a sample with a greater volume presents a greater absorbing thickness to the source and yields a higher counting rate in the external standard's channel. Nevertheless, if sample volume is held constant the method is valid and quite convenient.

External Standard Channels Ratio (ESCR)

Because of volume dependence, the ESC method generally has been replaced by the *external standard channels ratio* (ESCR) method. In the SCR method, two channels are used to determine the extent of attenuation of pulse spectra (shift to lower pulse heights), and the same principle can be applied to the spectra generated by an external γ-ray source. If a particular counting solution causes the β-ray spectrum to be shifted to lower values (quenching), the Compton electron spectrum for the same solution will be shifted by a comparable amount. In the ESCR method two channels are used for counting the pulses of the Compton electron spectrum with each channel covering a different portion of the spectrum. A change in the ratio of counts in the two channels reflects a change in the distribution of pulse heights that is related to the counting efficiency of the sample.

In practice, the sample vial is loaded into the counting chamber, the external standard source is moved into position next to the vial, counting occurs for a minute or less to yield the data needed for an ESCR value, then the external standard is moved back to its storage compartment while the sample's activity is counted as specified by the operator. For some instruments the ESCR measurement is made after the sample is counted.

Quench correcton curves are generated as described in the SCR method except that the coordinates of the graph are ESCR value and counting efficiency (for the activity in the sample, not that of the external standard). Although the γ-ray source is called an external standard the amount of activity in the source is not a critical parameter as long as sufficient counts are generated to yield a statistically valid ESCR value, and no use is made of the γ-source's actual activity in counting calculations. An example quench correction curve for ^{14}C is shown in Figure 9-5, and it was derived from the same samples used to generate Figures 9-2 and 9-3.

Most instruments have predetermined settings or dedicated channels for external standard measurements and the selection of the channel settings for ESCR values is not under operator control. Such was the case for the data shown in Figure 9-5. Normally, the ESCR counting mode is either invoked or not, and when it is selected, the ESCR values are printed out along with the sample's other counting data. Instruments with suitable data processing systems can assay a series of quenched standards and use that data to generate a "quench correction curve" that can be used to determine activities (dpm or dps) in unknowns.

A significant advantage of the ESCR method compared to the SCR method is that there is always a sufficient number of counts to produce a statistically valid ESCR value. In contrast, the validity of the SCR value is dependent on the sample's activity and length of counting time. Generally, it is not practical to count samples with activities close to background levels long enough to obtain a reliable SCR value, but if a sample contains enough activity to yield a reliable SCR value, then the SCR method is considered to be superior to the ESCR method because the spectral shifts being monitored by the ratios are more directly related to the sample's counting efficiency in the SCR method.

Compton Edge or H Number (H#)

The distribution of electron energies arising from interactions of γ-rays with an absorber by the Compton effect was discussed in Chapter 7. Figure 7-6 shows the theoretical shape of such a distribution and indicates that the maximum energy level for Compton electrons creates an upper limit, called the Compton edge. The position of the edge on an energy scale is determined by the energy of the γ-rays according to Equation 7-1. Cesium-137, which is the γ-ray source used in the H# method of quench correction,

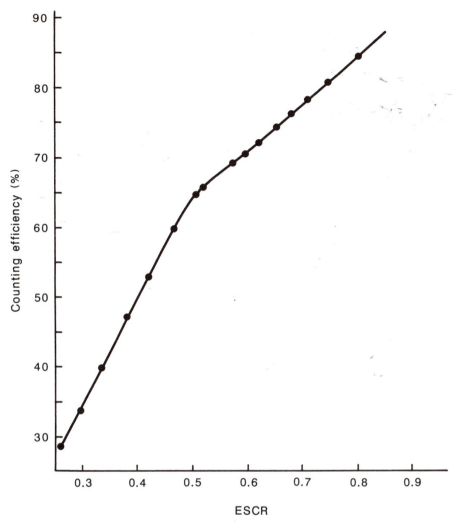

FIG. 9-5 A quench correction curve for ^{14}C samples using the external standard channels ratio (ESCR) method.

yields γ-rays of 662 keV and a Compton edge (maximum Compton electron energy) at 477 keV.

In a liquid scintillation solution the Compton electrons' energies are converted to light pulses that determine the magnitudes of the instrument's pulse heights. The scintillation efficiency of the solution represents the efficiency of the light conversion process, which is directly related to the level of quenching. Therefore on a pulse height scale, the position of the Compton edge for a particular source is related to the extent of quenching and, hence, is related to the sample's counting efficiency. A graphic or mathematical relationship between the pulse height of the Compton edge versus counting efficiency may be used for quench correction purposes. Quenching decreases the pulse height value of the Compton edge and of course decreases counting efficiency.

Unfortunately, actual pulse spectra do not produce as sharp a Compton edge as might be suggested by the theoretical spectrum shown in Figure 7-6, so the precise location of the edge is subject to interpretation. Logarithmic amplification causes the edge to appear sharper than does linear amplification, but without a precise definition based on actual spectra, reproducible locations cannot be accomplished. For that reason *H numbers* (H#) have been defined by their originator (Horrocks, 1974, 1977). The parameter most commonly used is the inflection point of the leading edge of a Compton electron spectrum. Using the inflection point location for an unquenched sealed sample as a reference, the position of another sample's inflection point is determined, and an H# is calculated by subtracting the position of the experimental sample from that of the reference sample.

To locate the inflection point of a sample, a series of measurements are made over a range of pulse heights surrounding the edge. This entails multiple counting periods, perhaps as short as a few seconds each, with narrow discriminator windows. Under such conditions, short counting periods and narrow windows, a source with an intense activity is needed. A multichannel analyzer simplifies the situation markedly because the data corresponding to a large number of narrow discriminator windows may be collected in one longer counting period. The instrumental settings necessary for determining H# are predetermined and normally not under operator control. If an operator selects that mode of quench correction the appropriate measurements with the sample vial exposed to the ^{137}Cs source are made in addition to the normal counting process of the sample, then the H# is printed out along with other counting data.

Using a series of quenched standards, a quench correction curve may be prepared as described for the previous methods except that the H# is used in place of other parameters such as ESCR and SCR. Data processing can reduce the operator's labor even more by using the data of standards to calculate the activity of unknowns directly.

For a sample with a given scintillation efficiency (level of quenching), the inflection point of a Compton electron spectrum is a unique parameter. Consequently, the H# of a sample does not depend on volume (at least within reasonable limits), level of sample activity, type of quenching agent, or cocktail formulations.

Automatic Pulse Selection Adjustments

In the Sample Channels Ratio section it was shown that for a given discriminator window, a lower counting efficiency due to quenching could be compensated for, to a large extent, by increasing the instrument's gain.

Increasing the gain causes a pulse spectrum that otherwise would lie partially below the lower limit of the window to be brought back into the window. The same effect can be achieved by lowering the window so the quenched spectrum lies within the window. Both type of adjustments, amount of gain and window position, can be performed automatically in response to a sample's quenching characteristics. For example, a quench parameter, such as ESCR, may be used. The degree of quenching in a sample is "measured" by comparing its ESCR value to that of an unquenched sample, the value of which may be stored electronically; then the instrument adjusts either the gain or the window so that the sample's pulse spectrum from the external standard shows the same pulse distribution (same ESCR value) as that expected for the unquenched condition.

As indicated previously, gain or window adjustment will not compensate for quenching entirely because a few of the weak particle pulses that produce countable pulses in the unquenched condition will not do so under quenched conditions. Such losses may not be statistically significant if the extent of quenching is small, but at higher levels the error becomes quite significant. Sample 1 in Figure 9-2 exhibited a counting efficiency of about 86%, whereas under the same conditions sample 10, which contained the same amount of activity, yielded an efficiency of about 65%. By increasing the gain to balance point conditions, the counting efficiency of sample 10 could be increased to slightly more than 80% (Figure 9-3).

Although automatic pulse selection adjustments are not adequate for quench correction purposes except under limited conditions, they do offer a significant advantage. Effectively, each sample regardless of its level of quenching is counted under balance point conditions (maximum counting efficiency for that sample). This improves the statistical reliability of the counting data and may permit shorter counting periods to be used. Its primary advantage is realized when a number of low activity (a few cpm above background) samples are being assayed.

Use of Multichannel Analyzers

In the methods of quench correction previously discussed two or three channel instruments are employed. Although these methods can be accomplished with instruments having multichannel analyzers, better and more convenient methods are possible. The multichannel analyzer permits a more complete characterization of a pulse spectrum than does a two- or three-channel analyzer because the spectrum is divided into smaller pulse height increments (channels or slots) that allow greater resolution of any changes in the distribution of pulses. Obviously, those quench correction methods that rely on changes in the sample's or an external standard's pulse spectrum can be used in instruments with multichannel analyzers, but the full capability of the multichannel analyzer is not utilized.

Sample Spectral Indexes (SSI)

The average energy (E_{av}) of a β^--particle pulse spectrum is a characteristic of the emitter, and the relationship between E_{av} and E_{max} is a function of the shape of the spectrum. The average energy represents the sum of the energies of all particles divided by the total number of particles.

A liquid scintillation electronic pulse spectrum differs from the energy spectrum, but the same logic applies. An average pulse height exists, and it represents the sum of all pulses heights (pulse height times the number of pulses having that height), divided by the total number of pulses. The average pulse height is a characteristic of the β-emitter involved, the scintillation efficiency of the counting solution, and any electronic transformation of pulse heights, such as logarithmic or linear amplification. For a given isotope and instrument, the average pulse height is an indicator of scintillation efficiency or quenching. As stated before, quenching shifts a scintillation pulse spectrum to lower pulse height values, which means that the average pulse height is shifted downward as well. The basic principle involved in this method is identical to that in the SCR method—the difference is the parameter used to "locate" the relative positions of pulse spectra.

In practice, a numerical spectral index or spectral parameter is computed and printed out, and this value is similar to the average pulse height. Computers can accomplish such tasks readily, but one method for determining the SSI is to identify the channel in the multichannel analyzer for which there as as many counts in the channels above it as there are in the channels below. Since the channels have a narrow, yet finite range of pulse heights (windows) the number above and below may not be perfectly equal, but there should be one channel where the numbers are more nearly equal than any other channel. This indicates that the precision of the determination will be affected by the channel width.

A quench correction curve may be generated by counting a series of quenched standards and plotting SSI versus counting efficiency. However, if an instrument has a multichannel analyzer it usually has sufficient computational power to store quench correction data and use it to perform the appropriate calculations needed to obtain the activity of a sample in dpm or dps units. Theoretically, the average pulse height is not the only parameter that could be used to "locate" the position of a pulse spectrum. Any point on the spectrum that can be identified precisely and located reproducibly should be a suitable index. After all, average pulse heights or spectral indexes are not E_{av} values because of the transformation of the β-ray spectrum in the liquid scintillation process. Also the average pulse height depends on the type of amplification used after the light conversion step. If logarithmic amplification is used, the average pulse height represents a lower energy scintillation event (fewer photons) than does the average pulse height when linear amplification is employed.

Because there is a better resolution of pulse heights in the SSI method as opposed to the SCR method, it appears that the SSI method is superior. One significant advantage is that separate quench correction curves are not needed for a wide range of quenching levels. Also the nature of the index calculation is such that shorter counting periods are needed for levels of precision comparable to those of the channels ratio method. The principal limitation of the SSI method is similar to that described for the SCR method. For a valid index, a fairly large number of counts must be collected, therefore samples that have counting rates close to background must be counted for extremely long periods of time. However, samples with counting rates of 500–1000 cpm can be assayed by the method with reasonable counting periods and acceptable accuracy.

External Standard Spectral Indexes (ESSI)

The basic principle involved in the ESSI methods is identical to that in the ESCR and H# methods. Quenching causes the pulse spectrum of Compton electrons to be shifted toward lower pulse heights, and a parameter that "measures" the extent of such shifts is dependent on the scintillation efficiency of the solution, and, hence, sample counting efficiency. ESCR values are appropriate parameters for determining spectral shifts and can be obtained with a few channels. However, a multichannel analyzer permits other parameters to be used as quench indicators, and because of their capability to characterize spectra more completely, such parameters should be superior to ESCR or to H# when determined without numerous channels.

The mechanical features described for counting the external standard in the ESC and ESCR sections are the same as those used in the ESSI methods, and quench correction curves are plotted similarly, ESSI versus counting efficiency. Because a large number of counts are generated by the external standard, the ESSI methods, like the ESCR and H# methods, are particularly valuable for correcting quenching in samples that have activities close to background.

Average Pulse Height Index.   The average pulse height index is completely analogous to the sample spectral index. However, the Compton electron spectrum generated by the external standard is used in place of the sample's spectrum.

Spectral End Point Index.   A number of indexes could be used to locate the position of a Compton electron spectrum, and one in use currently is called the spectral end point index. The index corresponds to a point on the spectrum where 99.5% of the total number of pulses (counts) have heights below it and 0.5% are above it. The actual index is the channel number that divides the spectrum as indicated.

H#.   The H# is an index and represents the inflection point of the Compton edge as described previously. However, this spectral feature can be located more precisely with a multichannel analyzer than with multiple counting periods in a few channels.

Mathematically Transformed Indexes.   As discussed previously, an instrument's electronic pulse spectrum differs from that of an energy spectrum for Compton electrons or β^--rays. Besides the changes caused by the scintillation process other factors may distort spectra such as volume effects, vial wall effects, and geometry within the counting chamber. With appropriate computer programs some of these distortions may be eliminated mathematically. Presumably, a "corrected" spectral index would extend the applicability of liquid scintillation counting to more unusual counting conditions such as very small volumes (25 μl) and unconventional vials such as microfuge tubes.

Efficiency Tracing

If integral counting (e.g., a discriminator window of 50–∞) is used, it is possible to predict the counting efficiency of a sample by counting it at several gain settings (Ishikawa et al., 1984). The inconvenience of counting a sample four or five times does not make the method attractive for an instrument without a multichannel analyzer, however the operating principle of the method is more easily presented if a single channel is considered. An example using ^{14}C samples follows.

A least quenched ^{14}C standard is counted with a constant window perhaps, 50–∞, and with various levels of gain such as 4, 5, 6, 8, 10, and 20%. The net cpm of the standard are plotted against the counting efficiency of the least quenched standard as shown in Figure 9-6. An unknown is counted under the same conditions and its data are plotted as net cpm versus the efficiency of the least quenched standard at the corresponding gain used to count the unknown. Projection of the plot to the "100%" counting efficiency axis yields the dpm of the unknown. Figure 9-6 shows data from two quenched samples with the same amount of activity as the least-quenched standard. Theoretically, the three lines should intersect at 23,500 dpm on the right (100%) axis. Using least-squares analyses for the two quenched samples yielded intercept values of 23,832 and 24,007 dpm with respective correlation coefficients (r) of 0.9999 and 0.9994.

The method appears to yield reasonable results when the maximum counting efficiency for the least-quenched standard is better than 90%, but it is not likely to receive widespread usage unless a multichannel analyzer is available.

The effect of counting samples at various level of gain can be reproduced with a multichannel analyzer by determining all counts above a series of

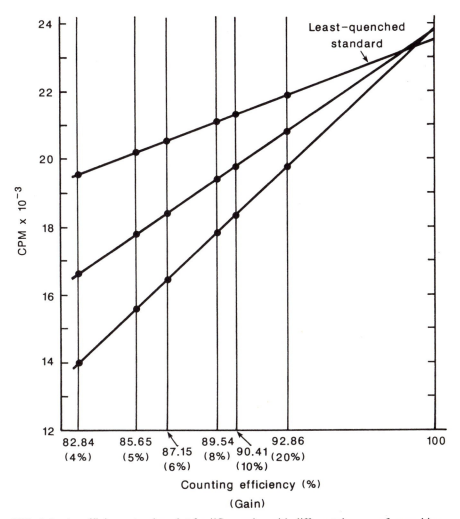

FIG. 9-6 An efficiency tracing plot for ^{14}C samples with different degrees of quenching.

lower channels. Suppose a 2000-channel analyzer is used and the entire pulse spectrum is covered by these channels, also the channels are numbered 1 to 2000 with 1 counting pulses with the lowest heights. The counting rate of a sample corresponds to the sum of the counts from a given channel to channel 2000. Therefore, the cpm obtained when 20% gain was used might correspond to the cpm in channels 1 through 2000, whereas for 10% gain the corresponding channels are 10 to 2000 and for 4% gain the analogous channels are 40 to 2000. With a multichannel analyzer many more plotting points are generated in one counting period because the cpm above each channel could be used as a data point, and with a computer linear regression of those points yields the intercept of a best fitted line on the 100% counting efficiency axis. This method is attractive because a series of quenched standards to generate a quench correction curve is not needed and dpms can be provided to the operator directly. The method appears quite attractive for samples with appreciable activities, but it has not received extensive usage yet.

COUNTING DUAL-LABELED SAMPLES

The pulse spectra of all β^--emitters overlap in the lower pulse height region because all exhibit continuous energy spectra from zero to a characteristic E_{max} value. However, if one emitter has a greater E_{max} than the other, the higher region of its spectrum will be free from pulses due to the presence of a lower energy emitter. When the ratio of E_{max} values is about 2.5, or higher, sufficient spectral difference exists for the two isotopes to be assayed in the presence of one another.

Counting ³H and ¹⁴C

In biological research, ³H and ¹⁴C are a commonly used isotope pair for dual-label studies. Their E_{max} values are 18.6 and 156 keV, respectively, which gives a ratio of about 8.4. Consequently, sufficient difference in their spectra exists to permit good assays of both activities. Figure 9-7 shows spectra for ³H and ¹⁴C individually and as a composite, in addition to hypothetical discriminator limits denoted by the letters p, q, r, and s.

In dual-label counting two channels are used and one has its discriminator adjusted to cover the lower region of pulse heights. This channel may be called the "lower channel." The "upper channel" has its discriminator adjusted to cover the higher region of pulse heights, and the precise discriminator limits may be indicated when referring to a particular channel.

From the data in Figure 9-7, if a p–r channel would be an appropriate lower channel, then the upper channel should be an r–s channel. Note that none of the ³H pulses would be counted by the r–s channel whereas the p–r channel would count most of the ³H pulses and many of the ¹⁴C pulses. If the counting efficiencies of each isotope are known in each channel, Equations 9-3 and 9-4 may be used to calculate the activities of ¹⁴C and ³H, respectively. Of course, in this example the counting efficiency of ³H in the r–s channel is zero.

$$\text{dpm}_c = \frac{\text{cpm}_{r-s}}{\text{CE}_{c,r-s}} \tag{9-3}$$

$$\text{dpm}_h = \frac{\text{cpm}_{p-r} - (\text{dpm}_c)(\text{CE}_{c,p-r})}{\text{CE}_{h,p-r}} \tag{9-4}$$

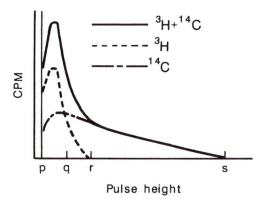

FIG. 9-7 Pulse spectra for ^{3}H, ^{14}C, and a composite of the two radionuclides.

In these equations the activities (dpm) or counting rates (cpm) and counting efficiencies (CE) of ^{14}C and ^{3}H are indicated by subscripts c and h, respectively; also where applicable the channel pulse height limits are given as subscripts.

For the spectra given in Figure 9-7, the counting efficiency of ^{14}C in the r–s channel ($CE_{c,r-s}$) is about 45%, which is large enough for reasonably accurate measurements of ^{14}C activity. There is more uncertainty in the calculation of ^{3}H activity because of the cumulative effects of additional mathematical terms, each of which contributes to the overall error. An analysis of statistical matters is presented in Chapter 13, but the conclusion is that errors can be quite large if the levels of ^{3}H and ^{14}C activities differ greatly.

Equation 9-3 is valid if the counting efficiency of ^{3}H in the upper channel (r–s) is zero, that is, none of the ^{3}H pulses exceeds the lower limit of that channel. However, if the counting efficiencies of both isotopes in both channels are greater than zero and are known, the activities of both isotopes can be determined by solving Equations 9-5 and 9-6 simultaneously. The same notation scheme is used for these equations as were used for Equations 9-3 and 9-4.

$$cpm_{p-q} = (dpm_c)(CE_{c,p-q}) + (dpm_h)(CE_{h,p-q}) \qquad (9\text{-}5)$$

$$cpm_{q-s} = (dpm_c)(CE_{c,q-s}) + (dpm_h)(CE_{h,q-s}) \qquad (9\text{-}6)$$

Solutions for Equations 9-5 and 9-6 are given as Equations 9-7 and 9-8, which yield the activity of ^{14}C and ^{3}H, respectively.

$$dpm_c = \frac{(cpm_{q-s})(CE_{h,p-q}) - (cpm_{p-q})(CE_{h,q-s})}{(CE_{c,q-s})(CE_{h,p-q}) - (CE_{c,p-q})(CE_{h,q-s})} \qquad (9\text{-}7)$$

$$dpm_h = \frac{(cpm_{q-s})(CE_{c,p-q}) - (cpm_{p-q})(CE_{c,q-s})}{(CE_{c,p-q})(CE_{h,q-s}) - (CE_{c,q-s})(CE_{h,p-q})} \qquad (9\text{-}8)$$

When the degree of quenching for all dual label samples is constant, the counting efficiencies of both isotopes in each channel can be determined by counting standards in comparable counting solutions. However, if different levels of quenching among experimental samples are anticipated, quench correction methods must be applied. Because of overlapping spectra, those methods that utilize the sample's spectrum in obtaining a quench parameter (e.g., SCR) are not appropriate. However, the methods that utilize an external standard (e.g., ESCR, H#, ESSI) are suitable for quench correction. Suppose the ESCR method was selected for a quench indicator. A series of variably quenched ^{3}H standards should be prepared and counted with a constant gain and discriminator windows of p to q and q to s. Then the counting efficiency of ^{3}H in each of the two channels ($CE_{h,p-q}$) and ($CE_{h,q-s}$) is plotted against ESCR to prepare two of the four quench correction curves needed. Likewise, a series of variably quenched ^{14}C standards is used to prepare similar quench correction curves for ^{14}C activity, i.e. ($CE_{c,p-q}$) and ($CE_{c,q-s}$) versus ESCR. When an experimental sample is assayed its ESCR value is used to determine the four counting efficiencies needed for Equations 9-7 and 9-8 from the four quench correction curves. Of course, many of the mathematical tasks involved in dual-label counting can be performed by microprocessors, therefore with appropriate setup procedures it is possible to obtain the activities of both isotopes from modern instruments directly. Without microprocessing capabilities, ^{14}C and ^{3}H are frequently assayed with the upper channel being free of ^{3}H pulses, that is, Equations 9-3 and 9-4 are employed, then only three quench correction curves are needed. When channel settings are judiciously selected, Equations 9-7 and 9-8 should yield slightly more accurate results under most conditions, but the calculations required are somewhat more laborious.

Determination of Channel Settings

Most instruments provide predetermined settings for channels to be used in dual-label assays of the common isotopes; however, settings may be developed by an operator with some instruments. An appropriate procedure depends on whether the instrument uses logarithmic or linear amplification and the ratio of the isotopes' E_{max} values.

Using ^{3}H and ^{14}C as an example, an instrument with a linear amplifier and adjustable gain controls may be set up by selecting a narrow, lower discriminator window, the width of which depends on the relative energies of the isotopes to be assayed. In the case of ^{3}H and ^{14}C, a window of about 20% of maximum is a good first approximation, and might be 50–250 on a pulse height scale of 0–1000. A ^{3}H standard is counted between gain adjustments until a maximum counting rate is obtained in the narrow channel. This establishes the gain to be used for both channels. In instruments where the gain is not under operator control (e.g., logarithmic amplifiers) the preceding step is not performed, and although the units used for discriminator window limits may differ, the following procedures are applicable to instruments with either type of amplification. Selection of discriminator settings may be performed in many ways, and most operator's manuals provide a suitable method; however, two possible methods follow.

1. Prepare a pulse spectrum using the ^{3}H standard covering the pulse height region from the lower limit, 50 in our example, to the pulse height where no ^{3}H pulses are recorded. A pulse spectrum for ^{14}C is prepared similarly using the complete range from the lower limit (50) to the maximum, 1000 in this example. The pulse height where the ^{3}H pulse spectrum reaches the abscissa (pulse height axis) becomes the upper limit for the lower channel and the lower limit for the upper channel. Suppose the value is 260, then the discriminator settings would be 50–260 and 260–1000. Such conditions will exclude ^{3}H pulses in the 260–1000 channel, and hence Equations 9-3 and 9-4 apply.

2. After the gain setting has been established, the upper limit of the lower narrow channel is varied between counting periods. Both ^{14}C and ^{3}H are counted with each discriminator setting, and appropriate windows to be tested may include 50–140, 50–160, 50–180, . . . , 50–360. From the counting rates, the counting efficiencies for each isotope may be calculated, and a graph, such as that shown in Figure 9-8, is prepared by plotting the counting efficiency of ^{3}H versus counting efficiency of ^{14}C for each window. The value to be selected as the lower limit of the upper channel and the upper limit of the lower channel is selected by viewing the graph.

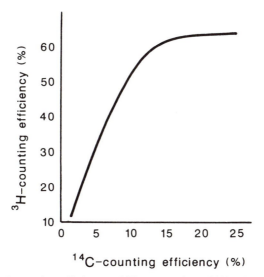

FIG. 9-8 A plot of counting efficiency of ^{3}H versus that of ^{14}C when the lower energy discriminator window width is varied.

Ideally, the window for the lower channel is one where the counting efficiency of ^{3}H is a maximum and that of ^{14}C a minimum. However, these are conflicting goals, and a compromise between them must be sought. The region of the curve (^{14}C efficiency around 10–15%) where it bends most sharply represents some windows suitable for such a compromise. In this region, ^{3}H is counted with high efficiency, close to the maximum, yet ^{14}C counting efficiency is reasonably low. One of the windows that provided data points in that region is selected as the lower window. Then the lower limit of the upper channel is set at the same value as the upper limit of the lower channel. Equations 9-7 and 9-8 are applicable for samples counted with settings established in this manner.

ADDITIONAL READING

Several excellent discussions about topics covered in this chapter are presented by others including Horrocks (1974), Kessler (1988), Kobayashi and Maudsley (1974), L'Annunziato (1987), and Wang et al. (1975).

REFERENCES

Horrocks, D. L. (1974). *Applications of Liquid Scintillation Counting*. Academic Press, New York.

Horrocks, D. L. (1977). *The H Number Concept*. Technical report 1095-NUC-77-IT, Beckman Instruments Inc., Fullerton, CA.

Ishikawa, H., Takiue, M., and Aburai, T. (1984). *Int. J. Appl. Radiat. Isot.* 35:463.

Kessler, M. J. (1988). *Liquid Scintillation Analysis, Science and Technology*. Publication No. 169-3052, Packard Instrument Co., Downers Grove, IL.

Knoche, H. W., Parkhurst, A. M., and Tam, S. W. (1979). *Int. J. Appl. Radiat. Isotopes* 30:45.

Kobayashi, Y., and Maudsley, D. V. (1974). *Biological Applications of Liquid Scintillation Counting*. Academic Press, New York.

L'Annunziata, M. F. (1987). *Radionuclide Tracers*. Academic Press, Orlando, FL.

Reed, D. W. (1984). *Int. J. Appl. Radiat. Isotopes* 35:367.

Wang, C. H., Willis, D. L., and Loveland, W. D. (1975). *Radiotracer Methodology in the Biological, Environmental and Physical Sciences*. Prentice-Hall, Englewood Cliffs, NJ.

CHAPTER 10

Liquid Scintillation:
Sample Preparation and Counting Atypical Emissions

Among topics concerning liquid scintillation counting there probably is the least amount of definitive technical directions, yet the most published information, about sample preparation methods. The reason for this situation is that the chemical and physical nature of samples from biological investigations varies widely and prevents the formulation of media that have acceptable properties for all kinds of samples.

AQUEOUS SAMPLES

Aqueous samples may form true solutions with certain cocktails and emulsions with others, but both types of liquid scintillation samples require certain conditions for valid assays.

True Solutions

Samples soluble in organic solvents such as toluene or xylene are ideal for liquid scintillation counting, because they are readily dissolved in cocktails that have high scintillation efficiencies. However, many biological samples contain water than is not miscible with such solvents. Cocktail formulations in which a portion of the aromatic solvent has been replaced by a solvent that is miscible with water can hold small volumes of water in a single homogeneous phase. Ethanol, methanol, and glycol derivatives are used for such purposes and are referred to as *secondary solvents*. Most

secondary solvents are quenching agents, and the performance characteristics of formulations that contain them are not as desirable as those with
aromatic solvents only. Dioxane, which may be considered as a primary or
secondary solvent depending on the proportion used, is miscible with water
and has acceptable energy transfer characteristics. The use of dioxane in
cocktails (see Chapter 9) is widespread, but safety concerns and problems
with purity have caused emulsion formulations to become more popular
for aqueous samples.

Emulsions

The presence of surfactants greatly increases the water-holding capacity of
liquid scintillation samples and offers another distinct advantage over
purely soluble systems. Salts and other molecules soluble in an aqueous
sample are less likely to precipitate in the presence of organic solvents.

Triton X-100 was mentioned in Chapter 9 as a popular surfactant. A
number of "Tritons" have been used successfully and their general structure is

$$CH_3-(CH_2)_n-\langle\bigcirc\rangle-(O-CH_2-CH_2)_n-OH$$

They differ in the lengths of their aliphatic and polyethoxy side chains. As
may be seen, the benzene ring and aliphatic chain written on the left side
of the ring is very hydrophobic and hence "soluble" in organic solvents
such as toluene, whereas the polyethoxy chain is "soluble" in water. Although cocktails that contain surfactants appear as solutions they actually
are emulsions that involve finely dispersed micelles. The organic solvent
may be any of the usual scintillation solvents, but toluene is a common one
and the micelles are spherical aggregations of the surfactant molecules in
which the hydrophobic ends are oriented outward into the toluene. When
water is present it "dissolves" in the interiors of the micelles and the sizes
of the micelles increase with increasing amounts of water. Thus, two
phases, aqueous and organic, exist with the surfactant molecules being at
the interface. The organic phase, which generally constitutes the bulk of
the emulsion, contains the fluor molecules, and hence is the phase where
scintillation events occur. A particular solute will partition between the two
phases, depending on its relative solubilities in the phases. For example,
[^{3}H]toluene will dissolve in the organic phase primarily and will be counted
with a higher efficiency than ^3H_2O, which would be located in the aqueous
phase. Of course emulsion type cocktails are not selected for radioactive
solutes that are soluble in organic solvents because a purely soluble system
would yield a higher counting efficiency. Consequently the usual condition

for emulsion cocktail use is where the radioactive sample is soluble in water and present in the interior portion of micelles.

The location of a radioactive atom within a micelle, the direction of its emission when it disintegrates, and the size of the micelle affect the proportion of the emitted particle's energy dissipated in the two phases. In general, energy is not transferred between phases, therefore the distance of a particle travels in the "aqueous" phase before entering the organic phase affects the scintillation yield for that particle. The energy dissipated in the "aqueous" phase is lost as far as its potential to excite fluor molecules; however, when fine dispersions exist, the loss will not be complete except for very weak particles. Effectively such energy losses behave, and can be treated as chemical quenching, where on the average the magnitude of all pulses are reduced. Consequently, the same methods described for quench corrections in Chapter 9 are applicable for emulsion-type counting.

The Triton compounds belong to the nonionic class of surfactants, but cationic and anionic surfactants offer advantages in cases where buffers and salts are present in the sample to be assayed. Quaternary amines may be used as cationic surfactants and sodium dodecylbenzene sulfonate is an example of an anionic surfactant.

TISSUE SOLUBILIZERS

Quaternary amines are strong bases and hence are effective in digesting small quantities of proteinaceous tissues. Hydroxide of Hyamine 10-X, a product of Rhom and Haas, is a classic tissue solubilizer. It is a 1.0 M solution of p-(diisobutylcreoxyethoxyethyl)dimethylbenzylammonium hydroxide in methanol and the ammonium ion's structure is

A number of commercial solubilizers have been developed and include BTS-450 (Beckman), NCS (Amersham), Protosol (New England Nuclear, Du Pont), Soluene (Packard), and TS-1 or TS-2 (Research Products International). These contain quaternary ammonium hydroxides or chlorides in concentrations of 0.5–1.0 M, but the structures of their active ingredients have not been disclosed.

Typically 20–100 mg of wet tissue is digested directly in a scintillation vial with 0.5–1.0 ml of the solubilizer, sometimes with additional water (10–100 μl) and heating at 40–50°C. Most proteinaceous tissues are dissolved in 1–2 hr and yield a slightly yellow colored solution that emulsifies in toluene-based scintillation cocktails. Bases such as the quaternary amines may cause chemiluminescence, and the color developed during digestion may cause color quenching; therefore treatment of solubilized samples with organic acids to neutralize the excess base and the use of peroxides to decolorize them prior to cocktail addition are common practices. Chemiluminescence and decolorization will be discussed later. The tissue solubilizers work fairly well for most animal tissues, but the carbohydrate polymers in plant tissues are resistant to digestion. Combustion methods (to be discussed later) are frequently used for plant samples.

ABSORPTION OF $^{14}CO_2$

In combustion methods for ^{14}C, CO_2 is trapped by a base, but the same principle is applied to the collection of $^{14}CO_2$ during metabolic reactions. For example, the amount of $^{14}CO_2$ respired by a small quantity of tissue may be measured using a small (20–50 ml) closed vessel. A small volume of an appropriate base is placed in a compartment within the vessel (perhaps a small glass cup or side arm) so that it is exposed to the gas phase inside the vessel, but does not mix with the medium containing the biological tissue. The CO_2 will be absorbed by the base, and after the completion of the incubation period an aliquot of the base may be mixed with cocktail to prepare the sample for counting.

Inorganic bases such as KOH and NaOH are effective absorbers, but are not very suitable for soluble liquid scintillation systems. Organic bases such as ethanolamine and its methoxy derivative are effective absorbers, and, with secondary solvents, are soluble in many types of cocktails. The quaternary amines used for tissue solubilization are excellent trapping agents and are easily incorporated into most cocktails, however, they are more expensive than the bases cited above.

SAMPLE COMBUSTION METHODS

Combustion of ^{3}H-, ^{14}C-, or ^{35}S-labeled samples to form $^{3}H_2O$, $^{14}CO_2$, or $^{35}SO_2$, respectively, has been performed since the advent of radiotracer methodology to improve detection sensitivity for certain types of samples containing one or two of these isotopes. In combustion methods, the tritium is converted to tritiated water, which is trapped by freezing or condensation. The products from a compound containing carbon and sulfur are the acidic gases, $^{14}CO_2$ and $^{35}SO_2$, and these may be absorbed in organic bases. A combustion method offers several distinct advantages:

1. It permits ^3H and ^{14}C to be separated before counting, hence avoiding some of the uncertainties of dual label counting.
2. It frequently yields samples with higher counting efficiencies by elimination of quenching agents.
3. It generally eliminates the potential problem of chemiluminescence.
4. It yields higher sensitivities in the assay of samples with low specific activities by increasing the amount of radioactivity from a biological sample that may be incorporated into a liquid scintillation sample.

To illustrate the latter advantage, suppose the specific activity of ^{14}C in a liver sample is about 0.5 dpm/mg. Using a tissue solubilizer (quaternary amine) about 100 mg (wet weight) of the sample may be incorporated into the liquid scintillation sample, which may yield a counting rate of about 40 cpm above background (50 dpm at 80% counting efficiency). Assuming the tissue is 65% water, a 2-g sample could be dried and would weigh 0.7 g. This quantity of dried liver could be combusted, the CO_2 collected, and incorporated into a liquid scintillation sample. The amount of radioactivity in such a sample would be about 1000 dpm and should yield a counting rate of about 800 cpm above background, assuming a counting efficiency of 80%. If the background was around 30 cpm, the solubilized sample would need to be counted for about 10.4 hr compared to 13.4 min for the combusted sample to achieve an acceptable degree of statistical accuracy. Obviously if a series of similar samples were to be assayed, 10-hr counting periods probably could not be tolerated whereas 15-min periods could.

Wet Oxidation

The complete oxidation of biological samples with strong acids and oxidizing agents, such as $HClO_4$, fuming sulfuric acid, periodate, hydrogen peroxide, and chromic acid has been used in the past, but rarely is used today except for small samples. Horrock's (1974) discussion of oxidation methods provides references to some older procedures.

Static Combustion

Oxygen bombs have been used to oxidize samples, and the "Shöniger flask" method has been the most widely used static method. Several variations in the physical arrangement have been used, but a typical flask is shown in Figure 10-1. The dry sample is placed in a platinum basket that is inserted into the flask after it has been thoroughly flushed with oxygen.

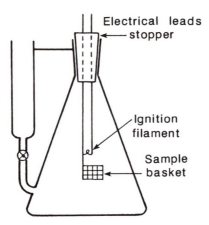

FIG. 10-1 A modified Schöniger flask for the combustion of ³H- or ¹⁴C-labeled samples for liquid scintillation counting.

The sample is ignited with an infrared source or by a hot electrical filament and allowed to burn. If tritium is to be assayed, the flask bottom is placed in a dry ice or liquid nitrogen bath for about 30 min to 1 hr to condense and freeze the water. The flask may be opened to add cocktail or cocktail may be introduced by several other methods, such as by a side arm with a stopcock or through a rubber septum. A negative pressure exists after cooling so liquids may be introduced easily. After dissolution and mixing, an aliquot of the sample is transferred to a liquid scintillation vial for counting. For ¹⁴C assay the flask is cooled to create a negative internal pressure and a CO_2 absorbing agent, such as ethanolamine, is introduced into the flask. After a period of time, cocktail is added, mixed, and an aliquot withdrawn for counting.

Obviously much labor is involved for each sample, and unless a number of flasks are processed concurrently, few samples can be processed in a day. The weight of dry material that may be combusted depends on the volume of oxygen available, but approximately 100 mg (dry sample) may be oxidized per liter of oxygen. One of the principal shortcomings in recovery of the isotope involves incomplete combustion. Unless the mass is compact and bound together well, small pieces may fall from the basket and not be completely burned.

Memory effects, the carry over of radioactivity from one combustion to another in the same flask, may be detectable when samples having high activities are combusted. Tritium, particularly ^3H_2O, is adsorbed or absorbed by every type of solid to a degree, and glass is no exception. In fact, glass contains bound water that is subject to exchange with water in its surroundings. Some of the water generated during a combustion will exchange with the water on and in the walls of the flask. Although the percentage of activity retained by the flask walls is so small that the recovery of ^3H_2O does not appear to be affected, the combustion of a high activity sample, followed by a blank, may yield detectable levels of tritium in the blank. Of course, water exchange occurs during normal washing procedures and generally this is sufficient to prevent noticeable memory effects, but soaking flasks in water for extended periods during washing may be necessary when very high and low activity samples are combusted sequentially.

Dynamic Combustion

In dynamic combustion, oxygen flows over the oxidizing sample, and the oxidized gases are trapped as they are vented. The process has been perfected and automated to the extent that commercial sample oxidizers have largely supplanted other combustion methods. A diagram of the general characteristics of a commerical oxidizer is shown in Figure 10-2.

The sample is introduced into the chamber, and oxygen flow begins. Via heated filament ignition occurs, and burning continues for a short period of time. The vapors from the combustion flow through a condenser and exchange column, which condenses the water while the CO_2 and other combustion gases pass on to a second section where the CO_2 is absorbed by an organic base. After the combustion is complete, N_2 is used to flush all remaining combustion products into the trapping sections, and steam is generated to aid in the exchange of ^3H_2O from the walls of the system so memory effects are alleviated. Most of the water is trapped by a condenser and it drips into a vial if tritium is to be assayed. Otherwise it is directed

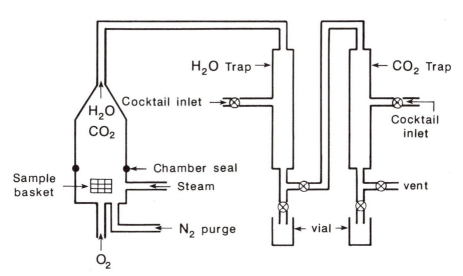

FIG. 10-2   A schematic diagram of a commercial sample oxidizer for the preparation of liquid scintillation samples.

to a waste receptacle. An exchange column is used to complete the H_2O trapping process, and assuming ^3H_2O is to be assayed, cocktail is used to wash the remaining H_2O from the trap into the vial.

If ^{14}C is to be assayed, the CO_2 is reacted with an organic amine that is automatically introduced into the reaction column. An exchange column is used to complete the trapping process, then cocktail is used to wash the products into the vial.

Samples containing either ^{14}C, ^3H, or both can be combusted by the method described and this permits the separation of the two isotopes if present in one sample. If one or the other of the isotopes is not present the appropriate trap's function is modified to avoid waste of reagents. Some instruments employ a catalyst in a train to ensure complete oxidation.

Dynamic oxidation instruments offer significant advantages over other methods by allowing much larger samples to be oxidized, generally affording more complete oxidation, and reducing labor requirements and sample preparation time.

HETEROGENEOUS SYSTEMS

Although emulsions are not true solutions their behaviors are similar to those of soluble systems and macroscopically they are homogeneous. When visible phases, two liquids or a solid and a liquid phase are present, a careful investigator should be concerned about obtaining meaningful results by liquid scintillation counting.

Two Liquid Phases

Generally the presence of two liquid phases in a scintillation vial can, and should, be avoided. Suppose a vial contains a toluene phase above a layer of water. The fluor molecules will be in the toluene phase, and hence scintillation events will take place only in this phase. Unless the radioactive sample is soluble in the organic phase a very low counting efficiency can be expected, because only the few emissions that enter the organic phase from the interface will produce counts. With surfactants, a homogeneous dispersion can be produced in most cases so a two-phase system should not be tolerated.

Solids in Liquids

Solids are subject to *self-absorption,* meaning that some or all of the energy of an emitted ray is absorbed by the solid matrix before the ray reaches the surface of the solid. For weak β^--emitters such as ^3H and ^{14}C, a millimeter thicknesses of most solids will completely absorb their emissions. Therefore if radioactive atoms are present in precipitates some self-absorption will occur, and the extent of the absorption depends on the energy of the rays, the size of the particle, and its density. It is impractical to determine the fraction of energy absorbed by the precipitate, hence impossible to ascertain the counting efficiency of such a system. Generally such situations should be avoided, but if conditions are controlled rigorously so that a reasonable level of reproducibility can be maintained, it may be possible to obtain *comparative* counting data between samples.

Consider the problem of counting solid particles of $Ba^{14}CO_3$ in a liquid scintillation medium. $BaCO_3$ is relatively dense so considerable self-absorption will occur for ^{14}C, but when surrounded by cocktail those emissions that reach the liquid are capable of being detected. Since the surface to mass ratio of particles increases as their diameter decreases smaller particle sizes yield higher counting efficiencies than larger ones. Obviously, any variation in particle size between samples decreases the reproducibility of counting results accordingly. Another factor that may affect results is whether all particles are completely surrounded by cocktail. In the case of $BaCO_3$ powders in toluene cocktails, settling will occur and the solid layer at the bottom is subject to self-absorption. Gelling agents, such as finely divided silica (Cab O'sil is a trade name) can be used to increase the viscosity of the medium so that solid particles remain suspended. Gels also may be produced by adding appropriate amounts of water to some of the commercial emulsion type cocktails mentioned in the Cocktail Formulations section of Chapter 9.

Filter Discs

One practice that has received widespread usage is the counting of precipitates on filter discs. Paper, glass, fiber, and membrane filters may be used, and normally they are placed face up on the bottom of a vial then cocktail is added without disturbing the disc. Although the method is popular and convenient it is not an acceptable method for the quantitative assay of radioactivity because it is virtually impossible to determine counting efficiencies. Nevertheless, assays that depend only on differences in counting rates may be performed satisfactorily under rigorously controlled conditions. Some of the factors that need to be considered as potential sources of variation between samples follow.

The precipitate itself is subject to self-absorption, and differences in the thickness of the precipitate layer are a source of variation. The quantity of precipitate and its distribution across the filter disc can affect layer thickness. In most successful assays the quantity of precipitate is very small and self-absorption by the precipitate is not great. However, self-absorption by the disc material may occur when small particles are trapped in interstitial spaces within the disc. This type of self-absorption is more pronounced with coarse paper and glass fiber filters because their pore sizes vary more than that for membrane filters, and the pores are larger. If the precipitate particles become dislodged from the filter the geometry of the sample is changed because cocktail may surround a free floating particle compared to one bound tightly to a disc. Also if the precipitate or quenching agents present on the disc are partially soluble in the cocktail, the counting rate may increase or decrease over a period of time until equilibrium is established. The ability of filters to transmit light from scintillation events may also affect counting rates. Cocktail may permeate interstitial spaces, and hence particles trapped in such spaces may produce scintillation events. If the light emitted is absorbed by the filter material such events would not be detected. Presumably higher light transmission is an advantage of glass fiber filters, but some membrane filters become transparent when exposed to cocktail as well. Obviously the position of the disc in the vial can affect counting results, so care in handling vials is necessary.

For protein and nucleic acid samples, self-absorption can be avoided by adding a small volume of tissue solubilizer to discs in counting vials and incubating them for 30–60 min before cocktail addition. However, tests should be performed to ascertain conditions necessary for complete dissolution of typical precipitates.

Electrophoretic Gels

The amount of radioactivity associated with electrophoretic bands may be judged from autoradiographs, but sometimes more precise measurements are desirable. The direct counting of gel pieces placed in cocktail should be avoided because of the absorption of radiation from within the gel.

If possible it is desirable to dissolve the protein, DNA, or RNA and prepare a homogeneous solution. This may be accomplished by several methods depending on the characteristics of the biological sample. Incubation of gel segments with tissue solubilizers is effective in freeing the material from the gel matrix, and while the matrix may remain intact, if the protein or nucleic acid is solubilized, normal quench correction methods may be applied to the counting results.

If the electrophoretic sample is soluble in water it may be leached from the gel by incubation with a small quantity of water (macerating the gel will hasten the process) that is subsequently incorporated into an appropriate cocktail. Polyacrylamide may be solubilized with strong oxidizing agents such as H_2O_2 and $HClO_4$, and with appropriate cocktails an entire gel slice may be incorporated into a liquid scintillation sample.

The counting of electrophoretic gel slices can and should be performed in a homogeneous system, and if the labeled material is solubilized, the liquid scintillation sample is considered to be homogeneous although pieces of gels may remain.

Thin-Layer Chromatogram (TLC) Absorbents

One practice that should be discouraged is the counting of absorbent scrapings or TLC segments in cocktails without testing to determine if the labeled material is eluted from the absorbent. Generally incubation of TLC absorbent with a solvent in which the labeled compound moves chromatographically will elute the compound from the absorbent quickly. When the eluate is mixed with a compatible cocktail a homogeneous system is obtained. The whole process may be conducted in a vial by adding a small volume of the elution solvent to the absorbent in the bottom of a vial. After a short period of agitation, cocktail is added and all suspended material is allowed to settle before counting.

The conclusion to be drawn is that precise measurements of radioactivity by liquid scintillation counting are not possible in heterogeneous systems. Generally methods are available to solubilize the labeled material and hence obtain suitable data. However, some heterogeneous systems may provide meaningful results under certain circumstances, but it is important to keep in mind that they usually show more variability than homogeneous systems. It is difficult to assess the statistical validity of counting heterogeneous samples without extensive measurements, but if differences between samples to be compared are less than about 20% it is unlikely a heterogeneous system will be appropriate.

SAMPLE PREPARATION PROBLEMS

Colored Samples

As indicated in Chapter 9 quench correction curves for chemical quenching differ from those for color quenching, and frequently it is desirable to avoid color quenching by decolorization of samples.

Hydrogen peroxide is probably the most widely used decolorizer, but other oxidizing agents, including $NaOCl$ and $HClO_4$, are effective. These agents are strong quenching agents and present solubility problems unless precautions are taken to decompose the decolorizer after the color disappears.

Chemiluminescence

Chemiluminescence is the phenomenon in which at least a portion of the energy of exothermic chemical reactions is released as light. A general form of such reactions is

$$A + B \longrightarrow C + D^*$$
$$D^* \longrightarrow D^\circ + h\nu$$

The chemical reaction yields a product (D*) that is in an electronically excited state, and the energy of excitation is dissipated by the emission of a photon of light. Therefore chemiluminescence is a single photon event unlike the multiple photon event produced by a β-particle. If the wavelength of the light emitted is in the range of the MP-tube's response spectrum, a single photon may produce a pulse; however, if the coincidence circuit is performing properly, a single photon will not yield a count. Then the question becomes, what is the chance of single photons striking both tubes simultaneously? Actually the pulses do not have to be exactly simultaneous because the coincidence circuits have a finite resolving time τ, which is in the order of 20–30 ns. This means if pulses from each tube are produced within a period of time equivalent to the resolving time, the pulses are processed as coincident pulses. If the pulse rates of the two MP-tubes are R_1 and R_2, the rate of coincident pulses (R_c) is given by Equation 10-1.

$$R_c = 2R_1R_2\tau \qquad (10\text{-}1)$$

When the rate of the chemical reaction is very high, perhaps 2×10^6 photons are released per minute, the probability that coincident pulses will be produced becomes significant. Assuming the number of photons striking each MP-tube is equally divided and the quantum efficiency of the photocathode is 25%, the pulse rate of each MP-tube is 2.5×10^5 cpm. If the resolving time is 21 ns (3.5×10^{-10} min/ct), the coincident pulse rate is

$$R_c = (2)(2.5 \times 10^5)(2.5 \times 10^5)(3.5 \times 10^{-10}) \text{ cpm}$$

$$R_c = 44 \text{ cpm}$$

As the photon release rate increases the number of coincident pulses increases dramatically. About 10,000 cpm is generated from a release rate of 3×10^7 photons per minute, and counting rates in the hundreds of thousands are not uncommon for freshly prepared samples that are subject to chemiluminescence.

Detection of Chemiluminescence.   Since single photon events are involved, the pulses are uniformly weak and can be eliminated by raising the lower discriminator limit. However, this excludes the weaker pulses from β-particles and yields lower counting efficiencies. Fortunately chemiluminescence can be eliminated under most conditions if samples subject to such chemical reactions are observed. Perhaps the first clue to an operator that chemiluminescence is occurring is an abnormally high counting rate, but two more definitive tests can be performed.

1. Monitor the counting rate over a period of time. The time course for most chemical reactions that cause chemiluminescence is usually short enough for the reaction to be complete in a few minutes to a few days, and the counting rate decreases with time, usually logarithmically. It is

not uncommon to observe initial counting rates around 10^5 cpm to decrease to 10^3 cpm in an hour; however, it will require several more hours for the chemiluminescence to be completely dissipiated.

2. Check quench correction parameters. Since the chemiluminescent pulses are very weak, the sample channels ratio (SCR) or sample spectral index (SSI) value will reflect the fact that a very high proportion of the pulses is in the low energy portion of the pulse spectrum. This means that the SCR or SSI value indicates more quenching than is actually present. Of course, one cannot ascertain the relative contributions of quenching and chemiluminescence, but if the SCR or SSI value changes with time, chemiluminescence is indicated.

Chemiluminescence is not subject to chemical quenching, so comparing SCR or SSI values to an external standard quench parameter such as the ESCR, ESSI, or H, can be diagnostic. For example, suppose the sample is moderately quenched. If chemiluminescence is occurring, the SCR value will reflect a very high degree of quenching while the ESCR value reflects a moderate degree of quenching. If the external standard quench parameter does not correspond to approximately the same degree of quenching as the SCR or SSI value, chemiluminescence is indicated. Normally the predicted counting efficiencies from the two quench correction methods should be within 1% of one another.

Sample Compositions Subject to Chemiluminescence.   The precise reactions that cause chemiluminescence in liquid scintillation samples are not understood well, consequently conditions that cause or eliminate the problem are not easily predicted. Nevertheless, oxygen appears to be the primary culprit. Excited oxygen molecules may undergo deexcitation by several processes, but the emission of a photon is one of its deexcitation modes, and this mode appears to be favored more when alkaline conditions exist. Normally dissolved oxygen is present in all samples prepared in air, but rarely causes detectable chemiluminescence, presumably because the oxygen concentration is not high enough or the media's composition favors other deexcitation modes for excited oxygen.

Oxygen may be removed by flushing solutions with N_2 or Ar, but unless the vials are flame sealed, air is likely to seep in around vial caps. Rarely are such precautions necessary to avoid chemiluminescence, but since oxygen is a strong quenching agent such practices may be used to obtain the highest possible counting efficiencies in critical studies.

Peroxides are used to decolorize samples and may be a component in oxidation mixtures. If excess peroxides are present, conditions may cause them to liberate oxygen and cause chemiluminescence.

Tissue solubilizers, such as the various quaternary amines, are strong bases; consequently samples that contain them frequently exhibit chemiluminescence. In contrast, acidic samples rarely cause chemiluminescence problems, but they do not appear to be immune.

Dioxane undergoes autooxidation in air and forms peroxides; therefore samples prepared with unpurified dioxane are more likely to exhibit chemiluminescence than samples prepared in toluene, xylene, or pseudocumene. The presence of oxidizers in dioxane-containing solutions may also generate oxygen causing chemiluminescence.

Probably the largest source of chemicals that promote chemiluminescence is the impurities introduced in the solvents and the chemicals used to treat a sample (solubilizers and decolorizers). Biological degradation products may contribute to chemiluminescence as well. Blood and urine samples frequently present chemiluminescence problems.

Methods of Reducing Chemiluminescence.   Since chemiluminescence is usually detected as a problem after a sample has been prepared, the most common method of dealing with the problem, at least for that sample, is to store the vial until its counting rate becomes stable. Generally 24 hr is sufficient in most cases. Warming liquid scintillation samples at 40–50°C accelerates the reactions associated with chemiluminescence and decreases the storage time, but care should be exercised to avoid vaporization of volatile solvents.

If chemiluminescence is recognized as a potential problem, steps may be taken when samples are being prepared to avoid or reduce the problem. Samples that contain bases, such as a tissue solubilizer, may be neutralized by addition of an organic acid such as acetic, benzoic, or ascorbic acid. HCl has been used successfully for neutralization as well when emulsions are employed. Generally acids do not cause problems, but their neutralization with amines before adding cocktail may be a worthwhile precaution. Oxidizing agents, such as peroxides, may be destroyed by the addition of suitable reducing agents. Regardless of the possibility of chemiluminescence, the use of high purity components for all components in a liquid scintillation sample is recommended.

Some of the instruments being manufactured today can identify chemiluminescence and correct for it.

Photoluminescence

In liquid scintillation counting, *photoluminescence* refers to phosphorescence that occurs in samples exposed to normal laboratory light. Fluorescence and phosphorescence are similar phemomena where light of one wavelength is absorbed by a molecule producing an excited state. The energy of excitation is subsequently reemitted at a slightly longer wavelength, and the distinction between fluorescence and phosphorescence is based on the lifetime of the excited state, which is the average length of time between excitation and deexcitation. In fluorescence, lifetimes are in the range of 10^{-8} to 10^{-4} s, whereas phosphorescent compounds have lifetimes that exceed 10^{-4} s, and may be as long as several minutes or hours. The fluors are fluorescent compounds, and fluorescence by the waveshifters is responsible for the typical purplish color in liquid scintillation samples. However, with a lifetime below 10^{-4} s, this fluorescence poses no problem because the small period of time the sample is subject to darkness in the counting chamber before counting begins is sufficient for fluorescence to decay completely.

Compounds subject to phosphrescence may be a problem if their lifetimes are sufficiently long. However, storage of samples in the dark for about 0.5 hr will circumvent many problems. Most automatic sample changers are enclosed and this eliminates or markedly reduces the possibility of phosphorescence interference.

Static Electricity

Electrical charges can build up on and within vials, and discharges that occur within the counting chamber may yield light that is detected by the MP-tubes. Instruments are designed to minimize spurious counts due to this phenomenon.

COUNTING VARIOUS TYPES OF EMITTERS

The discussion of liquid scintillation in Chapters 8 and 9 was devoted to detection of the commonly employed negatron emitters such as ^{3}H, ^{14}C, ^{35}S, and ^{32}P; however, other types of emitters may be detected efficiently by liquid scintillation.

Positron Emitters

For a given E_{max}, positrons are detected as efficiently as negatrons in liquid scintillation solutions, and the pulse spectra of β^+-emitters are similar to those of β^--emitters. Typically, a positron will dissipate its kinetic energy completely in the scintillation solution and on coming to rest undergo an annihilation reaction with an electron yielding two 511-keV rays. Because of the low density of the medium, most of the rays will escape the solution without an interaction, but a small fraction will interact by the Compton effect and a smaller fraction by the photoelectric effect. The dissipation of energy from the electrons arising from such interactions will contribute to the pulse heights of the positrons; however, quantitatively, the contribution of annihilation radiation to the pulse spectrum is negligible.

Electron Emitters

Two types of decay processes may eject electrons from the orbitals surrounding the decaying atom, and they are internal conversion and electron capture (see Chapter 3). Regardless of their origins, electrons dissipiate their kinetic energies in a liquid scintillation solution in the same manner as β^--particles. Consequently, scintillation events are initiated by such electrons.

In internal conversion, auger electrons as well as conversion electrons may be produced and both are emitted with discrete energies. Many radionuclides that decay by electron capture also yield auger electrons. The electronic rearrangements that occur in an atom after internal conversion or electron capture result in the emission of X-rays, and depending on the atomic number of the element involved, the energies of those rays may be low enough to be absorbed efficiently by small thicknesses of the scintillation medium. If the X-rays interact, they will yield electrons, primarily by the Compton effect.

The pulse spectra of radionuclides decaying by internal conversion or electron capture are quite different from those of β^--emitters in that most exhibit discernible peaks corresponding to certain emission energies. If produced, most pulses are stronger than the weaker pulses of β^--emitters so they generally are readily detected, but counting efficiency is dependent on the relative proportion of atoms that yield electrons directly or indirectly.

Several radionuclides that may be employed in biological and medical research and decay by electron capture are ^{55}Fe, ^{51}Cr, and ^{125}I. All may be counted with acceptable efficiencies by liquid scintillation.

α-Emitters

The energies of α-particles relative to most β^--particles are much higher and αs are emitted with discrete energies. Therefore, when dissolved in appropriate liquid scintillation solutions, α-emitters are counted with nearly 100% efficiency and produce pulse spectra with distinct pulse height peaks. Liquid scintillation is an effective assay method for α-emitters, but few α-emitters are useful tracers in biological and medical studies. Consequently, applications for liquid scintillation counting of α-emitters are not numerous and have not received much attention.

γ-Emitters

Except for very low energy γ-rays, the counting efficiency of γ-emitters is quite low because of the small fraction of rays that interacts with the liquid medium. However, some isotopes that emit other particles in addition to γ-rays can be counted effectively. Solid scintillation is the method of choice for most γ-emitters and ^{131}I is conveniently, and perhaps most often, counted by that method. Nevertheless, ^{131}I can be counted with a higher efficiency by liquid scintillation in many cases because this radionuclide decays by $β^-$-emission and the γ-rays arise from nuclear excited states of its daughter, ^{131}Xe. The E_{max} for 86% of the decays is about 607 keV, which means that a high percentage of its $β^-$-particles is detected by the liquid scintillation.

COUNTING ATYPICAL EMISSIONS

The radiation from bioluminescence and Cerenkov radiation are due to different processes than those involved in liquid scintillation, and discussion about counting such emissions with liquid scintillation topics is only justified because the same type of instrumentation is effective in assays involving these types of radiation.

Bioluminescence

Bioluminescence is not a counting inteference problem, rather it is a phenomenon that permits a liquid scintillation counter to be used to assay light-yielding biochemical reactions.

Luciferase, the so called firefly enzyme that catalyzes the production of greenish light from the hydrolysis of ATP, may be used to assay ATP concentrations in biological studies. The coincidence circuit should be deactivated so that the pulses from either or both MP-tubes are counted. Since the emissions occur one photon at a time, weak pulses are produced and the discriminator is not effective in distinguishing such pulses from most of the MP-tube noise pulses. This means that without the coincidence condition, the background (blank) counting rate will be much higher than those observed in normal radioactive assays. Nevertheless ATP assays are very sensitive, in the picomolar range. The assay of flavin and nicotinamide containing nucleotides has also been accomplished by similar methods.

Cerenkov Counting

In Chapter 5 the origin and nature of Cerenkov radiation were discussed briefly. Recall that Cerenkov radiation arises when the velocity of a charged particle exceeds the velocity of light in a transparent medium. The index of refraction (n) is equal to the velocity of light in a vacuum (c) divided by its velocity in a medium (c_m) as shown in Equation 10-2.

$$n = \frac{c}{c_m} \tag{10-2}$$

Since the velocity of a particle (v) must exceed c_m to cause Cerenkov radiation, the minimum velocity for a particle is readily calculated from the index of refraction of the medium. The minimum velocity is called the threshold velocity (v_t).

$$v_t = c_m = \frac{c}{n}$$

A common medium used for detection of β-particles is water, which has a fairly high index of refraction ($n = 1.332$), therefore the threshold velocity for a β-particle in such a medium is

$$v_t = \frac{2.9979 \times 10^{10} \text{ cm/s}}{1.332} = 2.2507 \times 10^{10} \text{ cm/s}$$

The kinetic energy of a particle having the threshold velocity may be calculated using Equation 2-4.

$$E_k = \frac{m_0 c^2}{\sqrt{1 - (v_t/c)^2}} - m_0 c^2$$

$$E_k = \frac{0.511 \text{ MeV}}{\sqrt{1 - (2.2507/2.9979)^2}} - 0.511 \text{ MeV}$$

$$E_k = 0.2626 \text{ MeV} \approx 263 \text{ keV}$$

The minimum kinetic energy is referred to as the threshold energy, and, as shown by the previous calculation, it decreases as the index of refraction increases.

Obviously the E_{max} values for ^{3}H, ^{14}C, and ^{35}S are below the threshold energy for water, but a large fraction of the β-particles from ^{32}P ($E_{max} = 1.71$ MeV) exceeds the threshold. Counting efficiencies in the range of 30–45% are common for ^{32}P in pure water. Although counting efficiencies much higher than this can be obtained by liquid scintillation counting, there are several advantages to counting ^{32}P in water.

1. The cost and time of sample preparation is reduced because no cocktail is required.
2. Since no cocktail is employed, the sample may be recovered for further studies.
3. For aqueous samples that have low specific activities, a larger sample may be counted.
4. Generally quench correction methods are not required, unless colored solutions are involved.

Other isotopes of biological interest may be assayed by Cerenkov counting and they include both negatron and positron emitters, such as several of the chlorine isotopes, ^{40}K, and ^{18}F.

Geometry of MP-Tubes. Normally the MP-tubes of a liquid scintillation counter are located opposite of each other (180°) and are operated in co-incidence. This arrangement is not ideal for detection of Cerenkov radiation due to its directional nature (see Chapter 5). The conical angle of emitted rays depends on the velocity of the particle and the index of refraction of the medium as shown by Equation 5-4.

$$\cos \theta = \frac{c}{nv} \tag{5-4}$$

E_{max} for ^{32}P is 1.71 MeV, which corresponds to a velocity of about 2.712×10^{10} cm/s (obtained from Equation 2-4), and $n = 1.332$ for water.

$$\cos \theta = \frac{2.9979}{(1.332)(2.712)} = 0.8299$$

$$\theta = 33.9°$$

$$2\theta \approx 68°$$

The maximum conical angle for ^{32}P in water is about 68° but few negatrons have that high of velocity; furthermore their velocities decrease along their paths, therefore smaller angles are more prevalent among individual Cerenkov events.

Operating a counter by deactivating the coincidence circuit improves counting efficiencies markedly, but it also increases the background counting rates. Generally the higher efficiency does not compensate for the background when the statistical accuracy of measurements is considered.

Wavelengths of Light. Although the light emitted by Cerenkov events is a continuous band that extends into the visible region, much of the light is in the near-UV region, and light of those wavelengths is not detected efficiently by the common MP-tubes. Glass, which may be present in the faces of MP-tubes and vials, absorb UV light, and the replacement of glass with quartz improves counting efficiency, but at a considerable expense. Because of less UV absorption, plastic vials offer better counting efficiencies than glass vials.

As indicated in previous chapters, most MP-tubes are not as sensitive to near-UV light as light in the visible region, therefore detection efficiency can be improved by employing a waveshifter. The waveshifters common to liquid scintillation cocktails are not used, instead a water-soluble compound that absorbs near-UV light and reemits it at a longer wavelength is desirable. Besides increasing the wavelengths of light, the inclusion of a waveshifter helps overcome the directional problem of Cerenkov radiation because emissions from waveshifters occur in all directions, thereby increasing the probability of light striking the MP-tubes.

A number of waveshifters have been evaluated for ^{32}P counting by van

Ginkel (1980), and some examples are quinine, esculin, and thymine. Sodium salicylate is a commonly employed waveshifter, but caution should be exercised in its use since it is sensitive to pH. At a concentration of 500 mg/liter, it yielded a counting efficiency of 66.8%, while at 1000 g/liter an efficiency of 85% was observed. The improved counting efficiency for the more concentrated sodium salicylate solution cannot be attributed solely to a greater waveshifting capacity, instead the refractive index of the solution is significantly higher than the less concentrated solution.

Refractive Index.   As mentioned earlier the energy threshold for Cerenkov radiation decreases as the refractive index increases, therefore the use of compounds, alone or in solutions, that have a higher refractive index increases the proportion of β-particles capable of causing a Cerenkov event. Both benzene and toluene have refractive indexes around 1.5, which translates to a threshold energy of 175 keV. Compared to water (threshold = 265 keV) these solvents should be much more efficient, but solubility problems with most ^{32}P samples and additional cost to prepare such solutions do not warrant their use generally.

The index of refraction for a solution is an additive function related to the molalities of the components; therefore aqueous solutions of compounds such as salicylic acid ($n = 1.565$) or glycerol ($n = 1.4756$) yield significantly higher refractive indexes than water alone, but again the simplistic advantages of Cerenkov counting are lost.

Sample Volume.   The efficiency in Cerenkov counting is affected by sample volume, unlike the situation with liquid scintillation counting. For reproducible results, the volumes of all samples must be kept constant. The optimum volume apparently depends on the energies of the particles; hence the radionuclide being assayed as well as the nature of the medium show volume effects, but usually the optimum volume is around 8–11 ml. Sample volume has a bearing on the geometry within the counting chamber, and it can affect particle pathlengths when very energetic particles are involved. Volume also affects the distribution of pulse heights, which explains why efficiencies may be affected.

Quenching.   Unlike the scintillation process, Cerenkov counting is not subject to chemical quenching, consequently unless contaminates are present at high enough concentrations that affect the index of refraction, they have no effect on counting efficiency, except if they are colored.

Color quenching occurs when compounds absorb light of the wavelengths emitted during Cerenkov events and do not reemit the light at appropriate wavelengths. Most of the quench correction methods suitable for color quenching in liquid scintillation counting are appropriate for correcting color quenching in Cerenkov counting. However, the problem frequently is more easily addressed by decolorizing samples, and one does not need to be concerned about causing chemical quenching if strong oxidizing agaents are used.

A good review and references on Cerenkov counting are offered by L'Annuziata (1987), and it is recommended reading for anybody contemplating measuring radioactivity by this method.

REFERENCES

L'Annunziata, M. F. (1987). *Radionuclide Tracers,* pp. 241–264. Academic Press, Orlando, FL.

van Ginkel, G. (1980). *Int. J. Appl. Radiat. Isotopes* 31, 307–312.

CHAPTER 11

Detection of Radioactivity by Semiconductors

SEMICONDUCTOR DIODE DETECTORS

The responsive component in the class of detectors to be discussed in this chapter involves a diode, a junction between two types of semiconductor materials. Commonly, such detectors are called *semiconductor* or *solid state detectors,* but neither term is restrictive in regard to a particular operating principle. For example, solid scintillation detectors could be, but generally are not, called solid state detectors. Here the term *semiconductor diode detectors* or more simply *diode detectors* will be used to identify this class of detectors.

Although more than adequate for many purposes, diode detectors are not particularly efficient in counting radioactive emissions, but they are superior to other types of commercially available detectors in spectroscopy, the measurement of ray energies. Consequently, diode detectors are used extensively for qualitative purposes and for quantitative purposes when mixtures of radionuclides are present.

Some general applications include the following:

1. Monitoring the presence of radionuclides in the environment. Since the types of emissions and their energies are known characteristics for most radionuclides, they can be identified and assayed in environmental samples.

2. Neutron activation analysis. The radioactivity induced in a sample by bombardment with neutrons is analyzed primarily by determining the

energies and half-lives of γ- and X-rays given off by activated samples, thus permitting elemental compositions to be ascertained.

3. Nuclear physics. The measurement of energies for particles or rays arising from particle interactions or from induced nuclear reactions aids in determining the reactions occurring.

Diode detectors are not used extensively in biological or medical research because quantitative (counting) assays of single known radionuclides is the principal kind of measurement needed. Nevertheless, knowledge of the operating principle and capabilities of diode detectors may lead to new approaches for old problems.

BAND THEORY

Band theory describes the electrical behavior of solids, how individual atoms participate in the movement of charges, and how their properties account for variations in conductivity.

Electron Energy Bands

In molecules some of the atomic orbitals combine to form molecular orbitals, which increase the number of allowed energy levels for the electrons involved in chemical bonds. Those electrons that normally participate in bonds are called *valence electrons,* and invariably they include electrons in the highest (outermost) energy shell.

When molecules crystallize, the close association between molecules permits molecular orbitals to overlap that effectively extends some orbitals throughout the crystal. Noncrystalline solids exhibit less definite but similar effects. The influence of neighboring atoms on a particular atom is easily visualized, but theoretically, all atoms in a solid have an effect on the potential energy levels of the valence electrons of a particular atom. If a solid piece of material contains 1 mol of atoms, the number of potential energy levels of an individual electron is increased by about 6×10^{23} times. Of course, the energy difference between levels is minuscule, and this creates a virtual continuous band of energies for valence electrons, called the *valence band*.

Consider a piece of metallic sodium. Each atom has a pair of electrons in its K shell (1s orbital), eight electrons in its L shell (2s and 2p orbitals), and a single electron in its M shell (3s orbital), which is sodium's valence electron. Since the valence electron is in the third principal quantum level (M shell), there are many unfilled sublevels in that shell, and the valence electron can be excited easily. That is, only a small increment of energy is needed to promote a valence electron to a slightly higher energy sublevel.

Thermal energy, such as that supplied by room temperature, is sufficient to excite many of the valence electrons so that their attraction to a particular atom is exceedingly weak; effectively they are free to move throughout the solid. Certainly, the energy given by the application of a small bias (voltage) will cause the electrons to flow in a particular direction. Electrons being conducted still have quantized energy levels, but the levels are essentially continuous and constitute a band of energies called the *conduction band*.

In our discussion, the valence electrons of elemental sodium were described as being in the valence band at one point, then later in the conduction band. This is the situation for good electrical conductors; their valence band energies overlap conduction band energies. This occurs when either of two situations, shown diagrammatically in Figure 11-1, exists. In Case 1, the valence electron's energy shell is unfilled, which allows electrons to acquire small amounts of energy and be promoted to higher sublevels. In Case 2, a principal shell is filled, but the highest sublevel in that level is higher on the energy scale than the lowest sublevel in the next higher shell. The point is, for good electrical conductors only small increments of energy are needed (if needed at all at room temperature) to "excite" electrons sufficiently for conduction, and this is possible when there are unfilled energy sublevels immediately above the valence electrons' energy levels. Metallic sodium corresponds to Case 1 in Figure 11-1, and it certainly qualifies as a conductor.

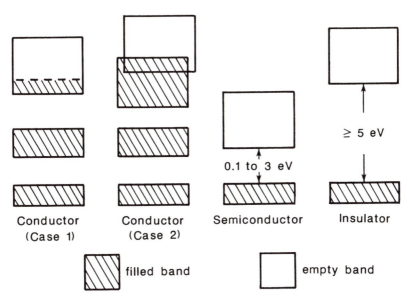

FIG. 11-1 Electron energy bands in various types of materials classified according to their ability to conduct electricity.

In insulators the valence electrons are in a filled shell, and considerable energy is required to promote them to the next higher level where they are free to move in the solid. Thus, there is an energy gap between the valence band and conduction band. The quantum mechanical nature of the atoms is such that electron energy levels between the valence and conduction bands are not allowed, hence this range of energies sometimes is called the *forbidden gap*.

Diamond is a good *insulator* and has a forbidden gap of 5.47 eV. The tetrahedral bonds of carbon are not easily depicted in one plane, but Figure 11-2 shows the bonding that occurs in diamond, disregarding proper space representation. Single bonds, indicated by the sharing of a pair of electrons (dots), connect each atom to four other atoms, and this creates *one* molecule the size of the piece of diamond. In carbon there is a pair of electrons in the 1s orbital and single electrons in each of the four hybridized sp orbitals of the L shell. The latter four are carbon's valence electrons, which are shared with four other atoms. Thus, eight electrons are involved in the bonding, and this fills the second principal quantum level (L shell). The energy levels of these eight electrons constitute the valence band. To excite a valence electron it must be promoted to the third principal quantum level, and the difference in energies between the levels represents the forbidden gap.

A diagram showing the relationship between bands in an insulator is shown in Figure 11-1 along with that for a *semiconductor*. The only difference between a semiconductor and an insulator is the size of the forbidden gap. Typically, insulators have gaps of 5 eV or more, whereas semiconductors have gaps in the range of 0.1–3 eV. Thermal excitation caused by normal temperatures becomes quite probable when energy gaps are below about 0.1 eV; hence, materials with very small gaps are classified as conductors.

The family of elements having a valence of four (C, Si, Ge, Sn and Pb) includes representatives of all three types of materials. Carbon in its diamond form is an insulator with a gap of 5.47 eV. Structurally, crystalline silicon is identical to diamond, but the gap is about 1.12 eV so it is classified

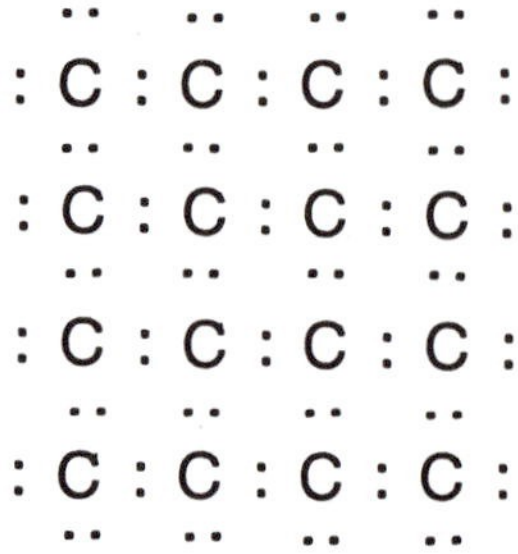

$$
\begin{array}{cccc}
\cdot\cdot & \cdot\cdot & \cdot\cdot & \cdot\cdot \\
: C : C : C : C : \\
\cdot\cdot & \cdot\cdot & \cdot\cdot & \cdot\cdot \\
: C : C : C : C : \\
\cdot\cdot & \cdot\cdot & \cdot\cdot & \cdot\cdot \\
: C : C : C : C : \\
\cdot\cdot & \cdot\cdot & \cdot\cdot & \cdot\cdot \\
: C : C : C : C : \\
\cdot\cdot & \cdot\cdot & \cdot\cdot & \cdot\cdot
\end{array}
$$

FIG. 11-2   The Lewis or electron dot structure of diamond.

as a semiconductor, as is Ge with a gap of 0.67 eV. Having a gap of only 0.08 eV places Sn on the borderline, but Pb has such a small gap it is considered a conductor, albeit not a good one.

Holes

The excitation of a valence electron to the conduction band level, and its subsequent movement, effectively creates a positive ion within the solid. The ion cannot move in the solid, but its positive charge can, and the positive charge created by the absence of an electron is called an *electron hole,* or, more simply, a *hole.* The hole is in the valence band, and it can move because a valence electron from a neighboring atom can fall into the hole, which neutralizes the positive charge, but also creates another hole (positive charge) in the donor atom. If a bias is applied, the holes and electrons move in opposite directions, but if there is no bias, both holes and electrons are free to wander. There is a probability that they may recombine, which yields energy, equivalent to the energy necessary to excite a valence electron to the conduction band, or the energy necessary for the creation of an *electron-hole pair.*

Conductivity of Solids

The movement of charges constitutes an electrical current, and both holes and electrons are charge carriers. Consider a solid to which electrodes are attached. When a negative charge is collected at one electrode a negative charge must be injected at the other electrode, likewise for positive charges. Furthermore, the movement of positive charges in one direction is equivalent to the movement of negative charges in the opposite direction. Therefore, the movement of the charge carriers, holes and electrons, do not have to be equal and opposite. If there are more electrons than holes, the electrons will be the primary carrier, and if holes outnumber electrons, the holes become the primary carrier. Of course, in many pure materials the number of holes and electrons will be equal, but this is not necessarily the case for all materials.

Conductivity is a measure of the availability of charge carriers, or more properly the charge carrier density. This is equivalent to the number of conduction electrons and holes for a given number of atoms. The probability of a valence electron gaining sufficient energy to be promoted to the conduction band may be calculated by the Fermi–Dirac distribution function, which is similar to the Maxwell–Boltzmann distribution function except that the quantized energy levels of electrons is taken into account. For a given material, the probability increases with temperature, which supports the observation that conductivity increases with temperature. For different materials at a given temperature, the probability is exponentially and inversely related to the materials' forbidden gaps. That is, for decreasing gap sizes, the number of conduction electrons and holes increases exponentially.

SEMICONDUCTORS AND JUNCTIONS

The primary semiconductors used for radiation detectors consist of either silicon (Si) or germanium (Ge), but particular "impurities" are introduced into them to alter their electrical behavior.

Intrinsic Semiconductors

The term *intrinsic* refers to the theoretical situation where a perfect crystal exists with no impurities.

Pure silicon has the same crystal lattice as diamond. The Si atoms are joined to one another by single covalent bonds creating one large three-dimensional molecule. The Lewis structure, which shows only the valence electrons as dots, is given for carbon in Figure 11-2, but a piece of silicon has an identical bonding pattern. Note that an octet of electrons surrounds each atom filling each valence shell. Since there are no energy levels immediately above the completed shell, electrons in silicon must be promoted to the N shell for excitation to occur, and the minimum energy necessary for such is about 1.12 eV. With such a band gap, about one electron out of 4.4×10^{10} valence electrons is in the conduction band at room temperature, and for each electron in the conduction band there is one hole in the valence band. Thus, in intrinsic semiconductors the number of positive and negative charge carriers is low, but equal.

A similar picture exists for pure Ge except that its forbidden gap (0.67 eV) is lower and the density of charge carriers is much higher; about 1 of 5×10^5 valence electrons is in the conduction band at room temperature. The availability of charge carriers is much larger for Ge than Si, which means its conductivity is greater. In semiconductor detectors a fairly low conductivity is desirable. Consequently, Ge detectors are operated at liquid nitrogen temperatures to reduce thermal excitation of valence electrons.

Doped Semiconductors

It is impossible to obtain perfectly pure crystals of Si or Ge, but certain elements may be introduced intentionally to alter the characteristics of the crystal. Such a practice is referred to as *doping,* and the "impurity" may be called a *dopant.*

n-Type Semiconductors. When Si or Ge atoms are replaced in a crystal lattice by atoms that have five valence electrons (N, P, As, Sb), negative charge conduction is enhanced. Figure 11-3a shows the Lewis structure for silicon doped with phosphorus. A single phosphorus atom is shown in the center of the lattice. Four of phosphorus' valence electrons are involved in bonds with surrounding Si atoms completing an octet in that shell. The unshared electron must occupy a higher energy level. Although this does not place the electron in the conduction band, it is only about 0.09 eV below it. The Fermi–Dirac distribution functions reveals that at room temperature virtually all of these electrons are in the conduction band because of thermal excitation. This affects the conductivity of the doped semiconductor. Suppose the concentration of phosphorus is as low as 100 parts per billion, then the proportion of electrons in the conduction band is about one per 10^7 atoms (Si and P). Recall that in intrinsic Si, the densities of holes and electrons are in the order of 1 per 10^{10}. The unbound electrons of phosphorus in the conduction band are not the result of hole creation in the valence band, and they greatly outnumber holes. Nevertheless, current can be carried by either type of carrier, and the total number of carriers has been increased dramatically, therefore the conductivity has been increased accordingly. This is reflected by a smaller forbidden gap. Of course, the gaps of individual atoms are not changed, but the gap in doped material is an average for all atoms, both Si and P. The conclusion to be drawn is that the effective forbidden gap can be regulated by controlling the concentration of the dopant.

The proportion of current carried by electrons and holes is directly related to their densities; therefore most of the current flowing through phosphorus-doped silicon will be carried by the *negative* charge carriers (electrons). Consequently, this type of semiconductor is classified as an *n-type*, and this term applies to all kinds of semiconductors if negative charge carriers predominate. Some obvious n-type semiconductors include Si and Ge doped with any of the Vb elements (N, P, As, Sb) in the periodic table. Dopants that readily donate electrons to the conduction band are sometimes called *donors*.

Lithium drifted Si and Ge are also examples of n-type semiconductors,

```
    ..   ..   ..   ..   ..                ..   ..   ..   ..   ..
  : Si : Si : Si : Si : Si :           : Si : Si : Si : Si : Si :
    ..   ..   ..   ..   ..                ..   ..   ..   ..   ..
  : Si : Si : Si : Si : Si :           : Si : Si : Si : Si : Si :
    ..   ..   .. .  ..   ..                ..   ..   ..   ..   ..
  : Si : Si : P  : Si : Si :           : Si : Si : B  . Si : Si :
    ..   ..   ..   ..   ..                ..   ..   ..   ..   ..
  : Si : Si : Si : Si : Si :           : Si : Si : Si : Si : Si :
    ..   ..   ..   ..   ..                ..   ..   ..   ..   ..
  : Si : Si : Si : Si : Si :           : Si : Si : Si : Si : Si :
    ..   ..   ..   ..   ..                ..   ..   ..   ..   ..

              a                                    b
```

FIG. 11-3 The Lewis or electron dot structure in silicon crystals containing (a) phosphorus and (b) boron impurities, as representatives of n- and p-type semiconductor materials, respectively.

but the crystalline structure is different from that just described. Lithium atoms are forced into interstitial spaces of the crystal lattice of Si or Ge and do not participate in the bonding pattern. However, lithium's single electron in the 2s level (its valence electron) is loosely bound and is easily excited to the conduction band level. This increases the number of electrons in the conduction band without creating holes in the valence band. Of course, other good conductors could be used in place of Li to accomplish the same effect, but lithium's utility is due to its relative size that makes it easier to infuse into pure crystals.

p-Type Semiconductors.   Doping Si or Ge with an element that has only three valence electrons (B, Al, Ga, In) creates a *p-type* semiconductor, one in which *positive* charge conduction is enhanced.

The bonding pattern of a boron atom in a silicon crystal is depicted in Figure 11-3b. Only seven instead of an octet of electrons surround the boron atom, and one of its silicon neighbors is similarly deficient. This represents a hole in the valence band because an electron is needed to complete the octet. The attraction between this type of hole and an electron is nearly as great as that between a hole and electron pair created by the promotion of a valence electron to the conduction band in pure silicon. That is, the energy necessary to create a hole in pure silicon is about 1.12 eV (the value of the forbidden gap), and the energy released when an electron fills the hole in the boron atom's valence shell is only slightly less. One of the valence electrons of silicon may be drawn to the boron's hole, which only rearranges the positions of holes. The net effect is similar to that described for n-type semiconductors except the charges are reversed; for each boron atom incorporated into the crystal lattice a hole is added without the promotion of an electron to the conduction band. Holes outnumber conduction electrons greatly, and the total number of charge carriers has been increased. As a result, conductivity is increased, the effective band gap is reduced, and holes are the primary charge carrier.

Those dopants that create holes in the valence band and readily accept electrons are called *acceptors*.

Compensated Semiconductors.   When a semiconductor material contains both types of dopants, electrons supplied by the donor atoms neutralize holes contributed by the acceptor atoms. If the concentration of acceptor and donor atoms is precisely equal, the material is described as *compensated*. In compensated material the densities of charge carriers are low, about that of intrinsic material. Although their properties are not equivalent, compensated regions in semiconductors are referred to as *intrinsic* regions and symbolized with an *i*.

Semiconductor Junctions

Some very useful electrical properties result when two slabs of different type (n and p) semiconductors are joined. If the electrical contact at the junction is sufficiently good for charge carrier transfers, a *depletion region, zone,* or *layer* is created in which the charge carriers have been depleted to near compensated levels. The depletion region is formed as indicated in Figure 11-4. Electrons from donor atoms diffuse across the junction and combine with the acceptor's holes. The transfer of an electron leaves an immobile positive ion in the n-material and forms an immobile negative ion in the p-material, and an electron and hole have been abolished.

Note that the positive ion is not a hole in the valence band, nor does the negative charge on the acceptor ion reflect a conduction electron. Recall that to create a hole in intrinsic silicon, about 1.12 eV is required (the energy necessary to promote a valence electron to the conduction band), and when an electron and hole combine, an equivalent amount of energy is given up. This energy represents the potential energy for electron–hole recombination and is directly related to the affinity of a valence electron to its atom. Using phosphorus as a donor example, the gap between the energy level of the nonbonding electron and the conduction band is only about 0.09 eV, and thermal energy effectively lower the gap to nearly zero. Thus, the attraction between the phosphorus ion and an electron is very slight and the potential energy for an ion–electron combination is vanishingly small.

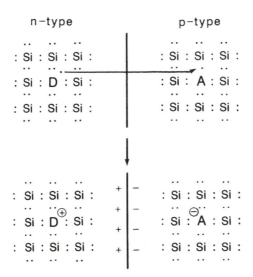

FIG. 11-4 The formation of a depletion region in a silicon semiconductor junction. The electron donor and acceptor are indicated by D and A, respectively.

In the case of the acceptor ion, the extra electron is tightly bound; nearly the same amount of energy is required to promote it to the conduction band as is needed for an Si atom.

The ions formed in the depletion region establish a *contact potential,* a potential difference between the n- and p-materials. Although the volume of the depletion region and contact potential become stable, a dynamic equilibrium is involved.

In the absence of an applied potential, the driving force for the movement of electrons from the n- to p-sides is due to the electrochemical potential of the dopants, that is, the potential for electron transfer between donor and acceptor atoms expressed in electrical units. The primary driving force for the movement of electrons from the p- to n-sides is the contact potential, which does exist if only a single charge difference exists. Since the rates of movement are proportional to the potentials, initially the *net* rate favors the n- to p-direction, but there is a low rate of movement from the p- to n-sides. As the depletion region grows, the potential difference between the two sides increases as does the rate of electron transfer from the p- to n-sides. Eventually, an equilibrium is established in which the rates of movement in each direction are equal. Under those conditions the potential between sides precisely counterbalances potential of the dopants.

As mentioned above, the electron transfers from the n- to p-sides are driven by the electrochemical potential of the dopants, which is a function of their concentrations (atoms per cm^3). Therefore the depletion volume (or width if thicknesses are equivalent) is larger if the concentrations of dopants are larger.

Assuming no applied potential and no effect by the nondopant atoms, the number of *donor ions* on the n-side must equal the number of *acceptor ions* on the p-side. This causes the depletion zones on each side of the junction to vary with the relative concentrations of the donor and acceptor atoms. Obviously, if the concentrations of both dopants are the same, the depletion region will be symmetrical about the junction. However, if the concentration of donor atoms is 10 times greater than the concentration of acceptor atoms, then the depletion zone on the p-side should be 10 times larger than the zone on the n-side. For several reasons that will become apparent later, nonsymmetrical depletion regions are common.

Suppose ionizing radiation interacts with the semiconductor material in the depleted region (Figure 11-5). Electron–hole pairs are created (ionization), which upsets their respective equilibrium concentrations. The holes are attracted to the p-side and electrons to the n-side, and some recombination may occur, but eventually, the "excess" electrons and holes are neutralized, and equilibrium is reestablished. The momentary movement of charges constitutes a current, which is an electrical signal that an ionizing event occurred in the depleted region. The signal is exceedingly weak because of the low potential difference (about 1 V) between the two sides;

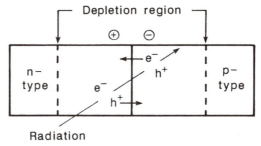

Radiation

FIG. 11-5   The movement of electrons (e^-) and holes (h^+) formed when ionizating radiation interacts in the depletion region of a semiconductor junction.

furthermore the responsive volume (depleted region) is small and contains relatively low atomic weight atoms so the probability of a ray interacting in that region is somewhat remote. Consequently, the detector just described is not a practical device, but the application of an electrical bias or voltage across the junction helps to overcome both of these problems.

Biasing Junctions.   Suppose good electrical contacts, *ohmic contacts,* are made to each slab of the semiconductor materials described in the preceding section, then consider the consequences when different biases are applied. Figure 11-6 shows the physical arrangements when zero, forward and reverse biases are applied.

The zero bias situation has already been discussed, and when the junction equilibrium is undisturbed, no current will flow, and the width of the depletion zone is stable.

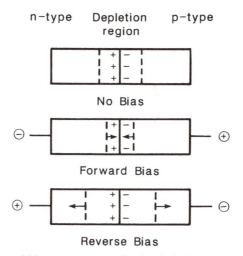

FIG. 11-6   The effect of bias arrangements in the depletion region of a semiconductor junction.

Current will flow if a forward bias is applied. Recall that the primary charge carrier in n-type material (outside of the depletion zone) is the negative type, which means electrons are readily injected on that side. The negative charges injected neutralize the positive charges of the donor ions causing the depletion zone to shrink toward the junction. On the p-side, holes are readily injected because that type of carrier predominates, and the holes combine with the "extra" electrons of the acceptor ions to cause the depletion region to shrink toward the junction. If the applied potential is sufficient, the depletion region and its original potential difference are detroyed, which allow electrons and holes to move across the junction. That is, current will flow.

When a reverse bias is applied, current will not flow. Outside the depletion zone it is difficult to inject electrons at the p-side and holes at the n-side because of the scarcity of appropriate charge carriers. However, any electron injected on the p-side can combine with the holes of the acceptor atoms creating acceptor ions; thus the depletion zone increases. Similarly, holes injected in n-material can accept electrons from donor atoms creating more donor ions and increasing the depletion zone. As the depletion zone grows, the potential across the junction increases until that potential counterbalances the applied potential *plus* the electrochemical potential of the dopants. Thus, with a reverse bias the width of the depletion region can be increased by increasing the applied potential.

The oversimplified description reflects the general actions of a rectifying diode in which current may flow in one direction only. However, some current does flow with a reverse bias. First, if a high enough potential *(breakdown voltage)* is applied the capacitance of the semiconductor can be overcome. Effectively, the voltage is sufficient to break the covalent bonds between Si or Ge atoms (valence electrons promoted to the conduction band), which allows large currents to flow and is likely to cause permanent damage to the crystal. Recall that a crystal represents one large molecule, and if a sufficient number of bonds are broken, the crystal is likely to fracture. Second, there is a very small current with lower reverse biases, and this is called *leakage current*. Figure 11-7 shows the general effects of applied voltage on current flow in a diode. A relatively low leakage current flows with a reverse bias, but with a forward bias, the current increases exponentially. Radiation detectors are always operated with a reverse bias, and in some applications the detector is cooled to reduce the leakage current.

RADIATION DETECTORS

The responsive portion of a biased diode detector is its depletion or intrinsic zone, and when radiation interacts with the atoms in that region, ions are created. In that respect, diode detectors are similar to gas-filled ion

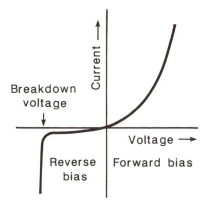

FIG. 11-7 The effect of bias voltage on current flow across semiconductor junctions. (From *Modern Physics for Applied Science,* by B. C. Robertson. Copyright © 1981 by John Wiley and Sons. Reprinted by permission)

chambers operating with a voltage that yields saturation currents. In both cases the ionization is the primary interaction detected, and the collection of resulting charges constitutes an electrical pulse, the magnitude of which depends on the extent of ionization. However, diode detectors offer several advantages.

Neither Si nor Ge has a high atomic weight, but as solids, their densities (2.33 and 5.373 g/cm², respectively) exceed that of air by about 2000–4000 times. Consequently, for a given size, diode detectors are more efficient because they are better absorbers, particularly for X- and γ-radiation.

In gas detectors, the electrical pulse is due to the collection of electrons primarily, since the movement of positive ions is very slow. In diode detectors, ions do not move, instead their positive charges move as holes, which have about the same mobility as electrons. This yields a shorter pulse duration and, hence, smaller dead time.

Perhaps the greatest advantage of diode detectors over gas-filled ones is the amount of energy required to produce an ion pair. Recall that in air about 34 eV per pair is needed. In Si at room temperature, the average energy required to create an electron–hole pair is about 3.62 eV, while it is 2.95 eV per pair in Ge at 80 K. This represents an approximate 10-fold increase in the number of ions formed from a given amount of energy dissipated in the absorber.

Diode detectors are subject to *radiation damage*. Ionization occurs primarily with the most prevalent atomic species, the Si or Ge atoms, thus the covalent bonds in the crystal lattice are broken when valence electrons are promoted to the conduction band. After the charges are collected, electrons and holes recombine (covalent bonds reformed), and equilibrium is reestablished. If the reformation of bonds restores the crystal lattice perfectly, no damage would result, but the absorption of a large amount of energy in a small area may cause atoms to be displaced and bonds to reform differently, that is, create crystal defects. The accumulation of defects will change the characteristics of the semiconductor, and, with time, render it useless as a detector. Susceptibility to radiation damage varies among the different types of detectors.

The primary use of diode detectors is in *radiation spectroscopy,* energy measurements of individual rays. Because the types of interactions for charged particles (α, β, protons, and heavy ions) are different from those of photons (X- and γ-rays), the basic designs of suitable diode detectors differ. Thus, there are two broad classes of detectors, *charged particle* and *photon* detectors, but within these classes designs differ to optimize measurements within certain energy ranges. Keep in mind that the responsive region of a diode detector is its depletion or intrinsic zone, and the zone must be flanked by unresponsive layers, either nondepletion regions of the crystal or the ohmic contacts. Such layers are called *dead layers.*

For charged particles, a thin window or dead layer is desirable to reduce absorption of energy from a particle before it enters the responsive region. However, the responsive region does not need to be exceptionally large because of the small ranges of charged particles in solids.

In contrast, a fairly thick dead layer or window is not a serious deficiency in photon detectors unless very low energy X-rays are being measured. However, a large and dense responsive volume is needed to efficiently absorb X- and γ-radiation.

Diffused Junction Detectors

Although more efficient designs exist for detection of charged particles and photons, the diffused junction detector is rugged and capable of operating at various temperatures. An n–p junction may be formed by diffusing a donor-type impurity into a p-type silicon crystal. With heating, one surface of the crystal is exposed to vaporized donor, and as the donor diffuses inward a concentration gradient is formed. At the surface, the donor concentration exceeds that of the original acceptor concentration forming an n-layer, but further inward the concentrations of dopants become equal, yielding a relatively thin compensated region, and beyond, the original p-type material remains. The n-surface is heavily doped so the depletion region is unsymmetrical and exists on the p-side of the junction primarily. Most of the n-layer is not sensitive to radiation and constitutes the dead layer or window that incoming rays must traverse. A diffused junction detector does not have a very large depletion zone, even with relatively high biases, so it is not an ideal detector for either class of radiation. However, it has good stability. High temperatures are necessary for diffusion of donor atoms, such as phosphorus, so when a crystal is cooled to normal laboratory temperatures, the donor atoms do not continue to migrate, which yields a stable detector composition. This type of detector is not as susceptible to radiation damage as other types and is particularly resistant to penetration of foreign contaminants, such as oil and fingerprints.

Surface Barrier Detectors

Surface barrier detectors are employed primarily as charged particle detectors, and variations in detector designs are used to optimize conditions for particular types and energies of charged particles.

A typical surface barrier detector is fabricated starting with a wafer of n-type silicon. The wafer's surface is etched by reactions that are not understood well, but apparently, oxides form at the surface and serve as electron acceptors. The suface layer is very thin and behaves as a heavily doped p-layer. By evaporation techniques, a thin metallic film (40 μg/cm^2, gold) is deposited on the surface and it serves as one contact. Frequently, aluminum is used for an ohmic contact on the nonetched surface.

Because of the relative concentrations of donors and acceptors, the depletion region is primarily on the n-side of the junction and is extended by the application of a reverse bias. Figure 11-8 shows the general appearance of the electric field strength within a cross section of a wafer when different biases are applied. The thickness of the wafer is indicated by the distance between the gold (Au) and aluminum (Al) contacts, and the depletion region (dr) is the depth or thickness in which an electric field exists. When the depletion depth does not extend to the Al contact, *partial depletion* exists, and when it just reaches the contact, *total depletion* is achieved. Although the depth of the depletion region cannot be increased further, the field strength can, and this condition is referred to as *overdepletion*. A given detector may operate under any of the conditions, but generally detectors are designed for a particular condition.

Detectors with a variety of depletion depths are available. A 0.1 mm

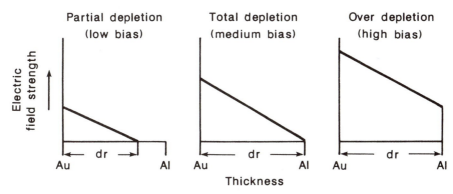

FIG. 11-8 Electric field strengths and depletion regions of a surface barrier detectors with different increasing reverse biases.

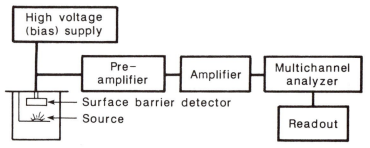

FIG. 11-9 A block diagram for a simple charged particle spectrometer.

thickness will absorb α-particles with energies up to about 15 MeV, whereas about 5.0 mm is needed to completely absorb 2.0 MeV negatrons.

Usually the gold contact serves as the window and is thin enough (equivalent to about 80 nm Si) so little window absorption occurs. However, light can penetrate such thicknesses, and the quantum energy of light is sufficient to cause ionization. Normally, charged particle spectroscopy is performed with the source and detector in a vacuum chamber. This eliminates the absorption of energy by air between the source and detector and also provides light protection. A typical instrumental block diagram for charged particle spectroscopy is shown in Figure 11-9, and it is quite similar to that for a solid scintillation spectrometer.

One quality factor in surface barrier detectors is their resolution of energies. For α-particles, the maximum resolution among different detector designs ranges from about 12 to 50 keV, and for β-particles, resolutions as low as 5 keV can be achieved.

Some detectors are designed to measure energy losses as charged particles pass through them. For this application, a thin totally depleted zone is desirable along with thin windows or contacts on both sides of the detector. Uniformity in wafer thickness is important in this case, but not so critical for nontransmission type detectors where the ranges of particles do not exceed the depletion depth.

The thin windows of most surface barrier detectors do not afford crystal protection from foreign materials; therefore they must be handled carefully. Also, this type of detector is subject to radiation damage and has a finite useful lifetime. However, unlike some detectors to be discussed later, surface barrier detectors can be operated and stored at normal laboratory temperatures. Also, generally at the expense of energy resolution, design features may include provisions for operating under harsh conditions.

Ion Implantation

Ion implantation is a method of producing doped semiconductor materials that may be used in different classes of detectors. In an accelerator, dopant ions accelerated to precise energy levels, bombard a homogeneous crystal that contains the opposite type of dopant. By regulating the kinetic energies of the dopants their depths of penetration into a crystal can be controlled precisely, thus producing a sharp junction boundary.

Boron, which has a relatively small size, is the most commonly used dopant ion. Since it is an acceptor, the crystal to be bombarded would be of the n-type. The boron ions pick up electrons readily and eventually are incorporated into the crystal lattice. A thermal annealing process, which provides sufficient energy for a low rate of bond breakage and reformation, aids in the process and in dissipating crystal imperfections caused by the bombardment. The surface bombarded has an excess of acceptors (boron) compared to the donors originally present, and the depth of penetration represents the junction, whereas n-type material makes up the remainder of the crystal. Ion implantation is used to fabricate both charged particle and photon detectors, but the design features differ accordingly.

Photon Detectors

Even in pure Si or Ge, some impurities remain, and they limit the amount the depletion region can be increased by the application of a high bias. Therefore, different designs are used to increase the sensitive portion of photon detectors, and a common feature of such detectors is the formation of a large "intrinsic" region between n- and p-contact surfaces. Sometimes such junctions are called n–i–p junctions.

Lithium-Drifted Detectors.   Lithium-drifted detectors are fabricated by starting with p-type Si or Ge, and are designated as Si(Li) and Ge(Li). One surface is exposed to Li at 25–60°C, and a reverse bias is applied. The ionization potential of Li is relatively low, and the atoms readily give their valence electrons to acceptor atoms as they diffuse into the crystal. The Li^+ ions are drawn into the crystal further by the bias. Instead of a concentration gradient being formed, the Li^+ ion migrates until the space charge is zero over the complete region. This yields a nearly perfect compensation of the acceptor atoms, and the picture is negative ions in the crystal lattice balanced by positive lithium ions in interstitial spaces. With no excess of holes or electrons, the drifted region behaves as intrinsic semiconductor material. Thus, a p–i–n junction is created with a large intrinsic (actually compensated) region between n- and p-contacts, the lithium surface and undrifted p-type material, respectively. The compensated regions in Si(Li) detectors may be as deep as 8–10 mm, but lithium is more mobile in germanium, and compensated depths of 15–20 mm are possible in Ge(Li) detectors.

With a larger intrinsic region, leakage current, and hence detector noise, increases. To reduce noise and increase energy resolution both Si(Li) and Ge(Li) detectors are operated at liquid nitrogen temperatures. For Ge(Li) detectors, storage at low temperatures is necessary to prevent Li migration, which destroys the intrinsic region. The lower mobility of lithium in silicon may allow such detectors to be warmed between uses.

Intrinsic Germanium Detectors.   Specially purified germanium, sometimes called high purity or hyper pure Ge, may be used to form the intrinsic region. Although the natural impurities in germanium are of the p-type (primarily Al), they are reduced to about 1 in 10^{12} atoms, which is low enough for large depletion regions to be formed without excessive biases. High purity Ge with n-type behavior is also produced. Lithium diffusion is used to form the n-layer, and the p-layer may be formed by boron implantation. Surface barriers (metallic contacts) are also used in some detectors.

Lithium diffusion creates a dead layer of about 600 μm, which is not an excessive thickness for high energy γ-rays, but is sufficiently thick to absorb X-rays with energies below about 25 keV. Ion implantation yields very thin layers, as low as 0.3 μm.

High purity germanium detectors are operated at liquid nitrogen temperatures to reduce leakage current and signal noise. However, unlike the lithium drifted detectors, these can be warmed and stored at room temperatures between uses without deleterious effects.

Detector Configurations.   As indicated in Figure 11-10, the detector and preamplifier is mounted on a Dewar containing liquid nitrogen, and an insulated cold finger extends from the detector to the liquid.

Detectors are manufactured in three basic forms: (1) planar, (2) coaxial, and (3) well. A planar detector is simply a cylindrical crystal with a height corresponding to the intrinsic region and contacts at each end of the cylinder (Figure 11-11). A coaxial design is also shown in Figure 11-11. The outer surface is n-type material, and the inner surface of the internal cylinder serves as the p-contact. For some designs, the contact surfaces may be reversed. A schematic well design is shown in Figure 11-12, and it is a modified coaxial arrangement, which functions much the same as a well type NaI (Tl) detector to give nearly 4π counting geometry.

Instrument Design.   A primary use of diode detectors is to measure energies of rays, and hence identify radionuclides; therefore, instrument designs are those common to spectrometers. By replacing the charged particle detector in Figure 11-9 with a photon detector, a simplified design is obtained.

Low Energy Photon Spectroscopy.   For the detection of low energy photons (X-rays), a thin window or dead layer is necessary, however, an exceptionally deep intrinsic zone is not necessary or desirable because leakage current increases with depth and that reduces resolution.

Si(Li) detectors are preferred for low energy photons, but Ge detectors

FIG. 11-10 A photon detector mounted in a cryostat that contains the detector and preamplifier along with electronics that aid in the protection of the detector. The cryostat is mounted on a liquid nitrogen Dewar for cooling. (Courtesy of EG&G Ortec, Oak Ridge, TN)

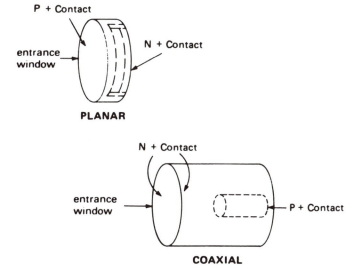

FIG. 11-11 Planar and coaxial detector configurations for photon detectors. (Courtesy of Canberra Industries, Inc., Meriden, CT)

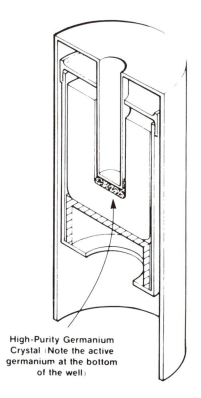

High-Purity Germanium
Crystal (Note the active
germanium at the bottom
of the well)

**Simplified Isometric View
of HPGe Well Detector Cryostat.**

FIG. 11-12 Well-type configuration for a high purity germanium detector. (Courtesy of EG&G Ortec, Oak Ridge, TN)

are designed for the same purpose. To optimize resolution and detection efficiency, detectors are designed for use in certain energy ranges, and some examples of actual product ranges are 1–60 keV, below 30 keV, and 2–200 keV. The resolution of photon detectors is quoted in terms of the full width (of the peak) at half maximum (height), or simply *FWHM,* and is expressed in energy units, eV or keV. Because resolution varies with the photon energy, the energy at which the resolution was determined is specified. For example, one model of a commercially available Si(Li) detector has a resolution specification of 165 eV at 5.9 keV.

γ-Ray Spectroscopy. For higher energy photons the thickness of the dead layer is not so important, but larger and more dense intrinsic zones are needed to improve detection efficiencies. Since germanium has a significantly higher density than silicon, germanium detectors are used for high energy spectroscopy primarily. Although such detectors are capable of detecting low energy photons, the resolution of the detectors is not as good as those designed especially for low energy radiation. The resolution of higher energy photon detectors is usually in the range of 2 keV at 1.33 MeV, and while 2 keV is about one order of magnitude higher than that for low energy photon detectors, the resolution relative to the photon energy is considerably better for the high energy detectors, e.g., 165 eV/5900 ev = 0.028 and 2 keV/1330 keV = 0.0015.

The advantages of diode detectors over ion chambers have been discussed, but for γ-ray spectroscopy, germanium detectors are remarkably superior to NaI(Tl) solid scintillation detectors. Typically, the resolution with NaI(Tl), 3 × 3 in. crystals is about 90–100 keV at 1.33 MeV. This advantage is illustrated in Figure 11-13, which shows ^{60}CO spectra obtained by solid scintillation and a high purity germanium detector. Because of the importance of the photopeak in identification of γ-emitting radionuclides, another performance parameter used for high energy photon detectors is the peak to Compton ratio. This is the ratio of the photopeak height relative to the average height of the Compton plateau near the edge, and it is an indicator of the proportion of rays that interacts by the photoelectric and Compton effects. Practically, it is an indicator of the ease and certainty in locating photopeaks. Also shown in the germanium detector's spectrum of

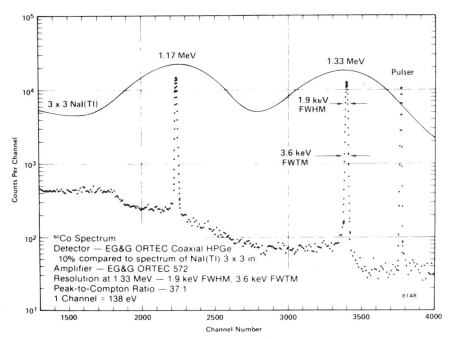

FIG. 11-13 ^{60}Co spectra obtained with a 3 X 3 in NaI(Tl) detector (solid line) and a high purity germanium diode detector. (From *Experiments in Nuclear Science AN34*, 3rd ed., 1984, p. 44. Courtesy of EG&G Ortec, Oak Ridge, TN)

^{60}CO is a peak labeled pulser. This peak is due to electronically generated pulses of known magnitude and is used to calibrate the multichannel analyzer's channels with respect to energy.

The excellent resolution of Ge detectors is also illustrated in Figure 11-14. Bullet lead, containing antimony and arsenic, was subjected to neutron activation resulting in the production of ^{122}Sb, ^{124}Sb, and ^{76}As. Photo-

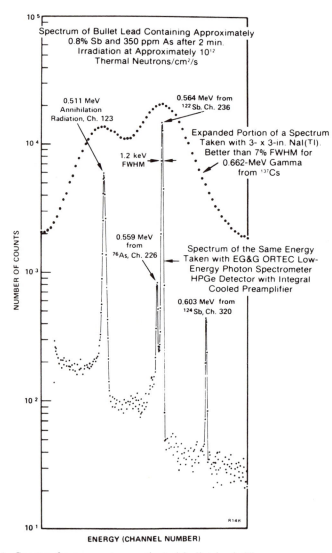

FIG. 11-14 Spectra from a neutron-activated bullet lead. The upper spectrum was obtained with a 3 × 3 in NaI(Tl) crystal, and the lower with a high purity germanium diode detector. (From *Experiments in Nuclear Science AN34,* 3rd ed., 1984, p. 48. Courtesy of EG&G Ortec, Oak Ridge, TN)

peaks for each radionuclide are readily apparent in the Ge detector's spectrum, but form one broad peak in the spectrum obtained by solid scintillation.

Although diode detectors are unexcelled for photon spectroscopy, counting efficiencies are better by solid scintillation methods, primarily because the responsive region, NaI(Tl), does not have the size limitations of diode detectors. Another advantage for counting γ-rays by solid scintillation is that the detector does not require cooling by liquid nitrogen, which contributes to a higher operating cost and inconvenience.

Compound Semiconductors

A search for semiconductor materials that can be used in diode detectors continues. Two desirable features sought are a high density, which would improve detection efficiency, and a capability of operating without the inconvenience of cooling the detector. To date the most promising materials are CdTe, HgI_2, and GeAs. Detectors made of some of these materials have been produced and used for special applications, but their replacement of silicon and germanium detectors does not seem imminent. For more information about such detector materials, the reader should consult Knoll (1979) and L'Annunziata (1987).

REFERENCES

Canberra (1981). Product Catalog. Canberra Industries, Meriden, CT.

EG&G Ortec (1984). *Experiments in Nuclear Science AN34*, 3rd ed. EG&G Ortec, Oak Ridge, TN.

EG&G Ortec (1989). Product Catalog. EG&G Ortec, Oak Ridge, TN.

Knoll, G. F. (1979). *Radiation Defection and Measurement*. Wiley, New York.

L'Annunziata, M. F. (1987). *Radionuclide Tracers*, pp. 145–148. Academic Press, Orlando, FL.

Robertson, B. C. (1981). *Modern Physics for Applied Science*. Wiley, New York.

CHAPTER 12

Autoradiographic Methods

RADIOGRAPHIC METHODS

For most people the medical or dental "X-ray" is a familiar radiographic procedure, whereby an image is recorded on a photographic film as a result of the differential absorption of X-rays directed at a portion of a person's body. However, radiography is not confined to animate specimens. Industrially, similar procedures are used to determine the structural integrity of materials such as that of metal castings. Because of the high densities of some materials, γ-ray sources may be used in place of X-rays in radiography.

In *autoradiography* the specimen is the radiation source, and the image results from the distribution of radiation emanating from the source. Although several methods for radionuclide imaging are used in medical diagnoses, the photographic emulsion is the most common detector in biological research.

THE PHOTOGRAPHIC PROCESS

Emulsions

A photographic emulsion consists of small silver halide *crystals* embedded in gelatin, and practically all of the emulsions used in common applications utilize silver bromide. When light or other forms of ionizing radiation strike a crystal, the crystal is "activated" and the development process converts it to a black, metallic *silver grain*. "Unactivated" crystals are dissolved during the fixing process, but the developed silver grains remain embedded in the gelatin matrix. Thus, areas on the film where many crystals were activated result in a higher density of silver grains after development and fixing, and the varying densities of silver grains throughout the film create a negative visual image.

The silver bromide crystals are produced in dilute gelatin solutions by the reaction of $AgNO_3$ with either KBr or NaBr. Crystal size depends on the rate of crystallization, which is affected by numerous conditions, including temperature, concentration of components, and rate of reagent addition. Precipitate digestion periods aid in obtaining more uniform crystal sizes, but they always exist in a range of sizes. This characteristic of manufactured emulsions is referred to as the film's grain size. Although the term grain may be used for both the AgBr and Ag^0 (metallic silver) particles, we will distinguish between them by calling the AgBr particles *crystals* and the Ag^0 deposits *grains*. Another characteristic of manufactured emulsions is the ratio of AgBr to gelatin. For X-ray film, crystal size is likely to be in the range of 0.2–3.0 μm with an AgBr to gelatin ratio of about 1:1. Higher ratios (5:1) and smaller crystals (0.02–0.5 μm) are used in nuclear emulsions, which are designed to record tracks of charged particles in emulsions.

Besides serving as a matrix for the AgBr, gelatin plays important roles in the photographic process. It prevents coalescence of crystals during crystallization and "chemically insulates" one crystal from another during the exposure, development, and fixing steps, yet it is permeable to the necessary reagents. Although mechanisms are poorly understood, gelatin participates in the formation of *sensitivity specks* in or on AgBr crystals. To define sensitivity specks the exposure process needs to be discussed.

Exposure

When a photon of electromagnetic radiation or a charged particle interacts with a crystal, all or a portion of the ray's energy is absorbed, resulting in the promotion of one or several valence electrons to the crystal's conduction band (see Band Theory, Chapter 11). About 2.5 eV is required per electron. Electrons in the conduction band and their corresponding holes are free to migrate throughout the crystal, but the electrons tend to collect at crystalline defect sites. A sharp edge represents a defect because ions on the very edge are not surrounded by a full complement of counterions. Foreign ions also create defect sites. It appears that gelatin supplies some sulfur that is incorporated as Ag_2S into crystals during the crystallization process, and a small quantity of I^- may be added to the crystallization mixture to increase crystalline defects in AgBr further. Apparently most crystals have several defect sites capable of trapping conduction electrons.

Sensitivity specks are sites in or on a crystal that make it susceptible to activation by radiation, and this process appears to involve the trapping of conduction electrons. Perhaps many, but not all, defect sites are sensitivity specks.

Current theory (Mees and James; 1966; Herz, 1969) suggests that conduction electrons combine with silver ions (Ag^+) yielding metallic silver (Ag^0) at sensitivity specks, and the speck becomes a *latent image center*. Depending on the energy dissipated in the crystal, the center may consist of only one, or perhaps 1000 silver atoms (Ag^0). That quantity of metallic silver is not visible, but when present, the development process converts all of the Ag^+ of the crystal to Ag^0, which then is sufficient to be seen microscopically.

Of course, the reduction of silver ions is a reversible process, consequently there is some probability that a silver atom may be oxidized and reincorporated into the crystal lattice. Thus, latent image centers are not perfectly stable over extended periods of time, and this accounts for the phenomenon called latent image fading. Certainly, the presence of oxidizing agents will promote latent image fading. The stability of latent image centers under normal conditions depends on the number of metallic silver atoms present. The average lifetime of one-atom centers is short, seconds or fractions thereof, whereas many two-atom centers appear to survive for days, and a 10-atom center is almost certain to survive until development.

Although it does not appear to be a critical feature of the photographic process, the fate of the holes created at the time of interaction is interesting. The holes seem to be trapped by bromine ions, which are converted to free bromine that diffuses out of the gelatin.

Development and Fixing

The development process is an amplification step whereby the large quantity of remaining silver ions in an *activated* crystal (crystal with a latent image center) is converted to metallic silver. Various reducing agents, such as hydroquinone, methyl *p*-aminophenol sulfate, and various phenylenediamines, catechol, and aminophenol derivatives may be used. Generally, a mixture of reducing agents is employed in an alkaline solution. In an activated crystal, development causes reduction of the remaining silver ions to metallic silver, and the bromine ions are released as HBr that diffuses out of the gelatin matrix. All crystalline silver ions are subject to reduction, but, fortunately, reaction kinetics permit almost complete reduction of the ions in a crystal containing a latent image center before any of the silver ions in nonactivated crystals are reduced. (This occurs because elemental silver acts as a catalyst in the reduction process.) Obviously, development time is a critical parameter. If left too long, all crystals are reduced to silver grains, whereas in an excessively brief development period, some crystals with latent image centers will not have begun to be reduced and others may be only partially reduced. Other factors affect development time as well: temperature, reagent concentrations, agitation of solutions, and the characteristics of the emulsion.

Within an activated crystal, reduction begins at a latent image center and the deposit of Ag^0 grows. At this stage the center is referred to as a development center. The deposition of Ag^0 does not seem to be governed by crystal shape, consequently, the center of mass of the resulting silver grain rarely coincides with that of the parent crystal, and the grains may be up to three times larger. Silver grains appear as convoluted ribbons under the electron microscope and as black dots under lower magnification. After the development period, unactivated AgBr crystals still exist in the gelatin matrix, consequently, the film is still sensitive to light.

The reduction reactions that occur in the developing solution are quenched in a stop bath. This is accomplished to a large extent by simply diluting the reagents. However, 1% acetic acid is used after many developers to neutralize the alkali. It also hardens the gelatin somewhat.

The fixer is a sodium thiosulfate or sodium ammonium thiosulfate solution, in which the thiosulfate complexes and solubilizes silver ions. Ammonia acclerates the process. Appropriate fixing periods vary among products, but if the period is long enough to dissolve all of the AgBr, time is not a critical factor in producing satisfactory images.

Physical Development

In the development process just described (called chemical development) the source of the metallic silver is the individual crystal's AgBr. In the physical development process silver ions present in the developer are deposited at latent image centers producing concentric layers of metallic silver. This results in a round, dense grain whose position is not likely to differ much from that of its parent crystal. Grain size is dependent on the development time period and reagent concentrations. Physical development is used more frequently for autoradiographic methods at the electron microscope level where slight differences in grain positions may affect the interpretation of an autoradiograph. Silver nitrate and silver sodium sulfite are used as sources of silver, and the deposition of silver may be performed before or after the fixing step (the dissolution of AgBr).

Sensitivity of Emulsions

Various parameters are used to measure emulsion sensitivity. Sensitivity is frequently referred to as film speed, but it relates the extent of visual blackening in developed film to the amount of radiation that constituted the exposure. Films are designed for specific applications by the manufacturer because most of the factors that affect sensitivity are controlled during the film's manufacturing process.

Radiation. The effectiveness of electromagnetic radiation in producing latent image centers varies with the wavelength (quantum energy) of the rays. Not only is there variation among types (visible, X and γ), but variation within types as well. The same is true for particulate radiation; response to electrons is different from that of α-particles, and there is variation among electrons with different energies as well.

Crystal Size and Density. Assuming ideal behavior, all the silver ions in a crystal with a latent image center are converted to metallic silver during chemical development, so more blackening is produced when larger crystals are employed. Some of the energy of the exposing rays may be absorbed by the gelatin instead of the crystals, therefore increasing the number of crystals per unit volume of emulsion (sometimes called density) increases the probability that radiant energy will be dissipated in crystals, thus increasing the number of latent image centers. Also, in the case of radiation that has a range greater than the thickness of the emulsion, energy absorption will be increased by increasing the thickness of the emulsion.

Hypersensitization and Latensification. The sensitivity of emulsions can be increased by treatment with certain reagents. For example, treatment of emulsions with triethanolamine appears to make them more sensitive to negatrons. Such treatments are sometimes referred to as *hypersensitization*.

Latensification is a term for processes that intensify latent images, that is, increase the percentage of fully developed silver grains from a given population of crystals with latent image centers. Gold latensification involves treatment of exposed films with gold salts, followed by a normal development process. It appears that gold ions deposit at the latent image centers, which increases the probability of the crystal becoming fully reduced.

Dual Exposure. Curiously, in light photography the image is enhanced if the same amount of light used in a single exposure is divided into two exposure periods. One method is to give a very short, low-intensity exposure followed by a longer exposure. It appears that *sublatent image centers* (centers that are not likely to become developable) are formed in the short exposure, then the second exposure produces highly stable and developable centers. As will be described later, this technique is used in autoradriography of chromatograms and electrophoresis gels and is referred to as *flashing*.

FACTORS AFFECTING RESOLUTION

Autoradiographic resolution reflects how well the developed emulsion reveals the positions of sources in a specimen. Different measurements of resolution have been used. For example, in macroautoradiography, it may be the minimum distance between specific sizes of sources that permits the sources to be distinguished from one another. In microscopic autoradiography, grain densities are measured, and one useful resolution parameter is the distance from a source that yields a grain density one-half of that directly over the source. Regardless of the measurement method, several factors affect resolution.

Distance Between Source and Emulsion

Resolution decreases as the distance between the emulsion and source increases. The effect is easily understood by considering the inverse square law (see Chapter 15). Since radiation is emitted in all directions, the paths of rays diverge from one another as they travel outward. As indicated in Figure 12-1, the area of the emulsion that subtends the rays' paths from a given sector of the source increases with the square of the distance from the source.

Thickness of the Source

A point source is illustrated in Figure 12-1, but if a source has appreciable thickness, then a portion of it lies a greater distance from the emulsion, which decreases resolution for the same geometric reasons mentioned above.

Range of the Source's Rays

As indicated in Figure 12-2 with source A, the size of a source's developed image is dependent on the range of its rays in the emulsion. Note that the effect is similar whether the range exceeds the thickness of the emulsion

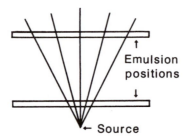

FIG. 12-1   The effect of distance between a radioactive source and photographic emulsion on image size and resolution capabilities.

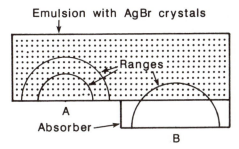

FIG. 12-2   The effect of particle range on image size and resolution capabilities.

or not. Consequently, X- and γ-ray sources are rarely used for critical auto-radiographic work, and since α-emitters are not useful tracers for most biological investigations, β-emitters predominate. Tritium, with its low E_{max} (18 keV), is often the isotope of choice when highest resolution is required.

Under certain circumstances, resolution may be improved by interposing an absorber between the specimen and emulsion. Two factors are involved: one, the reduction of the ray's range and two, a distance factor. The condition shown in Figure 12-2 for source B includes an absorber that has an absorption coefficient equal to that of the emulsion, but more efficient absorbers could be used. The minimum pathlength through the absorber is directly above the source. Therefore, if the absorber thickness is chosen appropriately, improved resolution is possible. For a strong β-emitter, such as ^{32}P, the placement of a plastic sheet between the specimen and emulsion may improve resolution, but, of course, sensitivity is reduced.

Emulsion Thickness

For β-emitters, a combination of factors causes resolution to be improved by using thinner emulsion layers. The density of silver grains in close proximity to the source is greater than that in areas further removed because of the geometric factor and the fact that the ranges of β-particles vary according to their initial kinetic energies. As shown in Figure 12-3, all emulsion thicknesses show a decreasing grain density as a function of distance from the point source, but the density decreases more rapidly in the thinnest emulsion layer. This effect may be visualized by superimposing successive layers of emulsions on the one closest to the source.

Size of the AgBr Crystals and Developed Silver Grains

A latent image center locates a point in the trajectory of a given particle or ray, but during development the entire crystal containing the image center is converted to a silver grain, and the center of the silver grain is usually displaced from that of its parent crystal. Also, if a ray strikes the outer

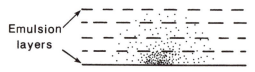

FIG. 12-3   The effect of emulsion layer thickness on image size and resolution capabilities.

edge of a crystal creating an image center there, the geometric center of the crystal differs from the image center by about the radius of the crystal. For these reasons, the errors in locating the trajectories of rays are smaller when smaller crystals are employed. Similarly, development processes that yield compact, as opposed to diffuse, silver grains provide better resolution.

Length of Exposure

The loss of resolution due to overexposure involves probabilities of AgBr crystals being struck by more than one ray or particle. Suppose the amount of activity and exposure time is such that every crystal within a distance of 10 μm has been hit at least once. Additional exposure time will not increase the silver grain density in that region of the emulsion if only one hit is required for grain development. However, further away where some unactivated crystals still exist, the grain density may be increased by continuing exposure. The probability of double hits decreases with distance from the source, but increases with time. Theoretically, resolution decreases from the inception of the exposure, but obviously, some exposure is necessary for visualization, and, furthermore, the probabilities of double hits do not increase rapidly until a significant fraction of the crystals has been hit once.

Obviously, many of the factors that improve resolution decrease detection efficiency. For example, more silver grains are likely to be produced from a 100-keV β-particle than a 10-keV particle. On the other hand reducing the distance, and hence the possibility of energy absorption between the source and emulsion, improves detection efficiency as well as resolution. Many variables are involved in the production of suitable autoradiographs, and conditions appropriate for a particular experiment depend on the specimen and autoradiographic objectives.

AUTORADIOGRAPHIC METHODS

The broad objective in autoradiography is to locate and distinguish various concentrations of labeled compounds in a planer section of a specimen, which may be as large as a sequencing gel or as small as a few cells on a microscope slide. The general procedure is the same. The specimen is placed in close contact to an emulsion (film), stored in darkness for a period sufficient for the formation of a latent image, then the film is developed for visualization. Although densiometric measurements on developed film may be performed, rarely is autoradiography used as a purely quantitative method.

There are two broad categories of autoradiographs, those whose images are viewed without magnification, called *macroautoradiographs,* and those viewed under a light or electron microscope, called *microautoradiographs.* The techniques and products suitable for the two forms of autoradiography differ considerably.

Radionuclides

The principal radionuclides used for biological experimentation, and, hence, for autoradiography are ^{3}H, ^{14}C, ^{35}S, ^{32}P, ^{125}I, and ^{131}I, but of course, this is not an exclusive list. Except for ^{125}I, all are β^--emitters although ^{131}I yields γ-rays as well. The E_{max} values for ^{3}H, ^{14}C, ^{35}S, ^{131}I, and ^{32}P are about 18, 156, 167, 606, and 1710 keV, respectively, therefore, their resolution capabilities vary accordingly. Although ^{125}I undergoes electron capture, a large fraction of the nuclear excitation energy (γ-rays) yields conversion electrons. Approximately 1.6 extranuclear electrons are emitted per disintegration and the majority have energies around 3 keV, but some are in the range of 20–35 keV.

Because of the energies of their emissions, ^{3}H and ^{125}I offer the best potential for high resolution autoradiography and are the isotopes of choice for studies at the electron microscope level. The energies of ^{14}C and ^{35}S β-particles are comparable to one another and suitable for work at the light microscope level as well as for macroscopic specimens. Neither ^{32}P nor ^{131}I is a desirable isotope for microautoradiography, but ^{32}P is used extensively for electrophoretic gel autoradiograms.

Nevertheless, the selection of radionuclides for autoradiography is not based on resolution capabilities only, and other factors such as detection efficiency, availability of suitable labeled compounds, convenience, and cost should be considered as well.

Chemography

Chemography refers to reactions in emulsions induced or inhibited by chemicals from the specimen that yield autoradiographic artifacts. With close contact, reducing agents in a specimen may diffuse into an emulsion and reduce silver ions to metallic silver. Effectively, latent image centers are created by chemical reactions rather than by radiation and lead to false images. Conversely, oxidizing agents may cause latent image fading. The normal development and fixing reactions may be inhibited by materials on the surface of an emulsions, such as a finger smudge or lipid from fatty tissue.

Two general methods are used to reduce the probability of chemography occurring: one, protect the emulsion from the specimen by using a thin barrier between them, and two, remove or inactivate possible interfering chemicals in the specimen. The most commonly encountered chemography problems involve water, and methods for dealing with it are discussed along with methods for specific types of specimens.

Macroautoradiography

Autoradiography has become a central method for molecular biological research because of its sensitive detection of small quantities of genetic materials and their products. Under usual conditions, autoradiography is more sensitive for visualizing spots or bands in electrophoresis gels than chemical methods and permits the recovery of components. But more importantly, components with specific binding properties can be distinguished from others, and such differentiation is not possible by chemical staining techniques. Chromatograms, either paper or thin-layer, are also used as macroautoradiographic specimens for similar reasons. The transport and distribution of labeled compounds in whole plants or animals may be studied by dissection and counting, but autoradiography provides visual images of distributions that may allow comparisons to be made between many anatomical features.

X-Ray Film.   The majority of macroautoradiography methods utilize X-ray films that are designed for medical radiography, but serve quite well for autoradiography when high resolution is not required. The usual film consists of a transparent base sheet of polyester, which usually has a bluish tint. Each side of the base sheet is coated with a 10- to 20-μm layer of emulsion covered by an antiscratch layer (1–2 μm) of hardened gelatin. As mentioned previously, crystal size is about 0.2–3 μm, and the AgBr to gelatin ratio is about 1:1. Having two layers of emulsion increases the film's sensitivity to X-rays, but for most emitters used in autoradiography, the radiation cannot penetrate the base sheet so only one emulsion layer is exposed by the specimen. Unfortunately, the unused layer contributes to the background haze, but the improvement in image quality by removal of that layer, which can be done after development, rarely warrants the inconvenience of doing so. Double sided film does relieve concern about which side to present to the specimen.

The antiscratch layer is sufficiently thick to absorb a sizable fraction of the β-particles from ^3H but not from ^{14}C, ^{35}S, or ^{32}P, consequently other products usually are used for locating tritium macroscopically. A special film distributed by LKB, called Ultrafilm ^{3}H, has no protective layer and is about 10 times more sensitive to tritium sources. To improve detection efficiency for tritium by ordinary X-ray film, a technique to be discussed later, called fluorography, is used.

"No Screen" X-ray film is usually preferred for macroautoradiography, and it is designed for the direct detection of X-rays. *Intensifying screens* are used with the screen type films, which are quite sensitive to visible light. The screens consist of dense fluorescent salts, such as calcium tungstate bound to a thin plastic or cardboard sheet, and a screen is placed on each side of the film prior to exposure. X-Rays are absorbed by the $CaWO_4$ of the screens and the resulting fluorescence exposes the film. Sensitivity is improved because the screens are more effective absorbers of X-rays than are the emulsion layers. Lower X-ray exposures to patients are possible with the screen type procedure, but some radiographic resolution is lost.

Although the sensitivities of X-ray films to β-particles do not differ widely (Scott and Bradwell, 1983), film speed and grain size are criteria to be considered in film selection. A more practical difference between commercial products concerns suitable development methods. For example, not all films should be developed in automatic developers, but some, such as Kodak X-Omat, are designed for that purpose. Of course, those films designed for automatic developers can be developed manually, but not vice versa.

Exposure. Because the physical nature of specimens, such as chromatograms, electrophoresis gels, plant materials, and animal sections, differs widely, exposure methods can be given only in general terms.

The exposure step is initiated in a darkroom, but a safelight is permissible. Generally, safelights contain a bulb of 15 W or less with a covering filter. Because the sensitivities of different emulsion types differ with respect to wavelength of light, the filter used should conform to the film supplier's recommendation.

The specimen is placed against the film's emulsion and secured, usually with tape, spring loaded clips, or perhaps staples so the positions of the film and specimen do not change relative to one another during handling. Also, some method to reorient the film and specimen after film development should be used. For some specimens this may be accomplished by snipping corners with scissors, but radioactive ink is useful as well. Such ink can be made by adding a quantity of radioactivity to waterproof (India) ink, and several specific activities may be prepared so the radioactivity in the ink is appropriate for the specimen's exposure time. Specific activities around 2.0, 0.5, and 0.05 μCi/ml may be appropriate for ^{32}P, but should be doubled or tripled if ^{14}C or ^{35}S is used. The ink may be applied to tape or gummed labels with a fiber tipped pen or an artists pen that has replaceable points. Nonradioactive markers, which employ phosphors with long decay times, are available commercially.

Glass or metal plates, the size of the film, may be used with clips, tape, or weights to apply gentle pressure so the film and specimen are in close contact. For most specimens an "X-ray exposure holder" (Figure 12-4) is a convenient item for protecting the film from light during the exposure period. Medical X-ray cassettes may be used as well, but they are more expensive.

Once protected from light, the holder may be stored, and the nature of the specimen may dictate appropriate storage conditions. Generally, emulsions should be kept dry, but desiccators are not necessary. Room temperature is satisfactory if the specimen is dry and not subject to deterioration. The exposure period varies with the amount of the radionuclide and its area of distribution. Because these parameters are rarely known with certainty, it is advisable to expose several identical specimens simultaneously then develop their films at various intervals.

According to studies by Tsuk et al. (1964) on activities of ^{14}C needed for images on chromatograms, Scott and Bradwell's (1983) investigations of X-ray film, and Herz's (1969) calculations involving stripping film, it appears that about 10^7 to 10^{10} β-particles per cm^2 are needed for good images, and detectable images may be created with as few as 10^5 to 10^6 β-particles per cm^2.

FIG. 12-4 X-ray holders that may be used for macroautoradiogram exposures.

Developing and Fixing.   Reagents and protocols for developing and fixing film are recommended by a film's supplier. Such recommendations almost always yield suitable autoradiographs, but the experienced autoradiographer frequently can improve images by experimentation with developing and fixing times or temperatures.

Increasing development time or bath temperature causes more silver reduction. Overdevelopment causes reduction of silver in crystals without latent images; on the other hand when underdeveloped, some crystals with latent image centers may not be converted to silver grains. Generally, conditions for the fixing step are not as critical, yet there must be sufficient time for dissolving all of the unconverted AgBr, and the film remains sensitive to light until it is removed. Usually safelights may be utilized during the developing and fixing steps.

Fluorography.   *Fluorography,* sometimes called scintillation autoradiography, was originated by Wilson (1958), who showed that the detection efficiency of ^{3}H on paper chromatograms could be increased significantly by impregnating the chromatograms with a scintillator before exposure to films. Theoretically, some of the β-particles reach the film but a majority are absorbed by the specimen's matrix. When scintillator molecules are in the matrix, they absorb some of the radiation energy and yield light that exposes the film. Compared to direct autoradiography, fluorography offers greater sensitivity advantages for weaker emitters, presumably because a higher proportion of the weaker particles is absorbed by the specimen's matrix. A considerable advantage is realized with ^{3}H, a lessor yet significant improved sensitivity is observed for ^{14}C and ^{35}S-labeled specimens, but there is little or no advantage in fluorography with ^{32}P. Resolution is better in direct autoradiographical methods than fluorography, but for macroscopic specimens, resolution frequently is not a critical factor.

The temperature during the fluorographic exposure step has a significant effect on sensitivity. For example, sensitivity to ^{3}H at $-70°$C is about 10 times better than at $-20°$C and about 60 times better than at room temperature (Bonner and Laskey, 1974). The reasons for the improved sensitivity appear to be 2-fold. First, scintillation efficiency is higher at lower temperatures, because among possible deexcitation processes, fluorescence is favored as temperatures decrease. Second, sublatent image centers are more stable at low temperatures. The energy of a single photon of visible light is sufficient to produce a latent image center in a AgBr crystal, but it consists of only one silver atom. Such a center, which may be considered to be a sublatent image center, has a probable lifetime of 1 s or less at room temperature, but at $-70°$C the probable lifetime may be in the order of minutes to hours. With a longer lifetime, the probability of a second photon striking a crystal with a sublatent image center becomes greater, which increases the probability of crystals becoming developable. This theory appears to have validity because flashing film increases fluorographic sensitivity as well.

Flashing of Films.   Laskey and Mills (1975) showed that preexposure of film to a low intensity flash of light increases film sensitivity in fluorography. They used an electronic photographic flash unit fitted with filters and regulated light intensities by varying the distance between the unit and film. Preexposure increases background fog, but the absorbance of areas exposed fluorographically increases more than enough to compensate for the induced fog. Optimum preexposure appears to yield a background fog absorbance of 0.1 with RP Royal X-Omat film.

Flashing of film does not appear to improve direct autoradiographic sensitivity, presumably because sufficient energy is absorbed by a crystal struck by a β-particle so that a stable and developable latent image center is produced.

Intensifying Screens.   The intensifying screens used in X-ray radiography may be employed with strong β-emitters such as ^{32}P to increase detection efficiency (Laskey and Mills, 1977; Swanstrom and Shank, 1978). For greatest sensitivity dual-sided screen type film is used with two intensifying screens, which have their coated (shiny) sides toward the emulsions. The specimen is placed on the back side of one screen. Presumably many of the β-particles will be stopped by the first screen, which yields a majority of the light that exposes the film, but very energetic particles may pass through the first screen and film to be absorbed by the second screen, yielding additional light. Exposure at $-70°C$ is necessary for maximum sensitivity, which may be about 10-fold better than direct autoradiography.

Because of the isotropic nature of fluorescence and radiation scattering, resolution capability is reduced when intensifying screens are employed.

Electrophoresis Gels.   Many molecular biology methods depend on electrophoretic separations of nucleic acids, which are visualized autoradiographically, and the tracer is usually ^{32}P or ^{35}S. Better resolution is achieved with ^{35}S than ^{32}P. Consequently, ^{35}S is preferred for sequencing gels, an example of which is shown in Figure 12-5. Tritium, ^{14}C, ^{35}S, and ^{125}I are common labels for proteins, but other isotopes are used as well. Because an investigator can control the amount of radioactivity applied to gels, exceptionally high specific activities are not required. However, the nature of research is such that short exposure periods (30 min to 48 hr) are highly desirable. Therefore, fairly high specific activities are normally used.

For exposure, X-ray film must be protected from the moisture in gels and this may be acomplished in several ways.

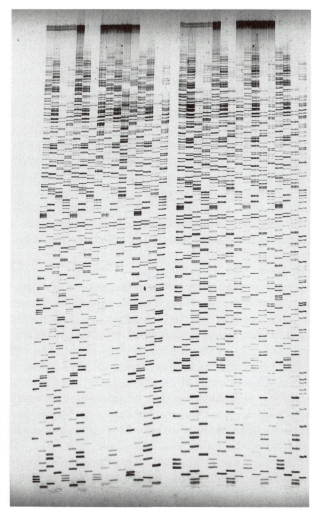

FIG. 12-5 An autoradiogram of a sequencing gel prepared by the dideoxy method with ^{35}S as the label. (Courtesy of Z. Chen and R. J. Spreitzer, University of Nebraska)

1. Dry the gels. The gel is transferred to a thick sheet of absorbent paper such as Whatman 3MM, and placed in a gel dryer which operates under vacuum at about 80°C, normally. Depending on the thickness of the gel, drying requires 1–2 hr. Gels can be dried sufficiently by blotting and allowing them to set on filter paper at room temperature with free air movement for 16–24 hr.

2. Wrap the gel in "plastic food wrap." Most brands of plastic wrap are not sufficiently impervious to water to avoid chemography, but Saran

Wrap® is satisfactory. Because of their emission energies, ^{32}P is detected efficiently when the specimen is wrapped, but very poor detection is observed with ^{35}S and ^{14}C, and ^{3}H is not detected at all.

3. Freeze the gel. Freezing the gels at -70 or -20°C immobilizes the water sufficiently so a gel may have direct contact with the emulsion, but care must be taken to avoid moisture condensation on the film. After exposure in a freezer, the specimen and film should be warmed before the holder or cassette is opened for film development. Alternatively, when removed from the freezer, the holder is taken to the dark room and the film is placed in the developer immediately. Some experimental adjustment of developing time may be necessary when this method is used.

Reagents, such as urea or sodium dodecyl sulfate (SDS), may be used to accomplish desired electrophortic separations, and they may cause chemography. Theoretically, complete dehydration or the presence of a perfectly impervious barrier should prevent the problem, but generally it is more practical to remove interfering reagents. This entails fixing the electrophoretic bands, then soaking the gels in several changes of water.

Intensifying screens may be used to increase detection efficiency of ^{32}P-labeled specimens, but direct autoradiography may be preferred because of resolution capabilities. However, if electrophoretic bands do not "light up" enough by the direct method, reexposure with intensifying screens may produce acceptable autoradiograms without an excessively long exposure period. Two of DuPont's Cronex Lightning-Plus screens with Kodak X-Omat AR film increases sensitivity to ^{32}P by about 8- to 10-fold (Swanstrom and Shank, 1978). Flashing film does not appear to increase sensitivity when intensifying screens are used.

Compared to direct autoradiography, fluorography of gels improves detection efficiencies of ^{14}C and ^{35}S by about 10-fold and ^{3}H by more than 500-fold, however, the improvement for ^{32}P may not warrant the extra labor involved in preparing gels for fluorography. Bonner and Laskey (1974) used the common liquid scintillation fluor, PPO, for fluorography of gels, and their general procedure follows. After electrophoresis and fixing or staining, the gel is soaked in two changes of dimethyl sulfoxide for 30 min each. This removes water, which is necessary for the penetration of the hydrophobic PPO. The gel is immersed in a 20% (w/v) solution of PPO in dimethyl sulfoxide for 3 hr, then the dimethyl sulfoxide is removed, leaving the PPO immobilized, by soaking in water for 1 hr or overnight. The vapor pressure of dimethyl sulfoxide is too low for efficient vacuum evaporation, and dimethyl sulfoxide causes chemography. Finally the gel is dried, preferably in a gel dryer, and exposure is conducted at -70°C. Flashing the film before exposure improves sensitivity several fold.

Chamberlain (1979) devised a more convenient fluorography method using the water-soluble fluor, sodium salicylate. Several convenient and effective products for fluorography of gels are available commercially from various suppliers, such as New England Nuclear (DuPont), Research Products International, and Amersham.

More specific details for various methods involving autoradiography of gels are given by Sambrook et al. (1989) and Davis et al. (1986).

Chromatograms. Autoradiography of chromatograms, both paper and thin-layer, is convenient compared to electrophoresis gels because chromatograms are dried easily and rarely exhibit chemography if developing solvents are completely removed. The surfaces of some thin-layer plates are friable, but, if necessary, powder movement can be reduced by spraying the plate with polyvinyl chloride or nitrocellulose.

For the sake of improved detection efficiency, fluorography is used commonly for tritium-labeled compounds and sometimes for ^{14}C and ^{35}S compounds. Following Wilson's (1958) report on fluorography of paper chromatograms, several methods for paper and thin-layer chromatograms have been developed. Randerath's (1970) method entails pouring a 7% solution of PPO in diethyl ether over chromatograms and allowing the ether to evaporate before exposure to film. Bonner and Steadman (1978) developed three methods, one of which differed from Randerath's in that chromatograms are dipped in the PPO-ether solution rather than pouring the solution on chromatograms. A second method involved immersing chromatograms in melted PPO and a third utilized 0.4% PPO in 2-methylnaphthalene, which is a solid at room temperature. Chromatograms may be dipped in the melted solution (PPO in 2-methylnaphthalene), or by adding 10% toluene, the fluor solution may be sprayed on chromatograms at room temperature. More recently Bochner and Ames (1983) found that 0.5% PPO in 1-methylnaphthalene and methyl anthranilate are efficient fluorographic materials and were easier to apply to chromatograms than 2-methylnaphthalene because they are liquids. As mentioned for electrophoresis gels, fluorography reagents, ready for use, are available commercially.

Plant Material. Although plant tissue may be sectioned, autoradiography is frequently performed with dry, pressed specimens, and except for the fragility of dried plants, the technique is relatively easy. Usually, the freshly harvested plant is arranged on a sheet of blotter paper, covered by a second sheet of paper, and with some method to supply gentle pressure, such as the weight of a glass plate, the tissue is frozen, then lyophilized. Generally, some rehydration of the plant material is necessary to avoid shattering, but exposure to air for 2 hr may be satisfactory. To immobilize the tissue on a support that can be handled, the tissue may be glued to cardboard at several points or pressed into a Styrofoam® sheet with the aid of glass or aluminum plates. Following mounting, the X-ray film is exposed to the plant tissue by direct contact. The possibility of chemography is reduced by using plastic wrap, but detection efficiency is reduced for ^{14}C and ^{35}S. X-Ray holders may be used for the exposure step or the film and specimen may be wrapped to exclude light.

When unsectioned tissue is autoradiographed, ^{3}H is not a useful tracer because of the limited range of its β-particles, but ^{14}C, ^{35}S, and ^{32}P labels work well.

For more specific details of suitable techniques autoradiography of plant materials the reader should consult Levi (1966) and Fiveland et al. (1972).

Animal Sections.   Animal autoradiography is performed with small animals when whole body sections are desired, but sections of organs from larger animals may be appropriate specimens as well. Sven Ullberg, the originator of whole-body autoradiography, with B. Larsson have summarized years of experience in their paper outlining recommended methods (Ullberg and Larsson, 1981). A person contemplating the use of whole-body autoradiography should consult their work, and that reviewed by Curtis et al. (1981). In brief, the general method is as follows.

Immediately after killing, the animal is frozen in a dry ice or liquid nitrogen bath, then placed in a solution containing sodium carboxymethyl cellulose, which is frozen quickly, yielding a reinforced frozen block. The block is mounted on a microtone stage and sagittal sections are made, keeping the microtone at $-20°C$. When sections containing the desired cross sections of the body are reached, tape is pressed on the surface of the block, and a section 5–200 μm thick is cut. The section adheres to the tape, which provides a means of handling sections. For best resolution, the thinnest sections possible are desirable, but 20-μm-thick sections are adequate for most purposes. Several sections to be used for autoradiography are collected and kept frozen. After lyophilization, which may require a day or so, the section is placed directly on the emulsion of the film and a sandwich containing the specimen and film, two layers of paper and finally two aluminum plates, is made. Spring loaded paper clips are used to apply gentle pressure to the aluminum plates. The specimen is wrapped or placed in a light tight box and stored at -10 to $-20°C$ for exposure.

The length of an appropriate exposure period depends on the film's sensitivity, the thickness of the section, the energies of the radionuclide's emissions, and the concentration of the radionuclide in the section. Carrier-free tracers are desirable to reduce exposure time, but if doses are to be physiologically appropriate, low concentrations may be unavoidable. Consequently, exposure periods in the order of a month or more are common. Appropriate exposure periods may be estimated by developing films from comparable sections at various intervals.

To aid in the interpretation of autoradiograms, some of the sections may be subjected to histological staining after film development. If one of the recommended types of tape is used, the sections may be stained without removal from the tape.

The radionuclide most commonly used for whole-body autoradiography is ^{14}C, and it is employed with a no screen X-ray film. Tritium-labeled specimens may be acceptable when X-ray film lacking the antiscratch layer, such as Ultrafilm ^{3}H, is employed. Stripping film, to be discussed later, has also been used for animal sections. Autoradiographs from a study involving the transfer of ^{14}C-labeled vitamin D_3 across the placenta of pregnant mice are shown in Figure 12-6.

Rogers (1979) discussed various experimental parameters involved in

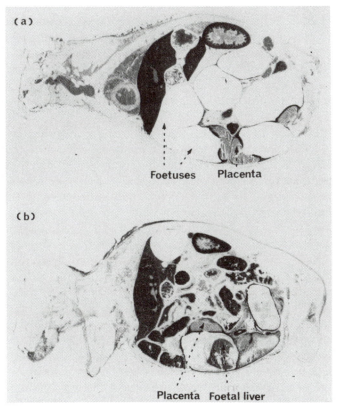

FIG. 12-6 Whole-body autoradiograms of pregnant mice that illustrate transplacental transfer of vitamin. D_3. Mice were injected intravenously with ^{14}C-labeled vitamin D_3 (28 ng, 2.5 μCi per animal) on day 17 of gestation. Animals were killed either (a) 7 hr or (b) 24 hr after injection. Only trace amounts are present in the fetal liver after 7 hr, but at 24 hr the level of isotope in fetal tissue is approximately equal to that of maternal tissue. (Kodirex X-ray film, 12 weeks exposure) (Reprinted with permission from *Whole Body Autoradiography*, by C. G. Curtis, S. A. M. Cross, R. J. McCulloch, and G. M. Powell. Copyright © 1981. Academic Press, Inc., San Diego, CA)

autoradiography of animal sections and additional information is given by Gahan (1972).

Microautoradiography

Although a line of demarcation for specimen size does not exist, microautoradiography involves the study of radionuclide distributions within cells or tissue sections that require a microscope for observation. Photographic images of labeled structures are not obtained, instead the relative densities of developed silver grains or particle tracks are used to correlate radionuclide distributions with cytological structures. An electron microscope autoradiograph of a neuromuscular junction is shown in Figure 12-7. As

FIG. 12-7   An electron microscope autoradiograph of a mouse neuromuscular junction showing the relative uniform distribution of newly inserted acetylcholine receptors at the top of the junctional folds in a denervated end plate. The isotopic tracer was ^{125}I-a-bungartoxin, which binds to acetylcholine receptors. Sch, schwann cells; JF, junctional folds; and M, muscle. Bar length, 1 mm. With this degree of magnification silver grains appear as convoluted ribbons rather than black dots that are observed with light microscopic levels of magnification. (Reproduced from Shyng, S.-L., and Salpeter, M. M. *Journal of Cell Biology*, 1989, 108:647, by copyright permission of the Rockefeller University Press, New York)

mentioned previously, silver grains appear as convoluted ribbons, and the high magnification of this autoradiograph reveals such structures.

Two types of photographic products are used in microautoradiography, stripping film and liquid emulsions, and somewhat different techniques are involved in their use.

Stripping Film.   Apparently the only stripping film currently available is Kodak AR-10, and it consists of a 5-μm layer of photographic emulsion on a 10-μm layer of gelatin, which is supported by a glass plate.

Generally, the specimen is mounted on a microscope slide and may or may not be prepared for histological observation before application of the film. A small sheet of film slightly wider than the width of the microscope slide is outlined on the plate with a scalpel then stripped from the plate. It may be applied directly to the specimen, but usually the film is placed in a flat dish containing distilled water or a 0.001% KBr solution. The film floats on the surface and should have its emulsion side down. The slide is submerged in the dish (specimen side up), moved under the film, and raised up at an angle so one edge of the film contacts the slide. As the slide is lifted, the film settles onto the slide. Obviously, it is important to avoid wrinkles and bubbles. The edges of the film are wrapped around the slide, and the slide is dried by standing vertically in a stream of air for about 1 hr. More complete drying is necessary and may be accomplished by storing in a desiccator at room temperature for 18–24 hr. After drying the slide is stored in a light-tight box at room temperature or in a freezer if a long exposure period is anticipated. Of course, the above procedures must be performed in a darkroom, but a safelight generally can be tolerated.

The film remains on the slide during development and fixation, and because of the permeability of gelatin, histological staining may be accomplished after autoradiogram development.

Liquid Emulsions.   Photographic emulsions "melt" at temperatures somewhat above room temperature, and in the liquid state an emulsion may be coated on a specimen that has been mounted on a microscope slide.

A bath or heating block at about 42°C is used to warm a small container of emulsion, and a slide with its mounted specimen is dipped in the emulsion. While being held vertically, the slide is withdrawn slowly and evenly. Tissue paper is used to wipe off the excess emulsion at the bottom and back side of the slide, and the slide is placed on a cooling block face up. After 20 min, the slide is transferred to a desiccator for overnight drying and then stored in a light-tight box at 4°C for exposure. As with stripping film, the autoradiogram is prepared and interpreted while remaining on the slide. Several factors affect the thickness of the emulsion layer, the temperatures of the emulsion and slide, the rate of slide removal from the emulsion, and the water content of the emulsion. Dilutions of emulsion may be used to obtain desirable layer thicknesses.

Nuclear track emulsions, which were developed to detect particle tracks in nuclear physics studies, are used as liquid emulsions. Products differ in grain (AgBr crystal) size and silver to gelatin ratio. An average crystal size of 0.2 μm and a ratio of 5:1 (AgBr to gelatin) are common. Some examples of emulsions are Kodak NTB, NTB2, and NTB3 as well as Ilford G, K, and L series. Kodak 129-01 has a very fine crystal size, about 0.06 μm.

Exposure and Development.   The length of suitable exposure periods is difficult to judge even with experience, and for that reason multiple specimens are prepared in most autoradiography studies. Periodically, individual autoradiograms are developed until optimum exposure is ascertained.

Proper conditions for development and fixation are also determined experimentally because the nature of the specimen and the thickness of emulsion affects the rates of the chemical reactions involved. Special developing procedures including physical development and latensification may be employed in microautoradiography.

Photographic Tracks.   The energy of an an individual β-particle is dissipated as it travels through a medium, such as a photographic emulsion, by interacting with the electrons or nuclei of molecules in the medium, and, as mentioned earlier, some of these interactions involve AgBr crystals, which are likely to become silver grains after development. Consequently, the track of an individual β-particle can be observed from its trail of silver grains, as shown in Figure 12-8. The initial energy of the particle shown in Figure 12-8 was about 190 keV, which exceeds E_{max} for ^{14}C and ^{35}S slightly and ^3H by a factor of more than 10. For comparison, the maximum path length for β-particles from ^3H is somewhat less than 3 μm, which is less than the thickness of the emulsion layer in stripping film. Nevertheless, silver grains may be produced in a volume of the emulsion, and when the focal plane of a microscope is adjusted through the emulsion layer, tracks may be observed. Of course, the trajectories of particles may be such to yield a track largely in one focal plane.

Interpretation of Autoradiographs.   The two most common measures of radionuclide distribution are called visual grain counting and track counting. The number of grains per unit volume of emulsion reflects the amount of radiation energy dissipated in that volume, but for a given thickness of emulsion, the number of grains per unit area, called the *grain density,* should be a proportional measurement. Of course, background grains will be distributed randomly throughout the autoradiogram, so evidence for the presence of radioactivity in a particular area of the specimen is based on a statistically significant difference in the grain densities of the area com-

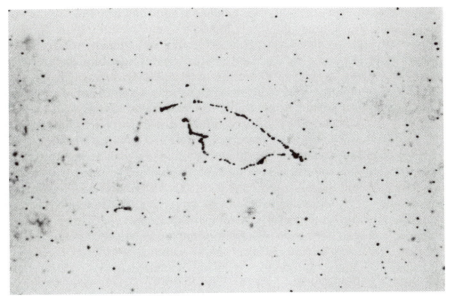

FIG. 12-8   A photomicrograph of a β-particle track recorded in Ilford G5 emulsion. The grains are smaller and more widely spaced at its origin than its terminus. Nuclear collisions are indicated by acute angles in the track. The initial energy of the particle was about 190 keV. X700. (From *Techniques of Autoradiography*, by A. W. Rogers. Copyright © by Elsevier Science Publishers BV, Amsterdam. Reprinted by permission)

pared to a "background only" area. Because many labeled compounds are present throughout a cell or tissue section, a differential distribution may be detected by comparing grain densities in areas corresponding to different cytological structures.

Grain counting is not as straightforward as it may appear because grains along tracks may lie on top of one another, be so close that individual grains cannot be discerned or may coalesce, and they range in size widely. Track counting, if performed properly, appears to be a better measurement since the number of tracks in a given volume of emulsion reflects the number of particles emitted. However, track counting requires decisions about what constitutes a track, and this is not a trivial matter because the lengths of tracks will vary depending on the angles and energies of emitted particles. An added benefit from observing tracks is gained, particularly at the electron microscope level, because trajectories of particles can indicate locations of their origins. For obvious reasons, thicker emulsion layers are used when track counting is to be performed.

With proper equipment photometeric, measurements of grain densities can be performed; also very sensitive optical density measurements are possible with the aid of microscopic optics. Rogers (1979) has presented an excellent discussion on nonvisual autoradiographic data collection.

Radionuclides. For the sake of resolution, tritium has been the most widely used tracer for microautoradiography, and for autoradiography at the electron microscope level, ^{3}H and ^{125}I are used almost exclusively. A critical facet in many studies is the availability of suitable labeled compounds. Generally, ^{3}H and ^{14}C are the most readily available tracers and suitable autoradiograms for light microscopy may be obtained with either. Higher specific activities are possible for ^{3}H compounds because of tritium's half-life, and high specific activities are necessary to avoid excessively long exposure periods. However, ^{3}H atoms are more labile than ^{14}C because of possible chemical exchange reactions as well as metabolic removal. Assuming suitable compounds are available, ^{35}S provides higher specific activities than ^{14}C and equivalent resolution. Although ^{32}P-labels may be used, resolution, sensitivity, and lability of the label make it less attractive than the other radionuclides mentioned above.

Additional information on microautoradiography has been presented by Williams (1977) and Gahan (1972).

DIFFUSION AND METABOLISM OF TRACERS

One of the assumptions generally accepted for radiotracer work is that the radioactivity detected is due to the presence of the labeled compound administered. Although this is a reasonable assumption in most cases, metabolism in biological tissues can lead to erroneous conclusions. In autoradiography, both macroscopic and microscopic, an additional assumption is necessary for meaningful results; the tracer is immobilized in its proper *in vivo* location during the exposure step, which means immobilization during specimen preparation as well. In most procedures, steps are taken to avoid tracer movement, such as freezing and lyophilization of specimens. Also histological fixation of tissues may be effective if labels are incorporated or bound tightly to cellular components that are fixed. Generally, small water-soluble compounds present the greatest problems, and possible movement of labels during manipulation of specimens should be evaluated carefully for all autoradiographic experiments.

NONPHOTOGRAPHIC RADIONUCLIDE IMAGING

A number of instruments are available that produce visual images analogous to those produced by autoradiographic methods.

Rectilinear scanners, which consist of a solid scintillation detector and a mechanism to systematically scan the whole body of a human, have been used for medical purposes for many years, and the "Anger" or gamma camera is another imaging device used in medical applications. The camera detector consists of a large-diameter (30–50 cm) NaI(Tl) crystal (0.6–1.3 cm thick) that has the faces of numerous MP-tubes coupled to it. The image is produced by the relative counting rates of the various MP-tubes, and is displayed with a cathode-ray tube, which may be photographed for a permanent record. Three-dimensional scanners are also available.

Recently, a number of instruments, popularly referred to as "gel scanners," have become available commercially, and they offer new technologies for producing images from planner-labeled specimens. The descriptions of one system that resembles autoradiography in some respects and two instrument types that utilize multiwire proportional detectors are presented in Chapter 17.

CONCLUSION

Although science is involved, autoradiography is an art. The general descriptions given in this chapter are not sufficient for a person to embark on autoradiograph preparations. The objective has been to provide background information that will expedite understanding and interpretations of the primary scientific literature dealing with autoradiography. L'Annunziata's (1987) chapter on radionuclide imaging is recommended reading for persons contemplating autoradiographic studies.

REFERENCES

Bochner, B. R., and Ames, B. M. (1983). *Anal. Biochem.* 131:510.

Bonner, W. M., and Laskey, R. A. (1974). *Eur. J. Biochem.* 46:83.

Bonner, W. M., and Steadman, J. D. (1978). *Anal. Biochem.* 89:247.

Chamberlain, J. P. (1979). *Anal. Biochem.* 98:132.

Curtis, C. G., Cross, S. A. M., McCulloch, R. J., and Powell, G. M. (1981). *Whole Body Autoradiography*. Academic Press, New York.

Davis, L. G., Dibner, M. D., and Battey, J. F. (1986). *Basic Methods in Molecular Biology*. Elsevier, New York.

Fiveland, F. J., Erickson, L. C., and Seely, C. I. (1972). *Weed Res.* 12:155.

Gahan, P. B. (1972). *Autoradiography for Biologists*. Academic Press, New York.

Herz, R. H. (1969). *The Photographic Action of Ionizing Radiations*. Wiley Interscience, New York.

L'Annunziata, M. F. (1987). *Radionuclide Tracers*, pp. 321–416. Academic Press, Orlando, FL.

Laskey, R. A., and Mills, A. D. (1975). *Eur. J. Biochem.* 56:335.

Laskey, R. A., and Mills, A. D. (1977). *FEBS Lett.* 82:314.

Levi, E. (1966). In Proceedings of the Symposium on Isotopes in Weed Research, International Atomic Energy Agency, Vienna, p. 189.

Mees, C. E. K., and James, T. H. (1966). *The Theory of the Photographic Process,* 3rd ed. Macmillan, New York.

Randerath, K. (1970). *Anal. Biochem.* 34:188.

Rogers, A. W. (1979). *Techniques of Autoradiography,* 3rd ed. Elsevier North Holland, New York.

Sambrook, J., Fritsch, E. F., and Maniatis, T. (1989). *Molecular Cloning: A Laboratory Manual,* Vol. I. Cold Spring Harbor Laboratory, Cold Spring Harbor, NY.

Scott, B. J., and Bradwell, A. R. (1983). *Int. J. Appl. Radiat. Isotopes.* 34:765.

Swanstrom, R., and Shank, P. R. (1978). *Anal. Biochem.* 86:184.

Tsuk, R. G., Castro, T., Laufer, L., and Schwarz, D. R. (1964). In *Proceeding of the Conference on Methods of Preparing and Storing Marked Molecules.* J. Sirchis, Ed. Euratom (EUR 1625e), Brussels.

Ullberg, S., and Larsson, B. (1981). *Methods in Enzymology.* W. B. Jakoby, Ed., p. 64. Academic Press, New York.

Williams, M. A. (1977). *Autoradiography and Immunocytochemistry.* Elsevier North Holland, New York.

Wilson, A. T. (1958). *Nature* (London) 182:524.

CHAPTER 13

Statistical Methods for Counting Data

RANDOM NATURE OF RADIOACTIVITY

For one atom of a particular radionuclide, one cannot predict when its nucleus may disintegrate, however for a large number of atoms the fraction that will decay in an interval of time may be calculated with reasonable accuracy. Of course, the number of atoms observed to decay will not agree with the calculated number precisely except in very rare instances because of uncertainties in the measurement of the number of atoms present and the random nature of radioactive decay. That radioactive decay is a random process can be observed easily by watching the irregular accumulation of counts by a scaler or by listening to the clicks of a survey meter with an audio output when its detector is placed close to a sample with low activity. Another method is to count a sample of a radionuclide with a long half-life (no perceptible decay during the course of observations) repeatedly for short intervals of time. For example, a sample of ^{14}C ($t_{1/2} = 5730$ yr) was counted by liquid scintillation 1318 times using 1-min counting periods. The number of counts observed varied about 20%, between 1228 and 1473, and most of the values were obtained more than once. A scatter plot, showing the number of times a particular count was observed, is shown in Figure 13-1. The plot approximates a *normal* or *gaussian distribution*.

NORMAL DISTRIBUTIONS

Figure 13-2 shows two normal distribution curves. Conceptually, a normal curve represents the frequency of various measured values for an infinite number of measurements. The general equation for a normal curve is given

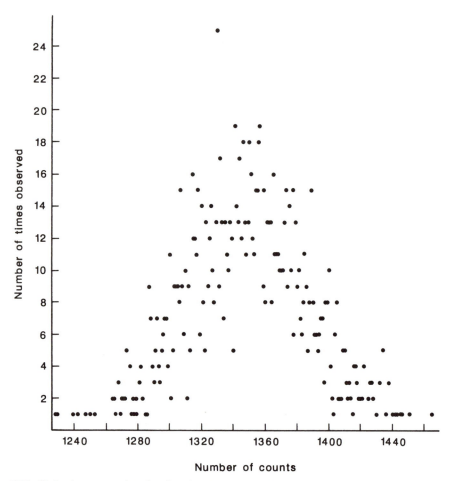

FIG. 13-1 A scatter plot showing the frequencies of various counting values obtained when a ^{14}C sample was counted repeatedly for 1-min periods.

as Equation 13-1, and for Figure 13-2; Y is the ordinate and X the abscissa.

$$Y = \frac{1}{\sigma\sqrt{2\pi}}\, e^{-(X-\mu)^2/2\sigma^2} \tag{13-1}$$

In this equation, both π and e have the usual mathematical meanings, while σ is a parameter called the *standard deviation*. The *true mean* is symbolized as μ, and it is the central value of an infinite number of measurements; practically speaking, it represents the actual value of the thing being measured.

A normal curve is fully defined by the two parameters, σ and μ, and the generation of such a curve may be visualized by the preparation of a histogram from data where measurements are made repeatedly, and the number of measurements whose values lie in a given interval (frequency) is

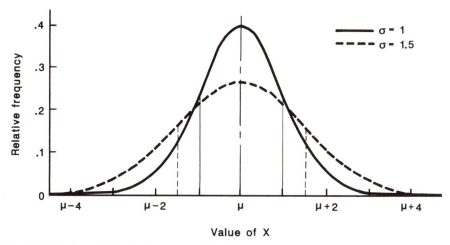

FIG. 13-2   Normal distribution curves with different standard deviations (σ), but the same true means (μ).

plotted against that interval. An example of such a histogram is shown in Figure 13-3. Using the centers of the bars' tops as points, a curve may be drawn that approximates a normal curve. Conceptually, a normal curve is identical except that the intervals (widths of bars) are infinitesimally small, and an infinite number of measurements is accumulated. As shown in Figure 13-2, a normal distribution curve is symmetrical about a vertical line through μ, and μ is at the maximum of the curve. The slope of the curve is a complex function, but when σ increases relative to μ, flatter curves result, indicating a greater spread among values of the individual observations (X_i). Regardless of the relative magnitude of σ and μ, the frequency only reaches zero at $+\infty$ and $-\infty$.

As indicated by Figure 13-3, the area under a curve can be ascertained by integration of $Y \cdot dX$ between specified limits. The limits $-\infty$ and $+\infty$ yield the total area, and all normal curves, regardless of their values of σ

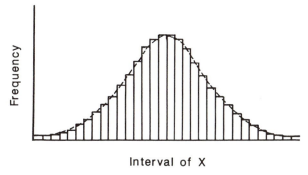

FIG. 13-3   A histogram that illustrates the conceptual basis of a normal curve.

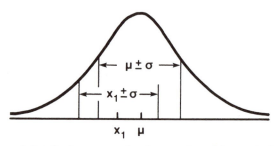

FIG. 13-4 A normal distribution curve showing overlap of intervals for μ and a single measurement X_1.

and μ, have the property that the area between $\mu - \sigma$ and $\mu + \sigma$ is about 68.3% of the total area. By extending the interval, the area is increased, and approximately 95.5% of the total area is included between $\mu \pm 2\sigma$, while 99.74% is between $\mu \pm 3\sigma$. The fractional area between particular limits represents the fraction of all observations between those same limits. Therefore, in an infinitely large series of measurements, 68.3% of them should be within $\mu - \sigma$ and $\mu + \sigma$, and 95.5% between $\mu \pm 2\sigma$.

The exact value of μ can never be ascertained, but a single measurement or observation provides a statistic that is an estimate of that population parameter, μ, and additional observations can improve our confidence about its probable value. If a single measurement (X_1) lies in the interval $\mu \pm \sigma$, then μ lies in the interval $X_1 \pm \sigma$ (Figure 13-4). Thus, if we can be 68.3% confident of X_1 being in the interval $\mu \pm \sigma$, we can be 68.3% confident that μ has a value between $X_1 - \sigma$ and $X_1 + \sigma$. Such intervals are referred to as *confidence intervals* or *confidence limits,* and they usually are expressed in units of σ. Obviously, greater confidence can be placed in larger intervals; for example, the assumption that μ is in the interval $X_i \pm 2\sigma$ is correct in about 955 cases out of 1000, and for the interval $X_i \pm 3\sigma$, correct in about 99.74% of the cases. The general form of confidence intervals is $X \pm ns$, where n is a factor that governs the range of the interval (for a given value of s), and hence the level of confidence of μ being in that interval. The symbol s indicates a *sample standard deviation* which is an approximation of σ. Table 13-1 provides data relating various intervals to levels of confidence.

TABLE 13-1   Confidence Interval Factors

Interval	Confidence level (%)
$\pm\ 0.6745\ s$	50.0
$\pm\ s$	68.3 (standard)
$\pm\ 1.96\ s$	95.0
$\pm\ 2\ s$	95.5
$\pm\ 2.58\ s$	99.0

For a more rigorous treatment of the normal distribution law the reader should consult a basic text on statistics, such as Steel and Torrie (1980).

Estimation and Use of the Standard Deviation

A common method of estimating the standard deviation of a normal distribution is to make a number (N) of repeated measurements that have values designated as $X_1, X_2, X_3, \ldots, X_N$. The mean value ($\overline{X}$) of these measurements is the arithmetic average value. This mean is referred to as the *sample mean* to distinguish it from the true mean, μ.

$$\overline{X} = \frac{X_1 + X_2 + X_3 + \cdots + X_N}{N} = \frac{\Sigma X_i}{N} \tag{13-2}$$

The *deviation* of each measurement ($\Delta_1, \Delta_2, \Delta_3, \ldots, \Delta_N$) is the difference between a particular measurement and the sample mean,

$$\Delta_1 = X_1 - \overline{X}, \Delta_2 = X_2 - \overline{X}, \ldots, \Delta_N = X_N - \overline{X}, \text{ or } \Delta_i = X_i - \overline{X}$$

and the sample standard deviation is given by Equation 13-3. To reiterate, the sample standard deviation (s) is an estimate of the true standard deviation (σ), and the term standard deviation in the remainder of this chapter will mean s, unless indicated otherwise.

$$s = \sqrt{\frac{\Delta_1^2 + \Delta_2^2 + \Delta_3^2 + \cdots + \Delta_N^2}{N - 1}} = \sqrt{\frac{\Sigma (X_i - \overline{X})^2}{N - 1}} \tag{13-3}$$

For ease in calculations the following algebraic identity may be used to ascertain the sum of the squares of the deviations.

$$\Sigma(X - \overline{X})^2 = \Sigma X_i^2 - \frac{(\Sigma X_i)^2}{N}$$

For the data presented in Figure 13-1, calculations were applied as follows.

$$\overline{X} = \frac{1,776,470 \text{ ct}}{1318} = 1347.85 \text{ ct}$$

The sum of the individual deviations squared was 1,770,325, which gave a standard deviation of 36.66 counts.

$$s = \sqrt{\frac{1,770,325}{1317}} = 36.66 \text{ ct}$$

Now, rounding figures to the nearest count and obtaining the interval, $\overline{X} \pm s$ or 1348 ± 37, one can predict that about 68.3% of the measurements should lie in similarly constructed intervals if a normal distribution is applicable. Actually, 65.8% of the measurements in our example are within the interval 1311 to 1385, and this appears to fit a normal distribution well if, indeed, $\overline{X}$ and s are close approximations of μ and σ. Calculation of the standard deviation by this method can be done readily with computers, but the time required to count large numbers of samples repeatedly, even just a few times each, would make many experiments impractical.

For many kinds of measurements, for example spectrometer readings, it is not possible to estimate σ by a single reading. Fortunately, the nature of radioactive decay and counting measurements do permit an estimation of σ from a single counting observation provided:

1. The rate of radioactive decay (λt) is small compared to the counting period (no perceptible decay during the measurement).

2. The total number of counts exceeds 100. These provisions are easily met in the majority of counting measurements, and hence Equation 13-4 is quite adequate for estimating σ.

$$\sigma = \sqrt{\mu}, \qquad \mu > 100 \qquad (13\text{-}4)$$

The derivation of Equation 13-4 is based on the binomial distribution law, but it is a good approximation of the normal distribution law under the conditions outlined above. For a rigorous derivation of Equation 13-4, see Frielander et al. (1981).

While the difference between a single measurement (X_i) and the true mean (μ) may be considerable, their square roots will not differ nearly as much, therefore as an approximation:

$$\sqrt{\mu} \approx \sqrt{X}$$

and as an approximation of σ,

$$s \approx \sqrt{X}, \qquad X > 100 \qquad (13\text{-}5)$$

The first measurement in the series used to generate the data shown in Figure 13-1 was 1349 counts, and using the approximation given by Equation 13-5, the sample standard deviation is

$$s = \sqrt{1349} = 36.73 \approx 37 \text{ ct}$$

This compares very favorably with the value of 36.66, obtained by repeating the measurement 1317 times and performing the calculations, but was this a fortuitous example? The two extreme values in the whole set of measurements were 1228 and 1473 counts.

$$\sqrt{1228} = 35.04 \approx 35$$
$$\sqrt{1473} = 38.38 \approx 38$$

These values are within two counts of the values of s obtained by the laborious method.

A Conclusion.   For radioactive counting data, Equation 13-5 provides a reasonable estimate of the standard deviation. Therefore, a single measurement permits μ and σ to be estimated. The counting measurement X may be expressed as $X \pm n\sqrt{X}$, where n corresponds to the desired level of confidence as indicated in Table 13-1. Thus, μ should lie in similarly constructed intervals $(X \pm \sqrt{X})$ about 68.3% of the times and in intervals of $X \pm 1.96\sqrt{X})$ about 95% of the times.

COUNTING STATISTICS

Counting Rates

Radioactivity is measured by counting rates (R), actually average counting rates, where X is the number of counts obtained in a counting period of t.

$$R = \frac{X}{t} \tag{13-6}$$

If 1800 counts were obtained in 10 min, the counting rate would be 180 cpm. What is the standard deviation of the counting rate? According to our previous discussion, the standard deviation in the measurement (X) is $\sqrt{X}$, but time is a measurement also. Fortunately, the error in measuring time by modern instrumentation is so small relative to the deviation in counts that t can be considered to be an errorless quantity. Therefore, the standard deviation of the counting rate $s(r)$ is

$$s(r) = \frac{\sqrt{X}}{t} \tag{13-7}$$

For our example, $s(r)$ is easily calculated,

$$s(r) = \frac{\sqrt{1800}}{10} = 4.24 \text{ cpm or 4 cpm}$$

The counting rate may be expressed as

$$\frac{X}{t} \pm \frac{n\sqrt{X}}{t}$$

The 68.3% confidence interval for our example is

$$\frac{1800}{10} \pm \frac{1 \cdot \sqrt{1800}}{10}$$

$$180 \pm 4 \text{ cpm}$$

Because $R = X/t$, another way of calculating the standard deviation of a counting rate is given by Equation 13-8.

$$s(r) = \sqrt{\frac{R}{t}} \tag{13-8}$$

and the counting rate may be expressed as

$$R \pm n\sqrt{\frac{R}{t}}$$

Relative Deviations

The *relative deviation* (RD) is a convenient method of expressing the size of the deviation with respect to the size of the measurement.

$$RD = \frac{\text{deviation value}}{\text{measurement value}} \tag{13-9}$$

Relative deviations may be given as fractional quantities or as percentages. For one counting measurement, the relative standard deviation (the relative deviation for the standard deviation) is given by Equation 13-10.

$$RD_s = \frac{s}{X} = \frac{\sqrt{X}}{X} = \frac{1}{\sqrt{X}} \tag{13-10}$$

Equation 13-11 gives the relative standard deviation for a counting rate.

$$RD_s = \frac{s}{X/t} = \frac{\sqrt{X}/t}{X/t} = \frac{1}{\sqrt{X}} \tag{13-11}$$

The relative standard deviations are the same in both cases, which shows that the relative uncertainty of a measurement depends on the *number of counts collected and is not affected by the time needed to collect them.*

At first glance, it may appear that increasing the counting time is not advantageous since the standard deviation increases with the square root of the number of counts (Equation 13-5). However, the magnitude of uncertainty means little unless it is compared to the magnitude of the measurement, and the relative deviation decreases as the number of counts increases (Equation 13-10).

Using our previous example, 1800 counts per 10 min, the relative standard deviation may be calculated.

$$RD_s = \frac{s}{X/t} = \frac{\sqrt{1800}/t}{1800/t} = \frac{1}{\sqrt{1800}} = 0.0236 \text{ or } 2.36\%$$

This means that the standard deviation is 2.36% of the counting rate. The counting rate can be expressed as $180 \pm 2.36\%$, which indicates that we can be 68.3% confident that the true counting rate is within the range 180 ± 0.0236 (180).

An uncertainty of about 2% may be acceptable in many experiments, but when the relative standard deviation is used, the level of confidence is not very comforting because there is about one chance in three that the true mean value differs from the experimental value by more than 2%. For the 95% confidence level what would the relative deviation be?

$$RD_{95} = \frac{1.96s}{X/t} = \frac{1.96}{\sqrt{X}} = \frac{1.96}{\sqrt{1800}} = 0.0462 \text{ or } 4.62\%$$

We can be 95% confident that the true counting rate is between $180 - 0.046\,(180)$ and $180 + 0.046\,(180)$ or (172 to 188).

How many counts should be accumulated so that the sample deviation is 2% of the counting rate at the 95.5% confidence level?

$$RD_{95.5} = 0.02 = \frac{2s}{X/t} = \frac{2\sqrt{X}/t}{X/t} = \frac{2}{\sqrt{X}}$$

$$X = \left(\frac{1}{0.01}\right)^2$$

$$X = 10,000 \text{ counts}$$

This is a convenient number to remember: counting a sample for a period long enough to accumulate 10,000 counts will yield a relative counting error of 2% with a level of confidence at 95.5%.

For a 1% relative deviation at the 95.5% confidence level the number of counts needed is 40,000, and at the 99% confidence level about 67,000 counts are needed.

Average Counting Rates

In our earlier discussion of the 1318 measurements with 1-min counting periods (data of Figure 13-1), we came to the conclusion that all the extra counting time did not greatly improve our estimation of σ compared to an estimate based on a single 1-min count. However, a comparison of the procedures to estimate the true mean (μ) was not addressed. Recall that

$$\overline{X} = \frac{\Sigma X_i}{N} = \frac{1,776,470 \text{ ct}}{1318} = 1347.85 \text{ ct}$$

$$s = \sqrt{\frac{\Sigma(X_i - \overline{X})^2}{N-1}} = \sqrt{\frac{1,770,325}{1317}} = 36.66 \text{ ct}$$

The standard deviation is for the distribution of individual measurements, that is, the standard deviation is the deviation expected for a single 1-min counting period. However, the sample mean, $\overline{X}$, is a much better estimate of μ than is any one of the single measurements. The *standard deviation of the sample mean*, $s(\overline{X})$, may be conceptualized as the deviation expected if the whole, multiple counting procedure were repeated many times, yielding multiple sample means $\overline{X}_1, \overline{X}_2, \overline{X}_3, \ldots$, and a mean for the sample means ($\overline{\overline{X}}$). The term *standard error of the mean* or just *standard error* is sometimes used for the standard deviation of the mean. Using Equations 13-2 and 13-3, the standard deviation of the sample means could be calculated, if the data were available. However, the standard deviation of the sample mean can be calculated readily from one set of data containing N measurements.

$$s(\overline{X}) = \frac{s}{\sqrt{N}} \tag{13-12}$$

Using the data from our one set we obtain:

$$s(\overline{X}) = \frac{36.66}{\sqrt{1318}} = 1.01$$

Therefore, the 68.3% confidence interval for the sample mean is

$$\overline{X} \pm \frac{s}{\sqrt{N}}$$

$$1347.85 \pm 1.01 \text{ ct}$$

or

$$1348 \pm 1 \text{ ct}$$

This indicates that if the whole set of measurements were repeated many times, about 68.3% of the times, μ should lie in similarly calculated intervals, and the interval widths would be approximately $\pm$ 1 count. Also, in 95.5% of the sets, the interval widths would be $\pm$ 2*s*, or about $\pm$ 2 counts.

Counting the sample repeatedly certainly narrowed the range of probable values for μ, but the improvement is not due to our choice of calculation methods. It is due to the consideration of a larger number of counts. The whole series of measurements could have been considered as one measurement since a total of 1,776,470 counts were obtained in 1318 min of counting time.

$$\frac{X}{t} \pm \frac{\sqrt{X}}{t} = \frac{1,776,470}{1318} \pm \frac{\sqrt{1,776,470}}{1318}$$

$$= 1347.85 \pm 1.01 \text{ cpm}$$

$$\approx 1348 \pm 1 \text{ cpm}$$

Obviously, the standard deviation of the counting rate is the same as the one calculated for the sample mean, but of course, the average counting rate (X/t) is the sample mean for the set of 1-min counts.

At the risk of belaboring the point, the accuracy of our estimate of the true counting rate is determined by the number of counts obtained, not by the time needed to collect them or the intervals of time used in their collection.

Net Counting Rates

So far our discussion has been limited to single counting measurements, but actual measurements involve net counting rates. The same statistical calculations discussed above apply to background counting rates, therefore the standard deviation of the background counts, X_b, obtained in t_b minutes is $s(b) = \sqrt{X_b}/t_b$. Let X_g be the counts obtained in t_g minutes when a sample is counted, then its standard deviation is $s(g) = \sqrt{X_g}/t_g$. The net counting rate is $(X_g/t_g) - (X_b/t_b)$, but what would the standard deviation of this difference be? One might be inclined to just add the standard deviations of both measurements, but intuitively that would not be correct because we could expect that many times when the gross counting rate is higher than its true mean the background counting rate may be lower than its true mean and vice versa. Therefore, to some extent errors should cancel, which would reduce the standard deviation of the difference $(X_g/t_g) - (X_b/t_b)$.

When numbers with associated independent uncertainties (standard deviations) are used in mathematical operations, such as subtraction and multiplication, the uncertainties are extended. The rules for handling such uncertainties are sometimes referred to as the law of propagation of errors. Equations 13-13 to 13-20 show how the standard deviation of values may be determined for various mathematical operations when two quantities (A and C) and their respective standard deviations, $s(a)$ and $s(c)$, are known.

For $A - C$,

$$s = \sqrt{[s(a)]^2 + ;s(c)]^2} \tag{13-13}$$

For $A + C$,

$$s = \sqrt{[s(a)]^2 + [s(c)]^2} \tag{13-14}$$

For $A \cdot C$,

$$s = A \cdot C \sqrt{\left[\frac{s(a)}{A}\right]^2 + \left[\frac{s(c)}{C}\right]^2} \tag{13-15}$$

For A/C,

$$s = A/C \sqrt{\left[\frac{s(a)}{A}\right]^2 + \left[\frac{s(c)}{C}\right]^2} \tag{13-16}$$

For A^Y (Y being errorless),

$$s = Y \cdot s(a) \cdot A^{Y-1} \tag{13-17}$$

For $\ln A$,

$$s = \frac{s(a)}{A} \tag{13-18}$$

For log A,

$$s = \frac{s(a)}{2.303A} \tag{13-19}$$

For e^A,

$$s = e^A \cdot s(a) \tag{13-20}$$

Equation 13-13 is appropriate for calculating the standard deviation of a net counting rate $s(n)$ and yields Equation 13-21.

$$s(n) = \sqrt{\left(\frac{\sqrt{X_g}}{t_g}\right)^2 + \left(\frac{\sqrt{X_b}}{t_b}\right)^2}$$

$$s(n) = \sqrt{\frac{X_g}{t_g^2} + \frac{X_b}{t_b^2}} \tag{13-21}$$

Since $R = X/t$,

$$s(n) = \sqrt{\frac{R_g}{t_g} + \frac{R_b}{t_b}} \tag{13-22}$$

Finally, the net counting rate and its associated deviation is given below, where n is the factor corresponding to the desired confidence interval.

$$\frac{X_g}{t_g} - \frac{X_b}{t_b} \pm n\sqrt{\frac{X_g}{t_g^2} + \frac{X_b}{t_b^2}}$$

Example Problem 13-1. A sample yielded 4783 counts in 5 min, and the background was 572 counts in 10 min. What are the standard deviations of the gross, background, and net counting rates? Also, calculate the relative standard deviations for each.

$$s(g) = \frac{\sqrt{X_g}}{t_g} = \frac{\sqrt{4783}}{5} = 13.83 \text{ or } 14 \text{ cpm}$$

$$RD_{s(g)} = \frac{13.83}{4783/5} = 0.0145 \text{ or } 1.45\%$$

$$s(b) = \frac{\sqrt{X_b}}{t_b} = \frac{\sqrt{572}}{10} = 2.39 \text{ or } 2 \text{ cpm}$$

$$RD_{s(b)} = \frac{2.39}{572/10} = 0.418 \text{ or } 4.18\%$$

$$s(n) = \sqrt{\frac{4783}{(5)^2} + \frac{572}{10^2}} = 14.04 \text{ or } 14 \text{ cpm}$$

$$RD_{s(n)} = \frac{14.04}{(4783/5) - (572/10)} = 0.0156 \text{ or } 1.56\%$$

Note that the relative deviation of the background counting rate is much higher than the gross rate, about 4% compared to 1.5%. However, the standard and relative deviations of the net counting rate are only slightly higher than those for the gross counting rate. Apparently, the uncertainty of the background does not contribute much to the uncertainty in the net counting rate in this example.

More counting time, hence more counts, will reduce relative deviations, but which of the measurements would have the greatest effect on the uncertainty of the net counting rate? Suppose the sample had been counted for 10 min rather than 5 min, then $X_g \approx 9566$ and $t_g = 10$ min.

$$RD_{s(n)} = \frac{\sqrt{[9566/(10)^2] + [572/(10)^2]}}{(9566/10) - (572/10)} = \frac{10.07}{899.4}$$

$$RD_{s(n)} = 0.0112 \text{ or } 1.12\%$$

The relative standard deviation of the net counting rate is about 0.4% lower than that when the sample was counted for only 5 min, a significant reduction. Suppose the counting time for the background was increased to 15 min and the sample counting time kept at 5 min, that is $X_g = 4783$, $t_g = 5$, $X_b \approx 858$, and $X_b = 15$.

$$RD_{s(n)} = \frac{\sqrt{[4783/(5)^2] + [858/(15)^2]}}{(4783/5) - (858/15)} = \frac{13.97}{899.4}$$

$$RD_{s(n)} = 0.0155 \text{ or } 1.55\%$$

The extra time spent counting the background reduced the relative standard deviation by only about 0.01%. The point is, when two measurements are combined by a sum or difference and one measurement is considerably larger than the other, the combined error will depend mostly on the larger component.

Rules of thumb:

1. If the background counting rate is less than $\frac{1}{10}$ the sample's gross counting rate, the relative deviation of the net counting rate will depend largely on the uncertainty of the gross counting rate. Therefore, count samples long enough to obtain at least 10,000 counts so the relative deviation of the net counting rate is about 2% or less at the 95% confidence level.

2. If the sample and background counting rates are within the same order of magnitude, the total counting time should be equally divided between the two measurements.

Example Problem 13-2. Using the same data presented in problem 13-1, (1) show the net counting rate and its associated 95% deviation and (2) calculate the relative 99% deviation for the net counting rate.

1. $\dfrac{X_g}{t_g} - \dfrac{X_b}{t_b} \pm 1.96 \sqrt{\dfrac{X_g}{t_g^2} + \dfrac{X_b}{t_b^2}}$

$\dfrac{4783}{5} - \dfrac{572}{10} \pm 1.96 \sqrt{\dfrac{4783}{25} + \dfrac{572}{100}}$

$899 \pm 1.96\,(14.04) = 899 \pm 28 \text{ cpm}$

2. From problem 13-1, $RD_{s(n)} = 0.0156$

$RD_{99(n)} = 2.58\,[RD_{s(n)}] = 2.58\,(0.0156)$

$RD_{99(n)} = 0.0402 \text{ or } 4.02\%$

Example Problem 13-3. The following counting data were obtained by liquid scintillation counting, and the counting efficiency of the sample is to be corrected by the sample channels ratio (SCR) method as described in Chapter 9. The SCR value to be used is the net cpm in channel A divided by that in channel B. What is the 95% confidence interval for the SCR value?

The counting data are

	Sample (counts/2 min)	Blank (counts/2 min)
Channel A	548	38
Channel B	1076	57

First, the net cpm and associated standard deviations are determined for both channels. Then the SCR value and its standard deviation is computed and adjusted for the 95% confidence level. Net cpm, channel A:

$$\dfrac{548}{2} - \dfrac{38}{2} \pm \sqrt{\dfrac{548}{2^2} + \dfrac{38}{2^2}}$$

$$255 \pm 12.1 \text{ cpm}$$

Net cpm, channel B:

$$\dfrac{1076}{2} - \dfrac{56}{2} \pm \sqrt{\dfrac{1076}{2^2} + \dfrac{56}{2^2}}$$

$$510 \pm 16.8 \text{ cpm}$$

SCR value:

$$SCR = \dfrac{A}{B} = \dfrac{255 \pm 12.1}{510 \pm 16.8}$$

$$s(\text{scr}) = \frac{255}{510} \pm \frac{255}{510} \sqrt{\left(\frac{12.1}{255}\right)^2 + \left(\frac{16.8}{510}\right)^2}$$

$$\text{SCR} = 0.500 \pm 0.029$$

95% confidence interval:

$$\text{SCR} = 0.500 \pm (1.96)(0.029)$$

$$\text{SCR} = 0.500 \pm 0.057$$

This answer means that we can be 95% confident that the SCR value is between 0.443 and 0.557, but this is a wide interval as shown by calculating the relative deviation of the value at the 95% confidence level.

$$\text{RD}_{95} = \frac{0.057}{0.5} \times 100 = 11.4\%$$

Such a relative deviation is unacceptably high for careful work. Another way of judging the reliability of the SCR value is to consider the range of counting efficiencies this may represent. Of course, there will be some error in every SCR quench correction curve but discounting that contribution, the range of SCR values (0.443 to 0.557) corresponds to counting efficiencies of 84.4–80.2% when the curve presented in Figure 9-2 is employed.

In Chapter 9 it was stressed that the SCR quench correction method is not a good one when low activity samples are involved, unless long counting times are used. The reason for such a statement is based on the number of counts that must be collected in each channel to obtain a reasonably reliable SCR value, and example problem 13-4 will address the apparent question.

Example Problem 13-4. How many counts must be collected to obtain an SCR value for which the relative 95% deviation is 2% or less, assuming the SCR value is about 0.5?

Notice that the background measurement is a minor component in the standard deviation of the net counting rates of each channel in example problem 13-3. For channel B, the standard deviation of the gross counting rate is

$$s(\text{b}) = \frac{\sqrt{548}}{2} = 11.7 \text{ cpm}$$

This is only slightly lower than the standard deviation for the net counting rate in that channel (12.1 cpm). Therefore, to simplify calculations, assume the standard deviation of the net counting rate is approximately the same as that for the gross counting rate. Let A and B be the total number of counts collected in channels A and B, respectively. Since an SCR value of 0.5 was assumed, $A = 0.5B$. Also let $s(\text{a})$, $s(\text{b})$, and $s(\text{scr})$ be the respective standard deviation of the counting rates in channels A and B, and the SCR value, A/B.

$$s(a) = \sqrt{A} = \sqrt{0.5B}$$

$$s(b) = \sqrt{B}$$

$$s(scr) = \frac{A}{B} \sqrt{\left(\frac{(0.5B)^{1/2}}{0.5B}\right)^2 + \left(\frac{(B)^{1/2}}{B}\right)^2}$$

$$s(scr) = (0.5) \sqrt{\frac{3}{B}}$$

$$RD_{95} = \frac{(1.96) \, s(scr)}{A/B} \times 100$$

$$2 = \frac{(1.96)(0.5)(3/B)^{1/2}(100)}{0.5}$$

$$B = \frac{(3)(1.96)^2(10^4)}{4}$$

$$B = 28,812 \text{ ct}$$

$$A = 0.5B = 14,406 \text{ ct}$$

Therefore, samples should be counted long enough to collect about 30,000 counts in the wide counting channel to obtain reasonably valid SCR values under the conditions of this example.

For the relative 95% deviation of the SCR value to be 1% or less, about 115,000 counts in the wide counting channel should be obtained. In preparing quench correction curves, the statistical validity of the counting data should be at least this good and can be obtained readily by using high activity standards (20,000 dpm) and sufficiently long counting periods (10 to 20 min).

Evaluating Optimal Counting Conditions

The statistical validity of a radioactivity measurement depends on the number of counts observed. Usually the time available for counting is limited, but frequently other factors that affect the number of counts recorded can be changed to improve the accuracy of results. Those factors may include the kind of radionuclide employed, type of detector (instrument), operating conditions of the instrument, sample preparation methods, and level of background radiation. These factors may be evaluated separately or together, but the criteria for such evaluation should be based on the relative standard deviations of measurements (Equation 13-9). For net counting rates:

$$RD_s = \frac{\sqrt{(X_g/t_g^2) + (X_b/t_b^2)}}{(X_g/t_g) - (X_b t_b)} = \frac{\sqrt{(R_g/t_g) + (R_b/t_b)}}{R_n}$$

Optimal conditions would be those that yield the lowest relative deviation with the least counting time. The reciprocal of the relative standard deviation (1/RD) is sometimes called a figure of merit (fm). Thus, one should strive to maximize the figure of merit.

$$fm = \frac{R_n}{\sqrt{(R_g/t_g) + (R_b/t_b)}}$$

Because automatic loading counters are used commonly, background is usually counted for the same amount of time as the samples, therefore, let $t_g = t_b$.

$$fm = \frac{R_n}{\sqrt{(R_g + R_b)/t_b}} = \frac{R_n}{\sqrt{(R_n + 2R_b)/t_b}}$$

The net counting rate (R_n) is a function of the activity of the sample (A) and counting efficiency (CE).

$$R_n = A \cdot CE$$

$$fm = \frac{A \cdot CE \cdot \sqrt{t_b}}{\sqrt{A \cdot CE + 2R_b}}$$

The figure of merit increases with increasing activity (amount of radioisotope employed), counting efficiency (affected by the type of radionuclide, type of detector, operating conditions of the instrument, and sample preparation), and length of time samples are counted ($t_g = t_b$). Higher background counting rates decrease the figure of merit. For optimizing instrumental performance, the counting efficiency and background counting rate are the values that may be changed because for a given sample, A is constant and t_b is subject to the time available for counting samples. For example, the counting efficiency and background vary for a liquid scintillation counter depending on pulse selection settings (amplifier and discriminator settings), and conditions that increase efficiency usually increase background as well. The question is, for a given amount of activity of a certain radionuclide that is to be counted for a given amount of time, what pulse selection settings would give the highest figure of merit? Many modern instruments have settings adjusted at the factory so the question may be one that a factory technician may have to address rather than a research scientist. Nonetheless, for this question, the figure of merit may be simplified by considering A and t_b as constants ($k = \sqrt{A \cdot t_b}$).

If $A \cdot CE = R_n \gg 2R_b$, one can ignore the background counting rate and seek to maximize efficiency because

$$\text{fm} = \frac{\sqrt{t_b} \cdot A \cdot \text{CE}}{\sqrt{A \cdot \text{CE}}} = \sqrt{t_b} \cdot A \cdot \text{CE}$$

$$\text{fm} = k \sqrt{\text{CE}}$$

However, if net counting rates (R_n) are low, $A \cdot \text{CE}$ is less than $0.1\,R_b$, and $k = \sqrt{t_b} \cdot A/\sqrt{2}$ then

$$\text{fm} = \frac{\sqrt{t_b} \cdot A \cdot \text{CE}}{\sqrt{2R_b}} = \frac{k \cdot \text{CE}}{\sqrt{R_b}}$$

In this case, one should try to maximize $\text{CE}/\sqrt{R_b}$. Frequently, the figure of merit is modified to highlight the term to be maximized.

$$\left(\frac{\text{fm}}{k}\right)^2 = \frac{\text{CE}^2}{R_b}$$

Certainly, if $(\text{CE})^2/R_b$ were maximized, the highest figure of merit, and hence lowest relative standard deviation would be achieved. Most commonly, the figure of merit is referered to as "efficiency squared over background" or E^2/B.

TREATMENT OF NONCOUNTING ERRORS

In assaying radioactive samples, measurements other than counting are involved frequently. For example, the preparation of a sample for counting may require pipetting or weighing procedures, and such measurements may contribute as much or more uncertainty to the desired measurement than the counting procedure. However, the overall standard deviation for an assay can be estimated using Equations 13-13 through 13-20.

Equations 13-13 and 13-14 can be combined and extended to include additional terms, likewise for Equations 13-15 and 13-16.

For $A - B + C + D$,

$$s = \sqrt{[s(a)]^2 + [s(b)]^2 + [s(c)]^2 + [s(d)]^2} \tag{13-23}$$

For $\dfrac{A \cdot B}{C \cdot D}$,

$$s = \frac{A \cdot B}{C \cdot D} \sqrt{\left(\frac{s(a)}{A}\right)^2 + \left(\frac{s(b)}{B}\right)^2 + \left(\frac{s(c)}{C}\right)^2 + \left(\frac{s(d)}{D}\right)^2} \tag{13-24}$$

Note that the square root term in Equation 13-24 is the relative standard deviation for the product and quotient, and the terms within the square root sign are the sums of the individual relative standard deviations squared.

Example Problem 13-5. The counting efficiency of ^{14}C in a liquid scintillation counter was determined by pipetting a 0.1-ml aliquot of a [^{14}C]toluene standard into a vial, adding scintillation cocktail and counting the sample. A blank for background was prepared similarly and yielded 124 counts in 5 min. The sample gave 23,562 counts in 1.0 min and the specific activity of the standard was listed as $3.995 \times 10^5 \pm 2.5\%$ dpm/ml. Accuracy specifications for the type of pipet used was 0.100 ± 0.005 ml. Calculate the counting efficiency and its associated standard deviation.

Let C, A, V, and E represent the net counting rate, specific activity of the standard, volume pipetted, and efficiency of counting, respectively, and let their associated standard deviation be $s(c)$, $s(s)$, $s(v)$, and $s(e)$.

$$E = \frac{C}{AV}$$

$$A = 3.995 \times 10^5 \pm (0.025)(3.995 \times 10^5)$$

$$= 3.995 \times 10^5 \pm 9.988 \times 10^3 \text{ dpm/ml.}$$

Equation 13-21 must be used to calculate the relative standard deviation of the net counting rate before it is used in the calculation of counting efficiency.

$$s(c) = \sqrt{\frac{23{,}562}{1^2} + \frac{124}{5^2}} = 153.5 \text{ cpm}$$

$$C = \frac{23{,}562}{1} - \frac{124}{5} = 23{,}537 \text{ cpm}$$

$$RD_{s(c)} = \frac{153.5}{23{,}537} = 0.0065 \text{ or } 0.65\%$$

$$E = \frac{23{,}537 \pm 153.5 \text{ cpm}}{(3.995 \times 10^5 \pm 9.988 \times 10^3 \text{ dpm/ml})(0.100 \pm 0.005 \text{ ml})}$$

$$E = 0.5892 \pm 0.5892 \ (RD_{s(e)}) \text{ cpm/dpm}$$

$$RD_{s(a)} = 0.025 \text{ or } 2.5\%$$

$$RD_{s(v)} = \frac{0.005}{0.100} = 0.050 \text{ or } 5.0\%$$

$$RD_{s(e)} = \sqrt{(RD_{s(c)})^2 + (RD_{s(a)})^2 + (RD_{s(v)})^2}$$

$$RD_{s(e)} = \sqrt{(0.0065)^2 + (0.025)^2 + (0.050)^2} = 0.0563 \text{ or } 5.63\%$$

$$E = 0.5892 \pm (0.5892)(0.0563) \text{ cpm/dpm}$$

$$E = 0.5892 \pm 0.0331 \text{ cpm/dpm}$$

or

$$E = 58.92 \pm 3.31\%$$

The answer indicates that we can be 68.3% confident that the counting efficiency is between 55.61 and 62.23%. For the 95% confidence level the interval is

$$E = 58.92\% \pm (1.96)(3.31)\%$$

$$E = 58.92\% \pm 6.49\%, \text{ or } 52.43 \text{ to } 65.41\%$$

The uncertainty in the counting efficiency is large—unacceptably large for careful workers. However, an examination of the relative deviations of the individual terms reveals why. The volume error ($RD_{s(v)} = 5\%$) contributes the most to the combined error ($RD_{s(e)} = 5.63\%$), while the contribution of the net counting rate uncertainty ($RD_{s(c)} = 0.65\%$) is almost insignificant by comparison. When terms with errors are multiplied or divided, the combined relative deviation must be higher than the highest individual relative deviation term. Such is not the case when terms are added or subtracted; note that the relative standard deviation for the background measurement is

$$1/\sqrt{124} \text{ or } 9.0\%$$

How could the uncertainty in the counting efficiency be reduced? Obviously, the volume measurement should be addressed first. Perhaps a more accurate pipet could be used, but pipetting small volumes has its limitations. Suppose 1 ml of the [^{14}C]toluene standard is transferred via pipet to a 100-ml volumetric flask. The standard is diluted to 100 ml with scintillation cocktail and then 10 ml of the dilution is transferred to a scintillation vial. Additional cocktail is added to give the final volume used for counting. The specifications of the pipets and volumetric flask are 1.000 ± 0.006 ml, 10.00 ± 0.02 ml, and 100.00 ± 0.08 ml, respectively, so the respective relative deviations are 0.6, 0.2, and 0.08%. The volume (V) of the undiluted standard transferred by this method is

$$V = \frac{(1.000 \pm 0.006 \text{ ml})(10.00 \pm 0.02)}{(100.00 \pm 0.08 \text{ ml})} = 0.1 \text{ ml}$$

and

$$RD_{s(v)} = \sqrt{(0.006)^2 + (0.002)^2 + (0.0008)^2}$$

$$RD_{s(v)} = 0.00637 \text{ or } 0.64\%$$

$$s(v) = 0.1000 (0.0064) = 0.0006 \text{ ml}$$

The combined volume measurement and its associated standard deviation is 0.1000 ± 0.0006 ml. Had this procedure been used in place of the one outlined in problem 13-3, the relative standard deviation of the counting efficiency would have been

$$RD_{s(e)} = \sqrt{(0.0065)^2 + (0.025)^2 + (0.006)^2}$$

$$RD_{s(e)} = 0.0265 \text{ or } 2.65\%$$

The relative deviation in the counting efficiency by this method is about one-half of the previous value, and now the largest source of uncertainty is due to the specific activity of the ^{14}C standard ($RD_{s(a)} = 2.5\%$). Little could be done to reduce the relative standard deviation further without obtaining a higher grade standard.

An important point is illustrated by example problem 13-5; counting errors generally are handled easily—just collect sufficient counts—but accuracies of measurements used to prepare a sample for counting, which all too often are ignored, frequently are the major source of uncertainty in radioactivity assays. Rapid transfer pipetters fulfill certain purposes well, but they have no place in the preparation of counting standards. Nothing less than the best quantitative analytical techniques should be tolerated.

In an earlier section, it was shown that there was no advantage in replicating counting measurements as opposed to using the equivalent length of counting time, but this is true when counting is the only source of error. Other experimental errors are not as easy to evaluate as indicated in example problem 13-5, and frequently the errors in certain steps cannot be ascertained readily. Therefore, there is a place for replicated radioactivity measurements. The combined uncertainty of a multiplestep procedure may be estimated by making a series of measurements in which each possible error contributing step is replicated, then application of Equations 13-2 and 13-3 will give the sample mean value and the standard deviation of the replicate values.

Example Problem 13-6. To measure photosynthesis rates by fixation of CO_2 ($NaHCO_3$), plant leaf discs were incubated with $NaH^{14}CO_3$ for an appropriate period of time, when the tissue was removed, washed, and digested so the $^{14}CO_2$ fixed could be assayed by liquid scintillation counting. The measurement was replicated five times, and all samples were counted for a period long enough to obtain at least 10,000 counts except for background.

The following data were obtained:

Sample	Net counting rate (cpm)
1	3775
2	4194
3	3697
4	4328
5	3986

Calculate the 95% confidence interval for the photosynthetic rate measurement.

Obviously, there are several measurements involved in such a procedure and each step should be performed with as much care as practical. However, a standard deviation well above 2% should be expected when experiments involve biological variations; therefore, the counting measurements were performed satisfactorily, and they do not need to be analyzed separately. Instead, the overall uncertainty will be evaluated.

(X_i)	$(X_i - \overline{X})$	$(X_i - \overline{X})^2$
3,775	−221	48,841
4,194	198	39,204
3,697	−299	89,401
4,328	332	110,224
3,986	−10	100
Total 19,980		287,770

$$\overline{X} = \frac{\Sigma X_i}{N} = \frac{19{,}980}{5} = 3996$$

$$s = \sqrt{\frac{\Sigma(X_i - \overline{X})^2}{N-1}} = \sqrt{\frac{287{,}770}{4}} = 268.2$$

For the sample mean, the 95% confidence interval is

$$\overline{X} \pm \frac{1.96s}{\sqrt{N}}$$

$$3996 \pm \frac{(1.96)(268.2)}{\sqrt{5}} = 3996 \pm 235 \text{ cpm}$$

Consequently, we can be 95% confident that the true value is between 3761 and 4231. The range is large, around ± 5.9%, but the data do not show unusual variation compared to many multistep biological experiments, and the counting errors contribute little to the overall deviation.

PROBLEMS

1. The following counting data were obtained for a sample.

> Sample: 6,924 counts in 10 min
> Background: 972 counts in 40 min

a. Determine the standard deviation of the gross counting rate of the sample.

b. Determine the relative standard deviation of the gross counting rate of the sample.

c. Determine the standard deviation of the background counting rate.

d. Determine the relative standard deviation of the background counting rate.

e. Determine the net counting rate and its associated standard deviation.

f. Calculate the relative 95% deviation for the net counting rate.

2. Verify mathematically that about 67,000 counts are needed for a 1% relative deviation at the 99% confidence level.

3. The following counting data were obtained for two samples.

Sample A: 9762 counts in 4 min
Sample B: 27,072 counts in 4 min
Background: 1161 counts in 40 min

a. Show the net counting rate of sample A (R_A) with its associated standard deviation.

b. Show the net counting rate of sample B (R_B) with its associated standard deviation.

c. Calculate the sum of the net counting rates for samples A and B ($R_A + R_B$) and the 95% deviation for the sum.

d. Calculate the relative 95% deviation for the sum of the counting rates, i.e., the relative deviation of the 95% error of ($R_A + R_B$).

e. Calculate the product of the net counting rates of samples A and B ($R_A \cdot R_B$) and the standard deviation of this product.

4. A solution containing ^{14}C was assayed by liquid scintillation counting. With a Mohr pipet, 1.00 ml of the solution was transferred to a scintillation vial, the appropriate volume of scintillation solution was added, and the sample was counted. The counting efficiency of the counter was known to be 0.741 ± 0.013 cpm/dpm where 0.013 represents the standard deviation of the counting efficiency. A catalog listed the accuracy specifications for the type of pipet used as 1.00 ± 0.02 ml, which means that the standard deviation of the volume delivered by the pipet is 0.02 ml. From the counting data given below, the error due to pipetting and the uncertainty of the counting efficiency, calculate the specific activity of the solution in units of dpm/ml and the standard deviation of this value.

Counting data:

Sample: 3240 ct/40 min
Background: 1480 ct/40 min

5. A radioimmunoassay for corticosterone production by rat adrenal cells was conducted using [^{3}H]corticosterone, appropriate antibodies, and

liquid scintillation counting. A suitable standard curve relating cpm to corticosterone concentration was prepared. Corticosterone production was determined by counting for replicates of a sample and using their mean counting rate, reading corticosterone concentration from the standard curve. Samples were counted for 1 min and background was 31 cpm. The four replicates yielded the following data:

Sample	Net cpm
1	10,801
2	10,372
3	10,035
4	10,640

Calculate the standard deviation of the sample mean, $s(\overline{X})$, and the interval for the mean at the 95% confidence level, using the method for replicated data (Example Problem 13-6).

REFERENCES

Frielander, G., Kennedy, J. W., Macias, E. S., and Miller, J. M. (1981). *Nuclear and Radiochemistry,* 3rd ed. Wiley, New York.

Steel, R. G. D., and Torrie, J. H. (1980). *Principles and Procedures of Statistics,* 2nd ed. McGraw-Hill, New York.

PART III

Biological Effects of Radiation and Radiation Safety

Undeniably, exposure to ionizing radiation constitutes a health hazard, but nuclear technology offers benefits to mankind as well. Although there are gaps in our knowledge concerning the effects of radiation on humans, our standards for protection and safety practices appear to have a sound logical basis that includes assessments of risks versus benefits. In Part III the biological effects of radiation, exposures due to radiation in our environment, protection standards, and radiation safety practices will be discussed. Hopefully, an understanding of these topics will allow persons who may be subject to occupational exposures to form a conclusion about the reasonableness of the rules and regulations that govern the use of radiation and radioactive materials in medical and scientific endeavors.

CHAPTER 14

Biological Effects of Ionizing Radiation

The effect of ionizing radiation on humans has been deduced largely by experiments with animals and observations following "radiation incidents," such as rare nuclear accidents and the two atomic bombs that were dropped on Japan at the close of World War II. The Japanese and United States governments continue to monitor the health of the survivors of the two bomb blasts. It appears that the recent Chernobyl nuclear power plant incident in Russia will be an important source of information concerning the effects of intermediate radiation levels on humans. A wealth of information is not available, but experiments with small animals, tissue cultures, and inanimate materials help provide an understanding of processes that occur at the molecular level and cause the gross effects observed in man. To be sure, extrapolation and deductions are involved.

RADIOLYSIS OF WATER

Since water is the predominate molecular species in living systems, radiolysis of water is an important process in the interactions of ionizing rays and living tissue. Radiolysis of water refers to the primary interactions between radiation and water and the subsequent reactions of the primary products with themselves and surrounding water molecules. There is a probability that all of the types of interactions between radiation and matter discussed in Chapter 5 occur, but the principal interactions for water are the ionization and excitation of water molecules.

The primary ions formed by the ionizing event, H_2O^+ and e^-, even in the absence of an electric field, do not recombine directly because of their reactivity with surrounding water molecules. In a span of about 10^{-8} s, secondary reactions lead to a variety of products: H_2, H_2O_2, OH^-, and H_3O^+, which appear to originate from a series of reactions that involve the *free radicals* of water, hydrogen (H·) and hydroxyl (HO·) radicals. The reactions and sequences are not understood completely, and they vary somewhat depending on the intensity of irradiation, purity of the water, and its pH. Although hydrogen ions may be symbolized as H^+, it is understood that these ions are largely solvated (H_3O^+), and electrons and free radicals may be solvated as well when they participate in the secondary reactions. Two reactions that account for the formation of hydrogen radicals are

$$e^- + H_2O \longrightarrow OH^- + H·$$

$$e^- + H_3O^+ \longrightarrow H_2O + H·$$

Hydroxyl radicals may be formed by other reactions.

$$H_2O^+ + H_2O \longrightarrow H_3O^+ + ·OH$$

$$H· + H_2O_2 \longrightarrow H_2O + ·OH$$

With sufficient energy, excited water molecules conceivably form both types of free radicals.

$$H_2O^* \longrightarrow H· + ·OH$$

Because of their unpaired electrons, free radicals are highly reactive and may form molecular hydrogen, hydrogen peroxide, and water.

$$H· + ·H \longrightarrow H_2$$

$$·OH + ·OH \longrightarrow H_2O_2$$

$$H· + ·OH \longrightarrow H_2O$$

If dissolved oxygen is present, the production of hydrogen peroxide is increased, presumably due to the formation of hydroperoxide radical ($HO_2·$).

$$H· + O_2 \longrightarrow HO_2·$$

$$H· + HO_2· \longrightarrow H_2O_2$$

In aqueous solutions, the free radicals and hydrogen peroxide may react with solute molecules, which extends the range of products. However, the radicals have short lifetimes and will not diffuse nearly as far in solutions as does H_2O_2.

RADIOLYSIS OF ORGANIC COMPOUNDS

Radiochemical reactions (reactions induced by radiation) yield many of the same products as thermochemical and photochemical reactions. This does not mean that the processes are identical, rather that an input of energy leads to similar compounds.

The primary interactions of radiation with organic molecules yield electrons (e^-), ions (M^+), excited molecules (M^*) and free radicals ($M\cdot$). When their kinetic energies are dissipated, electrons may combine with positive ions, but since an electron usually enters a vacant outer orbital, the newly neutralized molecule will be in an electronic excited state. An electron may also combine with a neutral molecule yielding an anion (M^-). Generally, the charge on an organic ion will be neutralized yielding an excited molecule, but prior to neutralization several types of reactions leading to different compounds may occur.

Charge migration within the molecule may induce rearrangements of atoms or functional groups. A charge may be transferred between two molecules, thus molecular ion formation is not restricted to those molecules involved in a primary ionizing event. Ionization also is a form of activation for reactions that may involve the transfer of atoms or groups between molecules.

Deexcitation processes for molecules have been discussed in several of the preceding chapters, and they include the emission of photons, intermolecular transfers of energy, and breakage of chemical bonds, that is decomposition. Excitation may also induce reactions with nonexcited molecules.

Free radicals are particularly reactive. The unpaired electron of a free radical may be transferred to another molecule forming the free radical of a different compound. Because of its unstable state, internal bonds in a free radical may be broken causing molecular decomposition. Two free radicals may combine forming a new compound, or the addition of a radical to a neutral molecule may yield a larger free radical, which may combine with another molecule to create an even larger radical. Continuation of this process leads to the formation of polymers.

When irradiated, essentially all compounds containing C, H, and O yield H_2, CO, CO_2, and CH_4, but the other products vary somewhat. Polymerization is more prevalent in unsaturated aliphatic and aromatic compounds than in saturated ones. Branching is induced in hydrocarbon chains, and products, both larger and smaller than the irradiated variety, are formed. Alcohols are oxidized to aldehydes and ketones and internal bonds next to functional groups are frequently broken. Carboxylic acids are decarboxylated readily forming CO_2 and hydrocarbons. Carbon–halogen bonds are relatively weak and are particularly susceptible to rupture.

Being organic compounds, the molecules that constitute the nonaqueous portion of cells are subject to similar radiolytic reactions, molecular rearrangements, loss of functional groups, disproportionation, and polymerization. In living tissues, an aqueous medium is involved, and the radiolytic products from water provide additional reactive species that may alter the structures of biological compounds by different reactions. The diversity of compounds involved makes the task of determining exact reactions exceedingly complex and inhibits progress in understanding the chemistry of such systems. Nevertheless, the precision in structure–activity relationships among biological molecules indicates that for many compounds a relatively minor modification in structure renders them biologically useless, if not deleterious.

RADIOLYSIS IN CELLS

In cells, which contain 60–70% water, the primary ionization and excitation events are most likely to occur with water molecules. Because of their short lifetimes free radicals will not diffuse far from their sites of origin. However, hydrogen peroxide, a strong oxidizing agent, may diffuse throughout the cell, and react with various functional groups, most notably sulfhydryl groups. By this and similar processes molecular damage may occur at a distance quite removed from the site where primary interactions occurred. This scenario is referred to as an "indirect hit" process.

A "direct hit" process is one in which a biological molecule is damaged by a primary interaction with radiation. Perhaps this could include interactions with the molecules of water immediately surrounding a biological molecule. Whether by a direct or indirect hit process, the impairment of cellular function is dependent on the function of the molecule altered. Damage to molecules that have precise structure activity relationships, such as enzymes, RNA, and DNA, will have a greater effect than damage to intermediates in a metabolic cycle.

Cellular Damage

Cells appear to have some protection against radiolytic processes. For example, catalase is an enzyme that dismutates hydrogen peroxide, and natural antioxidants may protect against free radicals. Whether altered by direct or indirect hits, some types of damaged molecules may be replaced, repaired, or eliminated by an irradiated cell.

It appears that for cell survival, damage to some molecules may be rather innocuous, but for others disastrous. For example, one or even several molecules of a particular enzyme may be damaged, but if the cell has its normal capacity for enzyme turnover, the defective molecules will be replaced and the cell should recover. On the other hand, if the molecule or molecules altered are involved in the control of that enzyme's synthesis or perhaps the genetic code for the enzyme has been altered, then repair may not be possible and cellular death results. Repair mechanisms for DNA are known so even the latter type of damage may not be permanent.

Assuming that among all of the biological compounds in a cell, there is a range of potentials for repair or replacement and a range of dispensabilities for cellular function, the probability for irreversible damage increases with radiation exposure. Certainly, at some higher level of exposure the extent of damage, even if not to a key process, is extensive enough to overwhelm a cell's capacity for repair and survival. Enzymes exhibit a range of sensitivities to radiation, but for complete inactivation doses in the range of millions of Grays (a dose unit to be discussed later, abbreviated as Gy) are required. Dry viruses require 3 to 5 $\times$ 10^3 Gy, bacteria 2 $\times$ 10^2 to 10^4 Gy, and human cells in culture around 1.0 Gy for complete killing.

Although more complex organisms are more sensitive to radiation, destruction of some cells may not cause permanent damage because many tissues are capable of regeneration; blood cells are turned over, and skin cells are replaced continuously.

Mutations caused by nonrepairable alterations of DNA have somewhat different ramifications. Most mutations are lethal and result in cellular death, but alteration of genetic control mechanisms may induce cancer in an organism. Also, if a germinal cell is involved, a mutant gene may be transmitted to succeeding generations; thus aberrations may be expressed as latent biological effects.

Bergonié and Tribondeau (1959) concluded that cells that have greater reproductive activity and are more differentiated show greater sensitivity to X-radiation. The conclusion has some general validity because most readily dividing tissues, such as bone marrow, are more sensitive than nondividing tissue, such as kidney and brain. However, lymphocytes, which do not undergo mitosis, are extremely radiosensitive and hence are a striking exception. Nonetheless, cellular division offers an amplification mechanism since a nonlethal genetic defect in a single cell could yield many identical defective daughter cells.

A number of compounds have been found to protect cell cultures (*in vitro*) against radiation effects. These include compounds with sulfhydryl groups (such as cysteine), free radical scavengers, and compounds that lower the concentration of dissolved oxygen in solutions. However, most are toxic to animals and hence are not practical protectants against radiation. Some compounds increase radiosensitivity when they are present during radiation exposures. Many of these are substances that alter normal nucleic acid metabolism.

The biological effects of radiation cannot be described by a precise set of chemical reactions because of the chemical complexity of living tissue.

However, general mechanisms for cellular damage, cell death, and genetic alterations are apparent. These mechanisms give some insight toward understanding the more complex situation involving whole living animals. Thus, the gross biological effects observed in animals depend on the level and type of radiation, the nature of the tissues exposed, and, at least to some extent, elements of chance.

RADIATION EXPOSURE AND DOSE UNITS

For radiation protection purposes, it is desirable to be able to make physical measurements that can be related to potential biological damage to humans, or better yet, to translate such measurements to quantitative health risks. The latter is a goal of NCRP, but much progress is needed before routine calculations of health risks can be made from physical data. In the meantime, health risks will need to be projected from units that involve physical as well as biological factors.

Unfortunately, the SI system of units has not replaced other systems, therefore knowledge in two systems is necessary to be able to understand current, recent and past literature. Table 14-1 provides a summary of units and general notations for certain measurements that will be used in this chapter.

Exposure Units

In the SI system, the unit for exposure (called the X-unit) is that quantity of X- or γ-radiation that produces in air ions carrying 1 coulomb (C) of charge of either sign per kilogram of dry air at standard temperature and pressure.

$$1 \text{ X-unit} = 1 \text{ C/kg} \qquad (14-1)$$

TABLE 14-1 Radiation Exposure and Dose Units

Type of unit	General symbol	Abbreviation (name)[a]	
		SI	cgs
Exposure	X	(X-unit)	R (roentgen)
Absorbed dose	D	Gy (Gray)	rad (radiation absorbed dose)
Dose equivalent	H	Sv (Sievert)	rem (roentgen equivalent man or mammal)

[a] 1 X-unit = 3876 R, 1.0 Gy = 100 rad, 1.0 Sv = 100 rem

Since the charge of an electron, or a unit positive ion, is 1.6029×10^{-19} C and on the average 34 eV is required to produce an ion pair, the energy dissipated per kilogram of air by 1 X-unit is

$$1 \text{ X-unit} = \frac{(1 \text{ C/kg})(34 \text{ eV/ion pair})}{1.6029 \times 10^{-19} \text{ C/ion pair}} = 2.12 \times 10^{20} \text{ eV/kg} \quad (14\text{-}2)$$

An exposure unit that has received much more common use, historically, is the *roentgen* (R). Originally, it was defined as that quantity of X- or γ-radiation that produces 1 esu (statcoulomb, SC) of charge of either sign in 1.0 cm³ of dry air at 0°C and 760 mm pressure. In equivalent units, $1 \text{ R} = 2.58 \times 10^{-4}$ C/kg of air since 1 cm³ of air weighs 1.2929×10^{-6} kg and $1 \text{ C} = 2.99793 \times 10^9$ SC,

$$1 \text{ R} = \frac{(1.0 \text{ SC/cm}^3)(1 \text{ C/2.9979} \times 10^9 \text{ SC})}{1.2929 \times 10^{-6} \text{ kg/cm}^3}$$

$$1 \text{ R} = 2.58 \times 10^{-4} \text{ C/kg} \quad (14\text{-}3)$$

The relationship between the X-unit and the roentgen is

$$1 \text{ X-unit} = (1 \text{ C/kg})(1 \text{ R/2.58} \times 10^{-4} \text{ C/kg})$$

$$1 \text{ X-unit} = 3876 \text{ R} \quad (14\text{-}4)$$

The amount of energy dissipated in 1 g of air by 1 R is

$$\frac{(2.58 \times 10^{-4} \text{ C/kg})(1 \text{ kg/10}^3 \text{ g})(34 \text{ eV/ion pair})}{1.6029 \times 10^{-19} \text{ C/ion pair}} = 5.47 \times 10^{13} \text{ eV/g}$$

$$(5.47 \times 10^{13} \text{ eV/g})(1.602 \times 10^{-12} \text{ erg/eV}) = 87.7 \text{ erg}$$

Although the amount of energy absorbed by air has been calculated for the exposure units, it is important to keep in mind that the basis for these units is the extent of ionization in air.

Absorbed Dose Units

Units based on the amount of energy absorbed are applicable for all types of rays and absorbers, not just air. However, for a given flux of rays, the amount of energy absorbed will depend on the type of rays, their energies, and the nature of the absorber. The symbol "D" will be used to indicate an absorbed dose, and it is the amount of energy absorbed per unit weight of the absorber. Appropriate units for D are the *Gray* and *rad*.

The SI unit for an absorbed dose is the Gray (Gy), which is defined as that quantity of any type of radiation that causes the absorption of 1 joule (J) of energy per kilogram of the absorber.

$$1 \text{ Gy} = 1 \text{ J/kg} \quad (14\text{-}5)$$

A radiation absorbed dose (rad) is that quantity of any type of radiation that causes the absorption of 100 erg of energy per gram of the absorber.

$$1 \text{ rad } = 100 \text{ erg/g} \tag{14-6}$$

Since 1 J is equivalent to 10^7 erg, the relationship between the rad and the Gy is

$$1 \text{ Gy } = 100 \text{ rad} \tag{14-7}$$

Relationship Between Exposure and Dose

The definition of exposure is based on the absorption of radiation by air, whereas dose definitions apply to all types of absorbers. However, for radiological protection, living tissue is of primary concern, and because of its greater density, tissue absorbs more energy per unit mass than does air. The extent of absorption in any absorber is dependent on the energies of the rays as well as the nature of the absorber, but the absorption coefficients for most soft tissues (not bone) are nearly equivalent for ray energies commonly encountered. The same can be said concerning ray absorption in air.

In tissue, 1 R of X- or γ-radiation dissipates about 96 erg/g, similarly about 34 J/kg is absorbed from 1 X-unit; therefore as an approximation:

$$1 \text{ R } \approx 0.96 \text{ rad } = 0.0096 \text{ Gy}$$

$$1 \text{ X-unit } \approx 34 \text{ Gy } = 3400 \text{ rad}$$

Because 1 R is approximately equivalent to 1 rad for X- or γ-radiation, these units are used interchangeably in many cases. Technically, such a practice is incorrect, but generally it is justified when the precision of measurements for radiation fields and biological effects is considered.

Absorbed Dose Rates

The biological damage created by the absorption of radiation is quite dependent on the intensity of energy deposition. Rates of exposure or dose rates are important factors in predicting potential biological effects. Therefore, dose rate units, such as rad/hr and Gy/da, are encountered frequently.

Biological Factors

The intensity of energy deposition has effects at microscopic levels because the damage created by rays or particles that lose their energies within a short distance is greater than those that lose an equivalent amount of energy over greater distances. These factors are correlated with the *Relative Biological Effectiveness* (RBE) of different types of radiation. The RBE depends on the type and energy of a particle or ray, the specific biological effect being considered, the species, and the specific experimental conditions. The term is useful in radiation biology, but for radiation protection

TABLE 14-2   Linear Energy Transfer (LET) Values and Corresponding Quality Factors (QF)

LET (keV/nm)	QF
3.5 or less	1
3.5–7.0	1–2
7.0–23.0	2–5
23.0–53.0	5–10
53.0–175.0	10–20

other factors are used and these are based on conservative upper limits for RBE values.

LET and QF Values.   *Quality factors* (QF) are used in the estimation of potential biological damage resulting from doses, and QF values have been correlated with *linear energy transfer* (LET) values for radiation as shown in Table 14-2. LET values are expressed in units of energy dissipated per unit pathlength, such as keV/nm.

The value of the LET depends on the type of particle or ray being considered and its energy. As discussed in Chapter 5, α-particles produce many more ions per unit pathlength than β-particles; therefore, the LET values and hence QF values are higher for α-particles. Also recall that charged particles cause more localization of ionization as their velocities decrease, therefore the dissipation of energy per unit pathlength is greater for a charged particle with a low velocity than the same type of particle with a higher velocity. As far as biological damage is concerned, the potential biological damage from 1 Gy of α-radiation is about equivalent to 20 Gy of γ-radiation. This is due to the differences in the LET values of the two types of radiation, and these differences are expressed in terms of QF values. Theoretically, a continuum of QF values exists, but for practical use, values have been assigned to various classes of radiation as shown in Table 14-3.

TABLE 14-3   Quality Factor (QF) Values Assigned to Various Classifications of Radiation

Radiation	QF
X-rays, γ-rays, negatrons, and positrons	1
Thermal neutrons[a]	5
Fast neutrons,[a] protons, α-particles, and heavy ions	20

[a]Thermal neutrons are low-velocity neutrons that have energies comparable to the energies of surrounding molecules or atoms (0.02–0.03 eV). In reactors, nuclear fission yields fast neutrons (in the MeV range), but their energies are reduced to thermal levels by passing through moderators such as graphite, water, and heavy water (D_2O).

Dose Equivalent

To relate relative biological damage to doses, dose equivalent units have been formulated. The general symbol "H" will be used to indicate dose equivalents.

$$H = (D)(QF)(DF) \tag{14-8}$$

The *distribution factor* (DF) has a value of one if the radiation field is perfectly uniform. For external fluxes it generally will be one, but for internal doses it may vary depending on the chemical nature of the radionuclide. For example, an internal dose of ^{131}I will not be uniformly distributed in the human body, instead it will be concentrated in the thyroid gland, and, consequently, the dose to the thyroid is much greater than that in other organs.

The *Sievert* (Sv) is an SI unit for dose equivalent (H) based on an absorbed dose expressed in Gy.

$$Sv = (Gy)(QF)(DF) \tag{14-9}$$

In the cgs system, the dose equivalent is expressed in units of *roentgen equivalent man* (rem) or *roentgen equivalent mammal,* which is related to the absorbed dose in units of rads.

$$rem = (rad)(QF)(DF) \tag{14-10}$$

The same quality and distribution factors discussed above apply to both Equations 14-9 and 14-10. Thus, a dose of 2 mGy from a uniform flux of fast neutrons would be a dose equivalent of 40 mSv according to Table 14-3.

$$H = (2 \text{ mGy})(20)(1) = 40 \text{ mSv}$$

DOSE–RESPONSE CURVES

Dose–response curves are simply graphs relating the magnitude of some biological response to different doses of radiation. The kinds of biological response or effects measured will depend on the nature of the experiment, but some examples are the number of animals that die after exposure, number of animals that develop tumors, frequency of cell transformations in tissue cultures, and capability of cell cultures to perform a given metabolic reaction.

Stochastic Effects

Stochastic effects are those governed by chance and as such are quite amenable to statistical analysis. A dose–response curve involving stochastic effects only should intersect at the origin of the graph axes, and the mag-

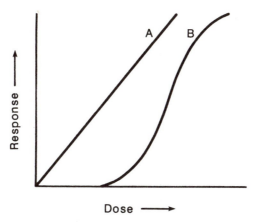

FIG. 14-1 Idealized dose–response curves for radiation exposures. A, stochastic effects; B, nonstochastic effects.

nitude of the response should be proportional to the dose (curve A in Figure 14-1). Thus, even for small doses there is some probability of observing effects. Cellular or organismal responses that do not involve repair mechanisms might be expected to exhibit such a response curve.

Nonstochastic Effects

Dose–response curves for nonstochastic effects have thresholds; that is, the dose must exceed a certain minimum value before any effects are observed. The magnitude of the response increases with the dose and sigmoid-shaped curves are observed. An idealized curve for nonstochastic effects is shown as curve B in Figure 14-1. One might expect single-cell killing rates to exhibit nonstochastic curves because of the possibility of cellular repair mechanisms.

The shapes of actual dose–response curves vary with the kind of biological effect being measured, the kind of radiation, and period of administration. In Figure 14-2, the fraction of surviving cells in cultures is shown as a function of radiation dose. Although the response axis is inverted compared with that in Figure 14-1, it is apparent that the shape of a response curve is affected by the nature of the radiation exposure. For X-rays administered at a low dose rate, and for neutron radiation, the survival curves appear to be linear; however, acute doses of X-rays produce a curvilinear response. Most radiation dose–response curves do not conform to either of the idealized shapes shown in Figure 14-1, but frequently thresholds seem to be present. Also, general forms are observed; for example, most curves relating dose to the induction of some form of cancer are similar to that of Figure 14-3, which shows the incidence of leukemia in mice exposed to various levels of X-radiation. At higher doses, the incidence of cancer

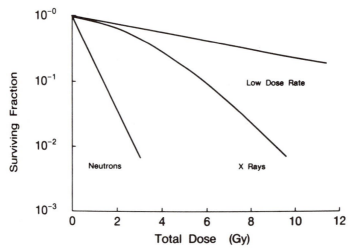

FIG. 14-2 Survival curves for cells in culture as a function of radiation dose. (From Rojas, A., and Denekamp, J. *Experientia* 45:43, 1989. Reprinted by permission of Birkäuser Verlag AG, Basel, Switzerland)

decreases because the radiation causes death by other mechanisms before cancers develop.

Compared to the idealized dose–response curves of Figure 14-1, another significant difference is apparent in the leukemia response curve (Figure 14-3); at zero dose the response is not zero. Of course, this reflects the natural incidence of leukemia, the incidence due to all other causes of this

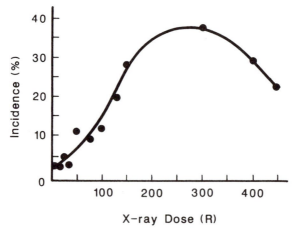

FIG. 14-3 Incidence of myeloid leukemia in RF male mice exposed to whole-body X-irradiation. (With permission from Upton, A. C. *Cancer Res.* 21:717, 1961, and The American Association for Cancer Research)

particular type of leukemia. The data points of the curve suggest there is a threshold above the natural incidence, but statistically that is difficult to prove or disprove in experiments of this type. The variability among individuals within "identical" groups confers a degree of uncertainty about the natural incidence and when coupled with the uncertainty of measuring a small response, a high degree of confidence cannot be placed on small apparent differences. Fortunately, in animal experiments larger numbers of animals may be used to improve the statistical reliability of data, but for human responses most of the data consists of observations following accidental exposures.

The reason for the apparent preoccupation with the low dose portion of radiation response curves is that the effect of low level radiation on humans is one area of knowledge that is severely lacking. What are the health risks? Obviously animal research is useful in projecting probable effects in man, but there is an amazing variation in the ability of animal species to withstand radiation exposure.

RADIATION EFFECTS IN ANIMALS

Acute Effects

Acute effects are those for which symptoms arise shortly after an exposure, perhaps within a few hours to several weeks; death is one example. The $^{30}LD_{50}$ is the dose that causes 50% of the irradiated animals to die within 30 days, but sometimes LD_{50} is used without specifying a 30-day period. For unknown reasons, there are wide differences in the LD_{50} values among animal species (Table 14-4), and, in general, the LD_{50} values show an inverse relationship with evolutionary development. The LD_{50} value for

TABLE 14-4   Radiation LD_{50} Values for Different Animal Species

Species	LD_{50} (Gy)[a]	Species	LD_{50} (Gy)
Sheep	1.55	Monkey	3.98
Burro	1.55	Rabbit	8.40
Swine	1.95	Mouse	9.00
Marmoset	2.00	Rat	9.00
Goat	2.30	Hampster	9.00
Guinea pig	2.55	Gerbil	10.59
Dog	2.65	Desert mouse	15.20
Man	2.25–2.70		

[a] 1 Gy = 100 rad

Source: Adapted from Bond (1969).

man is not clearly established, but it is probably in the range of 2.0–4.5 Gy (200–450 rad) for a single dose to the whole body. Below 1.0 Gy, survival is almost certain and probable up to 2.0 Gy. Survival is possible in the range of 2.0–4.5 Gy and virtually impossible above 5.0 Gy (Prasad, 1984, p. 99).

Whole-body exposure to large amounts of radiation causes radiation sickness, which is characterized by many of the following symptoms: nausea, vomiting, diarrhea, fatigue, fever, loss of hair, blood changes, lethargy, and convulsions. The radiation syndrome has been divided into three types: (1) the hemopoietic (bone marrow) syndrome, (2) the gastrointestinal syndrome, and (3) the central nervous system syndrome.

Most of the symptoms of the hemopoietic syndrome are due to bone marrow damage. The concentration of the white blood cells formed in marrow, called granulocytes, increases sharply for a day or so after exposure, then decreases for several weeks. Lymphocytes drop sharply and remain depressed for several months after exposure. Erythrocyte levels do not change much for about a week, then they decrease to a minimum somewhere between 1 and 2 months. The blood platelet count decreases for about a month and then recovers slowly over a period of several months. These blood changes make the body quite susceptible to infections, and death may occur 3 weeks to 2 months after exposure. The ability of the bone marrow to recover is a critical factor in survival, and bone marrow transplants increase survival rates. The threshold for symptoms of this syndrome is about 1.0 Gy and for death due to this syndrome about 2.0 Gy.

In the gastrointestinal syndrome, the primary cause of symptoms is the loss of the epithelial tissue in the small intestine. Nausea and vomiting occur within a few hours after exposure and diarrhea will occur somewhat later leading to dehydration. Since the threshold for this syndrome is about 5.0 Gy, the symptoms of the hemopoietic syndrome are observed also. Death usually occurs within 3 days to 2 weeks.

The brain is the primary organ affected in the central nervous system syndrome, and symptoms include lethargy and convulsions. About 20.0 Gy is necessary to cause the symptoms of this syndrome, and death occurs within 2 days.

There are some other acute effects that do not fall into the three syndromes. Skin erytherma (reddening of the skin) occurs after doses of about 3 Gy or more. Blistering, pigment changes, loss of hair, and necrosis due to infection are other possible skin effects.

The gonads are sensitive to radiation and may be irradiated without exposing the whole body. Permanent sterility is caused in males by 5.0–6.0 Gy and in females by 3.2–6.0 Gy. Doses of about 2.5 Gy in males and 1.7 Gy in females may cause temporary sterility with recovery in 1 to 3 years.

Eyes are radiosensitive, particularly the lens. Immediate as well as long-term effects are observed, including the induction of cataracts.

Latent Effects

Latent effects are those that appear some time, perhaps years, after an exposure. The dose does not have to be sufficiently high to cause acute symptoms; in fact, latent effects may appear when a person is exposed to low levels of radiation over a period of years. Cataracts have already been mentioned as a long-term effect, and life shortening due to processes that appear to be similar to aging occurs. Radiation-induced cancers of most organs have been observed, but leukemia and lung and bone cancers predominate.

The incidence of leukemia among the atomic bomb survivors in Hiroshima and Nagasaki began to exceed normal frequencies 3 to 4 years after the bombings and continues to be elevated. The threshold for an increased incidence of leukemia appears to be about 0.2 Gy.

The Beir report (1980) provided *estimates* for the excess mortality rate due to all forms of cancer. For a single dose of 0.1 Gy, the excess cases as a percentage of the normal rate was estimated to be 0.47%, and for a continuous dose rate of 0.01 Gy/yr and an accumulated dose of 0.75 Gy, the estimate was 2.8%.

Heritable effects have been observed for some lower forms of life, such as microorganisms, insects, plants, and mice. Irradiation of seeds has produced superior performing plants, but the induced mutations that were deleterious to individual seeds greatly exceeded those that were beneficial by several orders of magnitude. Probably, radiation has played an important role in evolution.

Highly reliable evidence concerning genetic effects in humans is not available. However, among the 35,000 children born to survivors of the Hiroshima and Nagasaki bomb blasts, there is no statistically significant difference in the incidence of infant abnormalities compared to the nonirradiated population. The doses may have been as high as 3.0 Sv (300 rem), but the average was around 0.25–0.35 Sv (Behling and Hildebrand, 1986). This suggests that the human threshold for heritable effects may be greater than 250–350 mGy, but the incidence rate may be just too small to be statistically significant with the number of subjects involved.

The above study involved effects on germinal tissue only, not fetuses. Compared to an adult, the human fetus is quite sensitive to low-dose radiation, and teratogenic effects are observed. Congenital defects include leukemia and mental retardation, which appears more likely to be induced when the dose is received during week 8 to 15 of pregnancy. Animal studies have shown congenital effects for doses as low as 50–100 mSv, but in humans less definitive data are available. According to the children born to women pregnant at the time of the bombing of Hiroshima and Nagasaki, it appears that about 250 mSv are required for a statistically significant increase in abnormalities. Smaller head size is an example of a teratogenic effect noted in these studies.

Two multiauthor reviews of the radiation effects on man and animals have been presented recently, one in *Experientia* (Fritz-Niggli, 1989), and a second that deals with the effects of low-level exposures (Jones and Southwood, 1987). A more detailed treatment of various subjects discussed here are given in those reviews.

FACTORS AFFECTING HEALTH RISKS

A single instantaneous exposure constitutes a greater health risk than an equivalent exposure administered at a low rate over a long period of time. Presumably this is largely due to body repair mechanisms. An acute effect such as radiation sickness requires a fairly high exposure (1.0 Gy) administered over a period of a few minutes or hours, but such an effect would never be observed if the same total exposure were administered evenly over a period of years. Thus, the total exposure level without an indication of the time over which the exposure was received can be quite misleading. For this reason radiation exposures or doses are frequently expressed as rates, such as R/hr, Gy/da, and Sv/yr. Rates are particularly important in assessing the probability of acute effects, but latent effects are better correlated with accumulated total dose.

The doses causing acute effects generally are reported as "whole-body" exposures, but effects, acute and latent, are different when irradiation is localized. For example, in treatment of cancer patients, doses in the range of 10–60 Gy may be given. At first glance, it appears that humans could not survive such treatments since the LD_{50} is much lower. However, the radiation is focused in the cancerous area, which is a small volume of the whole body. Although much of the immediately surrounding tissue receives a significant dose during the ablation of the cancerous tissue, the remainder of the body receives a much lower exposure. Also, the exposure may be administered at a rate of 2–3 Gy/day. Generally, mild radiation sickness accompanies radiation therapy.

Different tissues or organs exhibit different sensitivities to irradiation, and certain organs are more critical than others in maintaining life and reproductive processes. Red bone marrow is radiosensitive, and since it is responsible for replenishing formed elements in the blood, it is quite essential for life. For comparison, the risk of latent effects (primarily leukemia and bone cancer) arising from irradiation of red bone marrow is about four times higher than that from bone surfaces when whole-body exposures are involved. Other tissues that receive particular attention in assessing risks are the gonads, breast, lung, and thyroid.

For a worker engaged in an occupation where radiation exposure is possible, the greatest concern is the probability of suffering latent effects. Except for extremely rare nuclear accidents, occupational exposures are well below the threshold for acute effects. To assess risks from low doses of radiation the ICRP and NCRP have developed the concept called *Effective Dose Equivalent* (NCRP, 1987). The effective dose equivalent (H_E) takes into account the different mortality risks from cancer and severe hereditary effects associated with irradiation of various organs and tissues. Without considering the risks associated with the different sensitivities of various tissues, the dose equivalent to the whole body (H) is the sum of the dose equivalents (H_T) for all individual organs or tissues.

$$H = \Sigma\, H_T$$

or

$$H = H_{gonads} + H_{breast} + H_{lung} + H_{bone\ marrow} + etc.$$

However, weighting factors (W_T) have been assigned for various tissues and these represent the proportional risks associated with those tissues when the body is irradiated uniformly. The effective dose equivalent (H_E) is the sum of all tissues' dose equivalents multiplied by their respective weighting factors.

$$H_E = \Sigma(W_T \cdot H_T) \tag{14-11}$$

Table 14-5 gives the weighting factors for several tissues. These are current values that may change in the future. Nevertheless, the effective dose equivalent concept can be illustrated by the following.

Suppose a person has breathed natural radon gas and it has been determined that the dose equivalent to the lung is 10 μSv. Since this is the only tissue that receives exposure, the effective dose equivalent is easily calculated.

$$H_E = 0.12(10\ \mu Sv) = 1.2\ \mu Sv$$

This calculation indicates that a 10-μSv dose to the lungs only is a health risk equivalent to a uniform dose of 1.2 μSv to the whole body.

TABLE 14-5 Weighting Factors for Calculating Effective Dose Equivalents

Tissue	W_T
Gonads	0.25
Breast	0.15
Red bone marrow	0.12
Lung	0.12
Thyroid	0.03
Bone surfaces	0.03
Remainder	0.30
Total	1.00

If the quality of radiation is the same for all tissues, the total amount of radiation absorbed from a 10-μSv dose to the lungs is considerably less than a 1.2-μSv dose to the whole body, as shown by the following calculation. Assume that the quality factor is 1.0 so that 1 Sv = 1 Gy, which corresponds to the absorption of 1 J/kg of tissue. Also assume the body and lung weights are that of the "standard man," 70 and 1.0 kg, respectively.

$$\text{Whole body:} \quad 1.2 \ \mu\text{Sv} = (1.2 \ \mu\text{J/kg})(70 \ \text{kg}) = 84 \ \mu\text{J}$$

$$\text{Lungs:} \quad 10 \ \mu\text{Sv} = (10 \ \mu\text{J/kg})(1.0 \ \text{kg}) = 10 \ \mu\text{J}$$

Restating the situation in this example: if each of the tissues listed in Table 14-5 receives a dose of 1.2 μSv (the case of whole-body exposure), there will be a greater risk of latent effects appearing in each of these tissues. These combined risks are equivalent to the single risk resulting from a 10-μSv dose to lung tissues only.

CONCLUSION

Although the chemical and biological mechanisms for radiological damage are not understood clearly, ionizing radiation poses a life-shortening threat to humans. However, as we will see in the succeeding chapter, it is impossible to escape exposure entirely, but even elevated exposures, such as those that may be associated with particular occupations, do not involve unusual health risks. In fact, when guidelines are followed properly it appears that the nuclear industry is among the safest industries, certainly safer than the transportation, construction, agriculture, and mining industries.

Recall that according to the Beir report (1980), a very small, but statistically detectable increase in the probability of cancer is predicted for a single dose of 100 mGy (1 rad) or an annual dose of 10 mGy (0.1 rad) over a period of 75 years. Also recall, when the quality factor for the radiation is one, Gy = Sv and rad = rem. These figures may serve as benchmarks when discussing levels of background radiation and exposure limits for the public or radiation workers.

REFERENCES

Behling, U. H., and Hildebrand, J. E. (1986). *Radiation and Health Effects*. Report on the TMI-2 Accident and Related Health Studies. GPU Nuclear Corporation, Middletown, PA.

Beir, (1980). *Committee on the Biological Effects of Ionizations*. National Academy of Sciences, Washington, DC.

Bergonié, J., and Tribondeau, L. (1959). *Radiat. Res.* 11:587.

Bond, V. P. (1969). *Comparative Cellular and Species Radiosensitivity.* V. P. Bond and T. Sagahara, Eds. Igaku Shoin, Tokyo.

Fritz-Niggli, H. (1989). *Experentia* 45:1.

Jones, R. R., and Southwood, R. (1987). *Radiation and Health: The Biological Effects of Low-Level Exposure to Ionizing Radiation.* Wiley, New York.

NCRP (1987). *Recommendations on Limits for Exposure to Ionizing Radiation.* NCRP Report No. 91, National Council on Radiation Protection and Measurements, Bethesda, MD.

Prasad, K. N. (1984). *Handbook of Radiobiology.* CRC Press, Boca Raton, FL.

CHAPTER 15

Radiation Protection Standards and Dose Calculations

SOURCES OF IONIZING RADIATION

Since this chapter deals with standards for protection against ionizing radiation, it is appropriate to consider the sources of radiation to which the general public is subjected. First, there is the natural radiation in our environment, background radiation, and second, man-made sources, the principal component of which is medical X-rays.

Natural Background Radiation

The primary components of background radiation are cosmic rays, which originate from outer space; radiation from terrestrial sources, the natural radioactivity in the earth's crust; and radiation from internal sources, the naturally occurring radionuclides in the body.

Cosmic Radiation.   *Cosmic rays,* which arise from outer space, are a large source of background radiation in most areas of the world. Fortunately, our earth's atmosphere acts as a shield, but in addition to absorbing some of the radiation, it changes the nature of the radiation. Most of the *primary cosmic rays* that strike the outermost atmospheric levels are protons; helium nuclei constitute about 15%, electrons maybe 1%, and nuclei of heavier elements make up the remainder. The energies of the primary rays are very high, 20 to > 10,000 MeV, with an average of about 500 MeV per proton! The sun contributes significantly to the flux, but some rays are believed to originate in other galaxies.

At the top of the atmosphere, the primary rays interact with gaseous atoms, and these interactions involve very high energy nuclear reactions that result in the emission of many mesons and nucleons (n and p^+) that undergo subsequent nuclear reactions with other atmospheric atoms. The rays arising from these nuclear reactions are referred to as *secondary cosmic rays,* and they consist of very energetic photons, electrons, positrons (the soft component), and Mu mesons (the major type of rays in the hard component). The electromagnetic spectrum presented earlier (Figure 2-1) included "secondary cosmic rays" as a radiation type. This type refers to the photons of the soft component, and they, relative to γ-rays, have shorter wavelengths and much higher quantum energies.

The hard component penetrates to large depths in the earth's surface, whereas the soft component is readily absorbed by the atmosphere and almost totally absorbed by the outer few inches of the earth's crust. Background radiation levels are correlated with altitude because the extent of absorption of the soft component is dependent on the thickness of atmosphere involved. Cosmic radiation nearly doubles with an increase in altitude of 2000 m, and at supersonic transport altitudes (60,000 ft) the level is about 100 times higher than at sea level. The approximate annual dose equivalent at sea level is 240 μSv, at Denver, CO (elevation 1600 m) about 500 μSv, and at Leadville, CO (elevation 3200 m) about 1250 μSv. Airplane flights increase exposures; for a 1-hr flight at 39,000 ft a dose equivalent of about 5 μSv is received. A building, such as a home, provides some shielding, perhaps reducing exposure by 10%. Therefore, being indoors or outdoors has an effect on the exposure rate.

The NCRP (1987b) has estimated that the range of dose equivalents to the U.S. population is 150–5000 μSv, and considering location of residence, percentage of time spent outdoors, and some air travel time, they arrived at an average annual effective dose equivalent (H_E) of 270 μSv per person.

Terrestrial Sources.   Terrestrial sources of radioactivity contribute significantly to background radiation. All elements with atomic numbers above bismuth (83) have naturally occurring radioisotopes that belong to one of three decay series. These series start with either ^{238}U, ^{235}U, or ^{232}Th and undergo successive α- and β-decays eventually ending in the formation of one of the stable isotopes of lead (^{206}Pb, ^{207}Pb, and ^{208}Pb). In addition to members of these decay series, there are naturally occurring radionuclides for quite a few of the elements with atomic numbers below 83. Consequently, most mineral matter has a low level of natural radioactivity. The presence of higher concentrations of minerals containing members of the decay series, uranium for example, has a significant effect on the level of background radiation from terrestrial sources. Because of variations in mineral content at different geographic locations, the absorbed dose rate varies widely; it is about 230 μGy/yr in coastal plain areas, 460 μGy/yr in noncoastal plain areas, and 900 μGy/yr in the Colorado plateau. Other factors that affect exposure rates include the material used in home construction (bricks contain more radioactivity than wood) and time spent outdoors.

From this source of background radiation, the average annual effective dose equivalent (H_E) for the population of the United States and Canada is estimated to be about 280 μSv.

Although the majority of the inhaled radioactivity is from terrestrial sources, NCRP treats it separately. Most of the exposure is from "radon" (^{222}Rn, and its short-lived decay products). Radon is "exhaled" by the soil and arises from one of the natural decay series mentioned earlier. For that reason, the concentration of radon in outdoor air reflects the mineral content of the soil, and, as expected, it is variable. The average concentration in Colorado Springs is about 44 dps/m^3 and 4–5 dps/m^3 in New York.

The concentration of radon in indoor air has received much public attention recently, and the reason for concern is the observations that indoor concentrations frequently exceed outdoor concentrations markedly. Outdoors, radon is rapidly diluted as it is exhaled from the soil, but in a basement with little air exchange, dilution occurs slowly. The problem is reduced by increasing the air pressure against the surfaces of surrounding soil. Boosting ventilation rates may exchange air more rapidly, but can be counterproductive by allowing the soil to exhale more radon. Gaseous radon does not deposit in the lungs, and its contribution to the dose is small compared to its short-lived decay products (^{218}Po, ^{214}Po) that do deposit.

Among background sources, the inhaled component probably exhibits the greatest variation, however, the estimate of an average effective dose equivalent (H_E) to the population of the United States and Canada is 2.0 mSv/yr. Essentially all of this radiation exposure is to the lung tissue, and because of the risk of latent effects (cancer) the weighting factor (Table 14-5) for this tissue is about 10 times higher than its fraction of the body weight.

Internally Deposited Radionuclides. A number of radionuclides are deposited in the body, and their decay constitutes a significant source of background radiation. Some radionuclides are derived from the earth's crust and include members of the natural decay series, but quantitatively, ^{40}K, which is a nonseries nuclide, is the most important. Cosmogenic radionuclides are those formed by the interactions of cosmic rays with atoms in the atmosphere and earth surface. The four cosmological radionuclides that contribute most to exposures of humans are ^{14}C, ^{3}H, ^{22}Na, and ^{7}Be.

An example of cosmological formation is the production of ^{14}C from ^{14}N. An energetic neutron, generated by cosmic ray interactions, may strike and be absorbed by a ^{14}N nucleus that subsequently emits a proton yielding an atom of ^{14}C.

$$n + {}^{14}N \longrightarrow p^+ + {}^{14}C$$

The ^{14}C will be converted to atmospheric $^{14}CO_2$, which is incorporated into plants via photosynthesis and hence enters the food chain.

For the carbon in the biosphere, it appears that the production of ^{14}C is in equilibrium with its decay. This yields a constant specific activity of about 14 dpm/g of carbon. If carbonaceous material is prevented from recycling in the biosphere its activity will decrease with time, and this is the basis of Libby's (1955) carbon-14 dating method.

Most of the internally deposited radionuclides are taken into the body by the diet and drinking water, but some are inhaled. Because of uniform distributions, diet has little effect on ^{40}K, ^{14}C, and ^3H concentrations in the body, but concentrations of most members of the decay series do depend on food and water contents. The average effective dose equivalent for the population from internally deposited radionuclides is about 410 μSv/yr.

Total Background Radiation.   There is much variation in the natural background dose individuals receive. The geographic location of a person's residence has the largest effect on dose, but one's life-style has an effect as well. The total average dose for the population may be calculated by summing the four components cited above. In round numbers the total effective dose equivalent is about 3.0 mSv/yr, and radon is the largest single component when the linear, nonthreshold theory of doses is assumed.

Man-Made Exposures

Most of the exposure to the public by man-made sources is due to medical diagnosis and treatment. Occupational exposures, those to workers engaged in jobs that involve radiation, contribute to this category. Other components include consumer products, nuclear power, and fall-out.

X-Ray technology has improved dramatically over the past 25 yr, and exposure per film has decreased concomitantly. Although a significant component of the average annual dose, it has been reduced to about 18% of the total from all sources. The NCRP (1987c) estimates that the average effective dose equivalent (H_E) from medical X-rays is 390 μSv/yr and from nuclear medicine 140 μSv/yr, for a total medical component of 530 μSv/yr.

Political pressure has reduced the growth of the nuclear power industry, but the contribution due to the fuel cycle is only about 0.5 μSv/yr.

Consumer products and other sources include X-rays from airport luggage inspections, radioactive materials in luminous products, glass and ceramics containing uranium and thorium, tobacco products, and thorium-containing products, such as gas lantern mantles, camera lenses, and fluorescent lamp starters. The range of the annual effective dose equivalent for this component is 50–130 μSv.

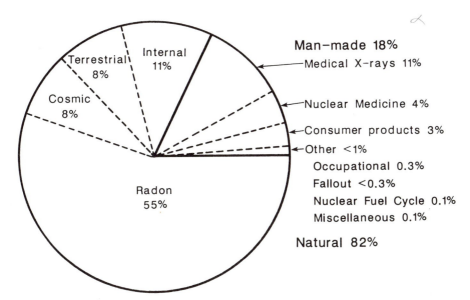

FIG. 15-1 The percentage contribution of various radiation sources to the total average effective dose equivalent in the U.S. population. (From NCRP, 1987c)

Fall-out has become a negligible exposure component in the United States.

Much more will be discussed about occupational exposures later, but the average effective dose equivalent for this category is 9 μSv/yr per person. Note, this is not the average dose for persons engaged in fields involving radiation, instead it is the combined total dose of all workers averaged over the whole population.

The average effective dose equivalent from all man-made sources is approximately 0.6 mSv. When the doses from natural and man-made sources are totaled and rounded, the average annual effective dose equivalent (H_E) is 3.6 mSv per person in the United States. The components of the average total annual dose are shown diagrammatically in Figure 15-1.

OCCUPATIONAL EXPOSURE LIMITS

The judicable question that must be addressed in establishing limits for exposures to man-made sources, or those sources mankind can control, is one of risk versus benefit. The benefits of medical X-rays are easily understood by individuals, consequently the small risks involved are accepted widely by the public. Nevertheless, radiation is not used indiscriminately by the health care professionals; instead approved guidelines, procedures and protocols must be followed when radiation exposures are administered. It appears there is little public controversy concerning the conclusion that health benefits more than compensate for possible risks when approved practices are followed.

The use of radiation benefits society in other ways as well; the use of radiotracers in medical research undoubtedly has contributed to longer life expectancies, the use of radionuclides in research has increased our capacity to feed the human race, and nuclear power as a substitute for fossil fuel may prolong our standard of living. Of course, individuals engaged in delivering health care (X-ray technicians, physicians, and nurses), generating nuclear power (mining, processing fuel, operating plants), and performing research involving radiation are likely to receive doses above background and usual medical levels. The benefits to be gained from such activities, except in the medical area, are not easily evaluated, and personal values frequently are important criteria in individual conclusions. Hence, controversy about the use of man-made radiation abounds. The position of NCRP in recommending occupational limits avoids the controversial benefit question.

The recommendations on occupational limits for exposure to ionizing radiation by the National Council on Radiation Protection and Measurements (NCRP) are based on the premise that health risks associated with occupations involving radiation should be no greater than those of "safe" occupations. Table 15-1 provides data relating the annual fatality rate among various occupations, and NCRP (1987a) defines "safe" industries as those with an average annual risk factor of or below 10^{-4} (1 fatality per 10^4 workers). NCRP's limits for occupational exposure appear to be "safe" because among Department of Energy (DOE) workers exposed to ionizing radiation, the annual fatality rate is about 2.5×10^{-5}.

Table 15-2 provides NCRP's recommended limits for occupational exposures. Recall that the effective dose equivalent, whether to one or sev-

TABLE 15-1 Annual Fatality Rates from Accidents in Different Occupations

Occupation	Number of workers 10^{-6}	Annual fatal accident rate per 10^4 workers
Trade	24	0.5
Manufacturing	19.9	0.6
Service	28.9	0.7
Government	15.9	0.9
Transportation and utilities	5.5	2.7
Construction	5.7	3.9
Agriculture	3.4	4.6
Mining, quarrying	1.0	6.0
All industries (U.S.)	104.3	1.1

Source: From NCRP (1987a).

TABLE 15-2 Recommendations by the National Council on Radiation Protection and Measurements for Exposure Limits

	Limits	
Occupational		
Effective does equivalent	50 mSv/yr	5 rem/yr
Dose equivalent for		
Lens of eye	150 mSv/yr	15 rem/yr
All other organs and tissues	500 mSv/yr	50 rem/yr
Cumulative exposure	10 mSv × age[a]	1 rem × age
Embryo–fetus, occupational (carried by a pregnant worker)		
Total dose equivalent to an embryo-fetus	5 mSv	0.5 rem
Dose equivalent	0.5 mSv/mo	0.05 rem/mo
Nonoccupational		
Effective dose equivalent	1 mSv/yr	0.1 rem/yr
Infrequent exposure[b]	5 mSv/yr	0.5 rem/yr

[a]Age in years

[b]Only applicable if the lifetime average of 1 mSv is not exceeded.

eral organs, is comparable in risk to a whole-body dose. Doses to individual organs or tissues may be higher than whole-body doses because they constitute a fraction of the whole body. The cumulative exposure takes precedence over the annual limit. For example, suppose a 30-year-old worker had received a cumulative dose of 280 μSv over a period of years, the limit for the succeeding year would be 30 μSv, not 50 μSv. When persons have accumulated exposures equal to 10 mSv times their ages, the annual limit becomes 10 mSv.

Should a person receive a dose in excess of limits inadvertently, their duties involving radiation exposure must be suspended, at least until their cumulative total is below the limit.

Pregnant Workers

Because of higher risks to an embryo or fetus, primarily mental retardation, the limits for pregnant workers are lower than those quoted as occupational. The total dose limit to the embryo–fetus is 5 mSv (0.5 rem) with no more than 0.5 mSv (0.05 rem) in any month (Table 15-2). These limits exclude medical exposures. Although the embryo–fetus is considered an entity separate from the mother, practically speaking these become the mother's limits if whole-body exposure is involved. Women capable of pregnancy should not be engaged in activities where exposures in excess of 5 mSv to an embryo could occur before pregnancy is detected. Internally deposited radionuclides are of special concern because it takes considerable time to rid the body, including the fetus, of them, therefore, unlike external sources, internal sources cannot be eliminated immediately after detection of pregnancy. A prudent rule followed in many organizations is to adopt limits for women of child-bearing age that correspond to the embryo–fetus limits.

NONOCCUPATIONAL EXPOSURE LIMITS

At first it may seem unnecessary to have nonoccupational limits, but it is virtually impossible to avoid some exposure to the public when radiation from man-made sources is used. For example, a medical X-ray unit must be accessible to the public, but a nonpatient passing in a hallway adjoining a room in which an X-ray unit is operating is subject to a possible exposure. Actually, construction requirements for X-ray facilities include shielding to protect the public. Frequently, lead sheets are incorporated into the interior walls where X-ray machines are located, but theoretically no absorber thickness is sufficient to absorb 100% of photonic radiation. The point is, radiation users must ensure that the public does not receive exposures above certain limits and absolutely zero exposure is impossible.

What philosophy is followed in establishing acceptable risks? The health risks faced by the public at large vary greatly and include such things as smoking, occupational accidents, overeating, automobile accidents, and accidental electrocutions. However, the usual range is about 1 mortality per 10^4 to 10^6 persons (risk factor of 10^{-4} to 10^{-6}), and these risks seem to be accepted by the public; people continue to drive automobiles. Bacground radiation represents a risk, but the annual effective dose equivalent along the Atlantic Seaboard is about 0.65 mSv compared to 1.25 mSv in Denver. People do not seem to be concerned about the differential risk associated with background radiation in different locations. With these factors in mind and risks of additional exposure, the following limits have been recommended by NCRP (1987a).

The annual limit for continuous exposure is 1 mSv (0.1 rem). An infrequent exposure of 5 mSv may be acceptable provided the average annual limit is not exceeded and the concept of *ALARA* is followed.

ALARA

The limits discussed above are precisely defined, but this does not mean that such exposures are acceptable if reasonable methods are available to reduce them. The acronym ALARA embodies the philosophy that all exposures, even those below limits, should be *As Low As Reasonably Achieveable*.

Suppose a scientist could employ a plexiglass shield while working with ^{32}P, and further, the shield would not be inordinately expensive or inconvenient and would reduce exposures by 50%. According to the ALARA concept, and in the eyes of most licensing agencies, it would be improper to not employ such a shield even though limits would not be exceeded without it. Consequently, acceptable exposures are those that are as low as reasonably achievable *and* do not exceed the limits cited above.

CALCULATIONS INVOLVING EXPOSURES AND DOSES

Exposure Rates from External γ-Ray Sources

Consider the task of calculating the exposure generated by a point source that emits γ-rays. Since the activities of sources are expressed as rates (Bq, dps, mCi, etc.), it is convenient to express the exposure from that source as a rate as well. The general notation for an exposure rate is $\dot{X}$, and various units are appropriate such as X-units/hr and R/min. To convert the activity of the source to an exposure rate, the amount of energy absorbed by a given quantity of air must be determined, and that is a function of the number of photons absorbed multiplied by their quantum energies. If the activity of the source and its mode of decay are known, the flux of photons emanating from the source may be calculated.

Using a hypothetical source, we will first determine certain constants that simplify calculations for known emitters. Suppose the source has an activity of 10^6 dps (1 MBq) and emits 1.00-MeV γ-rays. The fraction of disintegrations that yields photons of a given energy must be known. Many radionuclides emits γ-rays with different energies, and frequently only a small fraction of the distintegrations yields a γ-ray of a particular energy. For example, the percentages of disintegration that yield a given energy γ-ray for ^{131}I are 2.6%, 0.0802 MeV; 0.27%, 0.1773 MeV; 0.9%, 0.2725 MeV; 6%, 0.2843 MeV; 0.25%, 0.3257 MeV; 81%, 0.3645 MeV; and so on. Let f_i^p be the fraction of disintegrations that yield a photon of the ith energy level. In our example, there is only one energy, and we will assume for the sake of simplicity that the fraction is equal to one, that is $f_1^p = 1.0$.

When the activity of the source (A) is in dps, the flux will be in units of photons/s.

$$\text{flux} = (A)(f_1^p) = (10^6 \text{ d/s})(1.0 \text{ photon/d}) = 10^6 \text{ photon/s}$$

However, the flux density (Φ) is the number of photons striking one unit of surface area per unit time, and Φ is dependent on the distance of the surface from the source. The *inverse square law* must be utilized to calculate Φ.

Inverse Square Law. Because the radiation from a point source is emitted in all directions, the flux density levels are spherical in shape, with the source at the center (Figure 15-2). The flux divided by the surface area of the sphere ($4\pi r^2$) yields the flux density, and the radius (r) is the distance of the surface from the source.

$$\Phi = \frac{Af_1^p}{4\pi r^2}$$

The exposure rate is directly proportional to the flux density, therefore the exposure rate is inversely proportional to the square of the distance from the source. For radiation protection purposes, this is an important relationship because personal exposure rates can be reduced by 75% when distance from the source is doubled.

Let us assume the exposure rate is to be determined at a distance of 1 m from the source. For our example

$$\Phi = \frac{(10^6 \text{ d/s})(1.0 \text{ photon/d})}{4\pi \ (1 \text{ m})^2} = 10^6 \text{ photon/m}^2 \cdot \text{s}$$

The next step is to determine the proportion of the γ-rays to be absorbed when passing through a given amount of air.

Absorption of γ- and X-Rays by Air. The linear absorption coefficient, μ_ℓ, of air is fairly constant for γ- and X-rays having energies betweeen 60 keV and 2.0 MeV. As an approximation, we will assume $\mu_\ell = 3.46 \times 10^{-3}$/m for all γ-rays with energies in the range given above. The fraction of rays

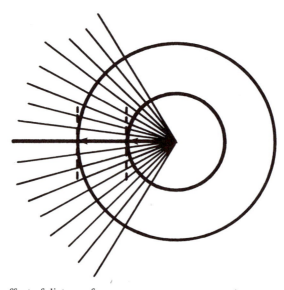

FIG. 15-2   The effect of distance from a source on exposure rates.

absorbed (f_a) by passing through a meter-thickness of air may be calculated using Equation 5-12.

$$I = I_0\, e^{-\mu_e X_e}$$

$$f_a = \frac{I_0 - I}{I_0} = 1 - e^{-\mu_e X_e}$$

$$f_a = 1 - e^{-(3.46\, \times\, 10^{-3}/m)(1\ m)} = 3.5 \times 10^{-3}$$

Since we specified a 1-m thickness, the fraction applies to each meter of thickness so our approximation becomes

$$f_a = 3.5 \times 10^{-3}/m$$

Now, the rate of energy absorption ($\dot{E}_a$) may be calculated by multiplying the flux density by the fraction of photons absorbed and the amount of energy per photon (E_i). Because there is only one γ-ray energy level, $E_i = E_1$.

$$\dot{E}_a = \Phi\, (f_a)(E_1)$$

$$\dot{E}_a = \frac{(A)(f_1^p)(f_a)(E_1)}{4\pi r^2}$$

For our example,

$$\dot{E}_a = \frac{(10^6\ d/s)(1.0\ photon/d)(3.5 \times 10^{-3}/m)(1.00\ MeV/photon)}{4\pi(1\ m)^2}$$

$$\dot{E}_a = 2.79 \times 10^2\ MeV/m^3 \cdot s$$

Note that with the flux being in photons/m$^2 \cdot$ s and the thickness being 1 m, $\dot{E}_a$ is the energy absorbed per cubic meter of air per second. Since the density of air is 1.293 kg/m^3 at standard temperature and pressure, there are 3600 s/hr and 2.12×10^{20} eV/kg per X-unit (Equation 14-2), the exposure in X-units/hr may be calculated.

$$\dot{X} = \frac{(\dot{E}_a)(10^6\ eV/MeV)(3600\ s/hr)}{(2.12 \times 10^{20}\ eV/kg \cdot X\text{-}unit)(1.293\ kg/m^3)}$$

$$\dot{X} = \frac{(2.79 \times 10^2\ MeV/m^3 \cdot s)(10^6\ eV/MeV)(3600\ s/hr)}{2.12 \times 10^{20}\ eV/kg \cdot X\text{-}unit)(1.293\ kg/m^3)}$$

$$\dot{X} = 3.66 \times 10^{-9}\ X\text{-}unit/hr \tag{15-1}$$

Using previous notations:

$$\dot{X} = \frac{(A)(f_1^p)(f_a)(E_1)(10^6\ eV/MeV)(3600\ s/hr)}{4\pi r^2\, (2.12 \times 10^{20}\ eV/kg \cdot X\text{-}unit)\, (1.293\ kg/m^3)}$$

When A is in units of MBq (10^6 dps), E_1 in MeV and r in meters, the X-unit exposure rate may be calculated according to Equation 15-2.

$$\dot{X} = \frac{3.66 \times 10^{-9}(A)(f_1^p)(E_1)\ X\text{-}unit \cdot m^2/hr \cdot MBq}{r^2} \tag{15-2}$$

Suppose a source emits γ-rays of i energy levels, Equation 15-2 may be transformed to 15-3.

$$\dot{X} = \frac{3.66 \times 10^{-9}(A)}{r^2} \, [(f_1^p)(E_1) + (f_2^p)(E_2)$$
$$+ \cdots + (f_i^p)(E_i)]\text{X-unit}\cdot\text{m}^2/\text{hr}\cdot\text{MBq}\cdot\text{MeV} \quad (15\text{-}3)$$

Specific γ-Ray Constant. Cobalt-60 emits two γ-rays per disintegration, and their energies are 1.173 and 1.332 MeV. What is the exposure rate from a 1 MBq source of ^{60}Co at a distance of 1 meter?

The appropriate fractions and corresponding energies for the γ-rays are $f_1^p = 1.0$, $E_1 = 1.173$ MeV, $f_2^p = 1.0$, $E_2 = 1.332$ MeV.

$$\dot{X} = \frac{(3.66 \times 10^{-9}\,\text{X-unit}\cdot\text{m}^2/\text{hr}\cdot\text{MBq}\cdot\text{MeV})(1\ \text{MBq})}{1\ \text{m}^2}$$

$$[(1.0)(1.173) + (1.0)(1.332)]\ \text{MeV}$$

$$\dot{X} = 9.17 \times 10^{-9}\ \text{X-unit/hr}$$

This particular value has special significance in radiation protection calculations, and it is called the *specific gamma ray constant* for ^{60}CO. Similar constants have been calculated for various γ-emitters and some are given in Table 15-3. The general symbol for a specific γ-ray constant is Γ, and its units are X-unit$\cdot$m^2/hr$\cdot$MBq or R$\cdot$m^2/hr$\cdot$Ci.

The exposure rate is calculated easily according to Equation 15-4 when Γ is known; the activity is in units of MBq or Ci and the distance is in meters.

$$\dot{X} = \frac{\Gamma A}{r^2} \quad (15\text{-}4)$$

TABLE 15-3 Specific γ-Ray Constants for Selected Radionuclides

	Γ	
Radionuclide	$\dfrac{\text{X-Unit}\cdot\text{m}^2}{\text{hr}\cdot\text{MBq}}$	$\dfrac{\text{R}\cdot\text{m}^2}{\text{hr}\cdot\text{Ci}}$
Cesium-137	2.30×10^{-9}	0.33
Chromium-51	1.11×10^{-10}	0.016
Cobalt-60	9.19×10^{-9}	1.321
Iodine-125	4.87×10^{-10}	0.070
Iodine-131	1.53×10^{-9}	0.22
Sodium-22	8.36×10^{-9}	1.20
Sodium-24	1.28×10^{-8}	1.84
Zinc-65	1.88×10^{-9}	0.27

Source: From Brodsky (1978).

For example, what is the exposure rate for a 10^7 dps source of ^{60}CO at a distance of 2.0 m?

$$A = \frac{10^7 \text{ dps}}{10^6 \text{ dps/MBq}} = 10 \text{ MBq}, \quad r = 2 \text{ m}$$

$$\dot{X} = \frac{(9.17 \times 10^{-9} \text{ X-unit·m}^2/\text{hr·MBq})(10 \text{ MBq})}{(2 \text{ m})^2} = 2.29 \times 10^{-8} \text{ X-unit/hr}$$

Example Problem 15-1. What is the exposure rate, in R/hr, from an ^{131}I source at a distance of 3 m if its activity is 5×10^{10} dpm?

$$A = \frac{5 \times 10^{10} \text{ dpm}}{2.22 \times 10^{12} \text{ dpm/Ci}} = 2.25 \times 10^{-2} \text{ Ci}$$

According to Table 15-3, the specific gamma ray constant for ^{131}I is 0.22 R·m²/hr·Ci.

$$\dot{X} = \frac{(0.22 \text{ R·m}^2/\text{hr·Ci})(2.25 \times 10^{-2} \text{ Ci})}{(3 \text{ m})^2} = 5.5 \times 10^{-4} \text{ R/hr}$$

Doses Involving External Sources of X- and γ-Rays

Dose calculations involving sources of X- and γ-radiation that are external to the human body may be accomplished by various methods, but one convenient method is to calculate exposure units and convert them to dose units. To be sure, there are simplifying assumptions in this method, but for most purposes the method is justified. The appropriate conversion factors, derived in Chapter 14, are 1.0 X-unit = 34 Gy and 1.0 R = 0.94 rad.

Example Problem 15-2. For a person, whose head is about 30 cm from a 10 mCi source of ^{137}Cs during a working period of 15 min, calculate the approximate total dose to the person's eyes. Of course, the entire face would receive a comparable dose, but because of the potential for cataracts, the eyes are of particular concern. Using SI units, Table 15-3 shows that $\Gamma = 2.3 \times 10^{-9}$ X-unit·m²/hr·MBq.

$A = 3.7 \times 10^2$ MBq and $r = 0.3$ m

$$\dot{X} = \frac{\Gamma A}{r^2}$$

$$\dot{X} = \frac{(2.3 \times 10^{-9} \text{ X-unit·m}^2/\text{hr·MBq}) \, 3.7 \times 10^2 \text{ MBq}}{(0.3 \text{ m})^2}$$

$$= 9.46 \times 10^{-6} \text{ X-unit/hr}$$

$$D = (9.46 \times 10^{-6} \text{ X-unit/hr})(34 \text{ Gy/X-unit})(15 \text{ min})(1 \text{ hr/60 min})$$

$$D = 8.04 \times 10^{-5} \text{ Gy}$$

Protection from External X- and γ-Ray Sources.   The two principal factors that may be used to reduce a person's exposure from external sources of X- and γ-radiation are *distance* and *shielding.*

As shown earlier, increasing the distance from a source reduces its flux density, hence it also reduces exposure and dose rates accordingly. Since exposure rates are inversely proportional to the square of the distance from the source, increasing one's distance from a source is an effective means of reducing the resulting dose. This principle may be used in various ways; for example, it is possible to reduce the exposure rate at the surface of a shipping package by securing a small source in the center of a large box. The use of remote handling equipment for conducting experiments involving high intensity sources is another example. Frequently, distance and shielding are used together to reduce exposure.

Equation 5-12 correctly predicts the attenuation of γ- or X-rays of a *specific wavelength* by a given absorber, but some types of interactions, such as the Compton and pair production processes, yield degraded rays (lower energy rays) that may be transmitted by the absorber. Thus, the radiation emanating from the back side of an absorber includes the attenuated flux of incident photons (correctly predicted by Equation 5-12) plus a portion of the degraded rays. Also in the derivation of Equation 5-12, the incident beam was assumed to consist of parallel rays and to strike the absorber perpendicularly. Other processes that produce scattered rays, such as normal scattering, which is identical to the phenomenon called X-ray diffraction, are affected by the angle of incidence. *Broad-beam* radiation refers to radiation that strikes an absorber with various angles of incidence, and the conclusion is that the attenuation of broad-beam radiation is overestimated by Equation 5-12.

In radiation protection, the attenuation of *exposure rates,* is needed, and for such calculations the term *half-value layer* (HVL) is used. HVL is the thickness of a specified absorber that reduces an exposure rate by one-half (NCRP, 1976). HVL is analogous to linear half-thickness ($X_{\ell\frac{1}{2}}$), but is larger. For example, using lead and ^{137}Cs γ-rays (0.662 MeV), HVL = 0.65 cm and $X_{\ell\frac{1}{2}}$ = 0.56 cm. As with $X_{\ell\frac{1}{2}}$, HVL values are dependent on the absorber and quantum energies of the incident photons. Equation 15-5 is applicable for calculating exposure rate reductions.

$$\dot{X} = \dot{X}_0 e^{-[(\ln 2)X]/\text{HVL}} \tag{15-5}$$

Where, $\dot{X}_0$ and $\dot{X}$ are the exposure rates at the front and back sides of an absorber and the thickness of the absorber is X.

Example Problem 15-3.   What is the minimum number of lead bricks (2 in thick) needed as a protective shield to reduce the exposure rate from a 0.5 Ci, ^{60}Co source to 10 mR/hr or less at a distance of 45 cm? Assume HVL is 1.20 cm.

According to Table 15-3, $\Gamma = 1.32$ R·m²/Ci·hr.

$$\dot{X}_0 = \frac{(1.32 \text{ R·m}^2/\text{Ci·hr})(0.5 \text{ Ci})}{(0.45 \text{ m})^2} = 3.26 \text{ R/hr}$$

Since the exposure is to be reduced to 10 mR/hr,

$$\dot{X} = 10^{-2} \text{ R/hr}$$

$$10^{-2} \text{ R/hr} = (3.26 \text{ R/hr}) \, e^{-[\ln 2)(X)]/1.2\text{cm}}$$

$$\ln(10^{-2}) - \ln(3.26) = -\frac{(\ln 2)(X)}{1.2 \text{ cm}}$$

$$X = 10.02 \text{ cm}$$

$$X = \frac{10.02 \text{ cm}}{2.54 \text{ cm/in}} = 3.94 \text{ in}$$

Since the bricks are 2 in. thick, a double wall of bricks would be sufficient to reduce the exposure rate to 10 mR/hr at a distance of 45 cm. Note that the inverse square law could be applied before or after the wall with the same effect. However, the wall should be built as close as practical to the source because that would attenuate a larger angle of the radiation emanating from the source and fewer bricks would be required.

External Sources of α- and β-Rays

Most α- and β-emitters do not represent a serious radiological hazard when outside the body. Under normal working conditions sources are usually located beyond their effective ranges in air, and source containers as well as clothing are effective absorbers for such emissions. Hands are the most likely portion of the body to receive significant exposure if the rays are capable of penetrating container walls.

With an E_{max} of 1.71 MeV, ^{32}P is about the only β-emitter commonly used in biological or medical research that may yield measurable exposures. For that reason the use of clear plastic shields are recommended when working with ^{32}P, primarily to protect exposed portions of the body such as the face and eyes. The energetic particles from ^{32}P will penetrate the paper coverings of film monitors (see Chapter 16) and when such monitors are worn as a ring, doses to the hands may be estimated.

INTERNAL SOURCES OF RADIATION

Consider the problem of determining the potential radiological damage caused by the intake of radioactive materials. Appropriate methods for calculating dose rates or dose equivalents depend on several factors, and these same factors are involved, in a more qualitative sense, in estimating the relative biological hazardousness of different radionuclides.

1. The energy of the emissions. The dose and dose equivalent are dependent on the amount of energy absorbed; therefore with other decay characteristics being equal, higher energy emissions yield greater doses. Table 15-4 gives the average energy absorbed per disintegration for a few radionuclides.

2. The type of emissions. Both α- and β-rays will be absorbed by the body if emitted internally, whereas γ-rays emitted inside the body have a good probability of escaping.

Recall that the QF values for β- and γ-rays are 1.0, whereas for α-particles it is 20. Thus, when other decay characteristics are equivalent, α-emitters are more hazardous than β-emitters because of their respective QF values, and β-emitters are more hazardous than γ-emitters because γ-rays are incompletely absorbed.

3. Distribution of the radionuclide within the body. If a radionuclide is distributed uniformly in the body, all organs would receive comparable doses. However, if the radionuclide concentrates in a particular organ then that organ would receive a greater dose than the other tissues.

Critical organ is a term that defines which portion of the body is most susceptible to radiation damage under a given set of conditions. The thyroid would be the critical organ if Na ^{131}I were taken internally because most of the resulting radiation dose would be received by the thyroid gland and the dose to the remainder of the body would be considerably smaller. This is due to the thyroid's nature to concentrate iodine. Because of such specific action, high doses of ^{131}I are used to oblate overactive thyroid glands without surgery.

Bone seekers are elements that may be natural components of bones or substitute for Ca in the bone's crystalline structure. Radionuclides of Sr,

TABLE 15-4 Biological Half-Lives and Average Absorption Energies of Selected Radionuclides

Radionuclide	Biological half-life, $t_{1/2b}$ (da)	Average energy absorbed per disintegration E_{av} (keV)
^{3}H	12	6
^{14}C	10	45
^{32}P	257	700
^{35}S	90	48.8
^{125}I	138	30
^{131}I	138	181.2

Ra, Ba, Ce, Pr, and Pu are examples of bone seekers. If accumulated in bone, they are not eliminated rapidly and lie close to bone marrow, which is responsible for generating blood cells.

With other decay characteristics being identical, those radioisotopes that concentrate in specific organs are more hazardous than those that are distributed uniformly.

4. Rate of elimination. After taken internally, the rate at which a particular radioisotope is excreted (its biological half-life) affects its hazardousness. The chemical compound in which the isotope is incorporated has an effect on the rate of elimination as does the chemical nature of the radionuclide itself. Most tritium-labeled compounds will exchange with H_2O and be metabolized readily. Therefore, most internal tritium will be converted to ^3HOH, which will be uniformly distributed throughout the body. Because of the large daily intake and excretion of water, the biological half-life of water is about 12 days. In contrast, minerals incorporated into bone are not likely to be excreted rapidly because of the slow turnover of bone minerals. Table 15-4 gives biological half-lives for some commonly used radionuclides. The half-lives are for total body elimination as opposed to elimination from specific tissues. Wang (1969) provides a complete list of total body half-lives for the elements and also includes half-lives in various tissues for some elements.

Elimination is also accomplished by radioactive decay. The total elimination rate is the sum of the rates for biological elimination and radioactive decay. Sometimes the sum is expressed as an *effective half-life,* or *combined half-life*. With other characteristics being equal, radionuclides with short effective half-lives represent less of a hazard than those with long effective half-lives.

Maximum Permissible Concentrations

Because of their differences in emissions, rates of elimination and propensity to accumulate in certain organs, various radionuclides represent different degrees of health hazards when they are taken into the body. Generally, the degree of hazard is reflected by values called *maximum permissible concentrations*. When a source of a particular radionuclide is released into the environment via air or water, there is a possibility that it may be ingested by humans. Therefore, concentration limits (maximum permissible concentrations) have been established for the release of each radionuclide via water and air.

Table 15-5 provides maximum permissible concentrations for a few selected radionuclides. Complete tables have been developed by the regulatory agencies and are included in their regulations.

TABLE 15-5 Maximum Permissible Concentrations for Selected Radionuclides

Radionuclide	Physical form[a]	Restricted area Air (μCi/ml)	Water	Unrestricted area Air (μCi/ml)	Water
^{3}H	S	5×10^{-6}	1×10^{-1}	2×10^{-7}	3×10^{-3}
	I	5×10^{-6}	1×10^{-1}	2×10^{-7}	3×10^{-3}
^{14}C	S	4×10^{-6}	2×10^{-2}	1×10^{-7}	8×10^{-4}
	CO_2	5×10^{-5}	—	1×10^{-6}	—
^{22}Na	S	2×10^{-7}	1×10^{-3}	6×10^{-9}	4×10^{-5}
	I	9×10^{-9}	9×10^{-4}	3×10^{-10}	3×10^{-5}
^{32}P	S	7×10^{-8}	5×10^{-4}	2×10^{-9}	2×10^{-5}
	I	8×10^{-8}	7×10^{-4}	3×10^{-9}	2×10^{-5}
^{35}S	S	3×10^{-7}	2×10^{-3}	9×10^{-9}	6×10^{-5}
	I	3×10^{-3}	8×10^{-3}	9×10^{-9}	3×10^{-4}
^{125}I	S	5×10^{-9}	4×10^{-5}	8×10^{-11}	2×10^{-7}
	I	2×10^{-7}	6×10^{-3}	6×10^{-9}	2×10^{-4}
^{131}I	S	9×10^{-9}	6×10^{-5}	1×10^{-10}	3×10^{-7}
	I	3×10^{-7}	2×10^{-3}	1×10^{-8}	6×10^{-5}

[a]S, soluble forms; I, insoluble or particulate forms.

Body Burden

The amount of radioactivity or amount of a given radionuclide in a person's whole body is referred to as a *body burden*. To calculate doses from internal concentrations of radionuclides, body burdens must be ascertained.

Calculations Involving Internal Sources of α- and β-Rays

Example Problem 15-4. Suppose a person's only source of drinking water has a level of tritium equal to the maximum permissible concentration (MPC) for an unrestricted area. What dose and dose equivalent would a 70 kg man receive if he lived continuously with such a source of water? Tritium would not concentrate in his body, but would approach the concentration of tritium in his drinking water, which according to Table 15-5 is 3×10^{-3} μCi/ml. Assuming 60% of a person's body weight is water, the body burden may be calculated.

Burden = $(70 \text{ kg})(0.60)(10^3 \text{ g/kg})(1 \text{ ml/g})(3 \times 10^{-3} \text{ μCi/ml})$ = 126 μCi

As indicated in Table 15-4, the average energy of the β-particles emitted by tritium is 6 keV, and all of that energy would be absorbed. Thus, the average energy absorbed per disintegration (E_{av}) is 6 keV. The instantaneous dose rate (dose per second or minute) can be calculated by determining the rate of energy deposition and dividing that value by the weight of the man.

$$\dot{D} = \frac{(126 \ \mu\text{Ci})(2.22 \times 10^6 \ \text{d/min}/\mu\text{Ci})(6 \ \text{keV/d})(1.6 \times 10^{-16} \ \text{J/keV})}{70 \ \text{kg}}$$

$$\dot{D} = 3.84 \times 10^{-9} \ \text{J/min·kg}$$

$$= 3.84 \times 10^{-9} \ \text{Gy/min or } 3.84 \times 10^{-7} \ \text{rad/min}$$

If the concentration of tritium remained constant, the annual absorbed dose would be

$$D = (3.84 \times 10^{-9} \ \text{Gy/min})(5.256 \times 10^5 \ \text{min/yr}) = 2.02 \ \text{mGy or } 202 \ \text{mrad}$$

Since the emission is a β^--particle and the distribution factor is one, the dose equivalent is

$$H = (2.02 \ \text{mGy})(1)(1) = 2.02 \ \text{mSv or } 202 \ \text{mrem}$$

This dose equivalent is one-fifth the limit for a continuous occupation exposure (10 mSv/yr) and about equivalent to two-thirds of the average background dose. Although the dose is small, it should not be tolerated willingly. The quantity of water on this earth and the rapidity of its movement through drainage and evaporation cycles make it inconceivable that a person could attain a body burden within several orders of magnitude of this example even if water were released at the maximum permissible concentration. Therefore, it appears that the maximum permissible concentrations for tritium release are reasonable and conservative, at least according to the exposure limits given in Table 15-2.

If internal contamination with radioactivity occurs, it is most likely to happen by accidental ingestion, after which the radionuclide will be eliminated. The mathematical treatment of doses caused by ingestion of a quantity of radioactivity is illustrated in the following example problems.

Example Problem 15-5. Assume that 1.0 mCi of ^{32}P is uniformly distributed inside a 70 kg man. Calculate the instantaneous, initial dose rate this person would receive.

The approach is to calculate the number of atoms that decay during a small period of time (Ao, which is the initial disintegration rate or body burden), and multiply that by the average energy dissipated per distintegration (E_{av}, see Table 15-4), then divide by the weight of the organ (body) receiving the exposure. Let $\dot{D}_0$ represent the initial, instantaneous dose rate and W the weight of the organ, in this case the whole body.

$$\dot{D}_0 = \frac{(Ao)(E_{av})}{W} \tag{15-6}$$

$$\dot{D}_0 = \frac{(1 \ \text{mCi})(3.7 \times 10^7 \ \text{d/s·mCi})(0.70 \ \text{MeV/d})(1.6 \times 10^{-13} \ \text{J/MeV})}{70 \ \text{kg}}$$

$\dot{D}_0 = 5.92 \times 10^{-8}$ J/s·kg

$\quad = 5.92 \times 10^{-8}$ Gy/s or 5.92×10^{-6} rad/s

Example Problem 15-6. Determine the absorbed dose that the man described in example problem 15-5 would receive from the internal ^{32}P (1) during the first 30 days of exposure and (2) during his lifetime.

The initial dose rate, calculated in the previous problem, would decrease with time since ^{32}P decays and phosphorus is excreted; therefore mathematics dealing with the combination of biological elimination and radioactive decay must be developed.

Most compounds, including ^{32}P, are eliminated by excretion, expiration, etc. according to first-order kinetics. Since radioactive decay also obeys first-order kinetics, the equations for elimination are completely analogous to those used for radioactive decay. To distinguish between elimination and decay the following symbols are defined.

$t_{\frac{1}{2}b}$ = biological half-life of the compound, time required for half of the radioactive atoms to be eliminated from the body

λ_b = biological elimination constant, analogous to the decay constant and equal to $(\ln 2)/t_{\frac{1}{2}b}$

$t_{\frac{1}{2}r}$ = half-life for radioactive decay

λ_r = radiological decay constant

N = number of radioactive atoms present *inside* the body at any time

N_0 = number of radioactive atoms present inside the body at $t = 0$

The rate of loss of radioactive atoms due to biological elimination is

$$\left(\frac{dN}{dt}\right)_b = -\lambda_b N$$

And the rate of loss of radioactive atoms due to radioactive decay is

$$\left(\frac{dN}{dt}\right)_r = -\lambda_r N$$

The effective or combined rate of loss is

$$\left(\frac{dN}{dt}\right)_c = -\lambda_b N - \lambda_r N = -(\lambda_b + \lambda_r)N$$

Let $\lambda_c = \lambda_b + \lambda_r$

$$\left(\frac{dN}{dt}\right)_c = -\lambda_c N$$

This equation is identical to the one used to derive radioactive decay equations except the combined decay constant (λ_c) has replaced the radiological decay constant. Therefore, by rearranging the equation, integrating and evaluating the constant of integration we obtain

$$N = N_0 e^{-\lambda_c t} \tag{15-7}$$

From the above equations, relationships between half-lives are easily derived.

$$\lambda_c = \lambda_b + \lambda_r$$

$$\frac{\ln 2}{t_{\frac{1}{2}c}} = \frac{\ln 2}{t_{\frac{1}{2}b}} + \frac{\ln 2}{t_{\frac{1}{2}r}}$$

$$t_{\frac{1}{2}c} = \frac{t_{\frac{1}{2}b} \cdot t_{\frac{1}{2}r}}{t_{\frac{1}{2}b} + t_{\frac{1}{2}r}} \tag{15-8}$$

To calculate the dose we must calculate the body burden and determine the number of radioactive atoms that decay *inside* the body. If we sum the increments of decay inside the body over a specified period, the total number of atoms disintegrated inside the body would be obtained.

Let $-S$ = the sum of atoms decaying inside the body from $t = 0$ to t = an unspecified time. The negative sign for S is needed to indicate decay.

$$-S = \int_{t=0}^{t=t} -\left(\frac{dN}{dt}\right)_r \cdot dt \quad \text{since} \quad \left(\frac{dN}{dt}\right)_r \cdot dt = -\lambda_r N$$

$$-S = \int_{t=0}^{t=t} \lambda_r N dt$$

Since $N = N_0 e^{-\lambda_c t}$

$$-S = \int_{t=0}^{t=t} \lambda_r N_0 e^{-\lambda_c t} \cdot dt$$

$$-S = \lambda_r \cdot N_0 \int_0^t e^{-\lambda_c t} \cdot dt$$

$$-S = \frac{\lambda_r \cdot N_0}{-\lambda_c} \left(e^{-\lambda_c t}\right)\Big|_0^t$$

$$-S = \frac{\lambda_r \cdot N_0}{-\lambda_c} \left(e^{-\lambda_c t} - 1\right)$$

$$-S = \frac{\lambda_r \cdot N_0}{\lambda_c} \left(1 - e^{-\lambda_c t}\right)$$

The absorbed dose would be

$$D = \frac{-S \cdot E_{av}}{W}$$

$$D = \frac{\lambda_r \cdot N_0 \cdot E_{av}}{\lambda_c \cdot W}\left(1 - e^{-\lambda_c t}\right) \tag{15-9}$$

Compare Equation 15-9 to 15-6. It is obvious from Equation 4-14 that $\lambda_r N_0$ is the initial disintegration rate or body burden, which when multiplied by E_{av}/W is the initial instantaneous dose rate; therefore

$$D = \frac{\dot{D}_0}{\lambda_c}\left(1 - e^{-\lambda_c t}\right) \tag{15-10}$$

After determining λ_c for ^{32}P, Equation 15-10 can be used to calculate the dose for the man in this example problem. According to Table 15-4, $t_{1/2b} = 257$ da.

$$t = \frac{(t_{1/2b})(t_{1/2r})}{t_{1/2b} + t_{1/2r}} = \frac{(257 \text{ da})(14.28 \text{ da})}{(257 \text{ da}) + (14.28 \text{ da})} = 13.53 \text{ da}$$

$$\lambda_c = \frac{\ln 2}{13.53 \text{ da}}$$

Notice that the effective half-life for ^{32}P is less than either the radioactive decay or biological elimination half-life. Now the dose for the 30-day period can be calculated by use of Equation 15-10, and an answer for example problem 15-6 may be formulated,

$$D = \frac{5.92 \times 10^{-8} \text{ Gy/s}}{\ln 2/[(13.53 \text{ da})(8.64 \times 10^4 \text{ s/da})]}\left(1 - e^{-[(\ln 2) (30 \text{ da})]/13.53 \text{ da}}\right)$$

The time units in D_0 and λ_c must cancel, and of course the time units in the exponent component must cancel as well.

$$D = 9.984 \times 10^{-2} \text{ Gy } [1 - e^{-1.537}]$$

$$D = 7.8 \times 10^{-2} \text{ Gy or 7.8 rad}$$

The maximum dose a person should receive is easily calculated from Equation 15-10 by using an infinite time for elimination of ^{32}P.

$$D = \frac{\dot{D}_0}{\lambda_c}\left(1 - e^{\lambda_c \infty}\right)$$

$$D = \frac{\dot{D}_0}{\lambda_c}[1 - 0]$$

$$D = \frac{\dot{D}_0}{\lambda_c} \tag{15-11}$$

$$D = \frac{5.92 \times 10^{-8} \text{ Gy/s}}{\ln 2/[(13.53 \text{ da})(8.64 \times 10^4 \text{ s/da})]} = 9.984 \times 10^{-2} \text{ Gy}$$

Reviewing Equations 15-10 and 15-11 it may be seen that the $(\dot{D}_0/\lambda_c)$ term represents the maximum dose and the $(1 - e^{-\lambda_c t})$ term is the fraction of the maximum corresponding to a time period specified as t.

Obviously, the man described in this example problem will not live an infinite amount of time so he will not receive the maximum dose. However, he would receive about 78.5% of the maximum in 30 days, therefore after 10 effective half-lives, his absorbed dose would approach the maximum closely because the concentration of ^{32}P would have been reduced to nearly zero.

Example Problem 15-7.   A technician performed an experiment with curie amounts of tritium gas. Although an approved hood was used, an accident occurred, and there was a possibility that she could have inhaled some tritium. At the conclusion of the experiment (several hours later) she found that the counting rate of 1 ml of her urine was 11.3 cpm above background. Long counting times were used so the counting data were quite accurate, statistically, and the counting efficiency of the counter was determined to be 43.6% for that sample. Calculate the maximum absorbed dose she can expect to receive from the internal tritium. Assume her measurement reflected the concentration of tritium in her body water, she weighed 54 kg and 60% of a person's body weight is due to water. The half-life for radioactive decay is 12.26 yr, and according to Table 15-4, $E_{av} = 6$ keV and $t_{\frac{1}{2}b} = 12$ da. Although the body burden is equal to $\lambda_r N_0$, it can be calculated more simply from the data given:

$$A = \frac{(11.3 \text{ cpm/ml})(54 \text{ kg})(0.60)(10^3 \text{ g/kg})}{(0.436 \text{ cpm/dpm})(1 \text{ g/ml})} = 8.398 \times 10^5 \text{ dpm}$$

Apparently, the technician had somewhat less than a microcurie of tritium as a body burden. The instantaneous dose rate may be calculated.

$$\dot{D}_0 = \frac{E_{av} \cdot \lambda_r \cdot N_0}{W} = \frac{E_{av} \cdot A}{W}$$

$$\dot{D}_0 = \frac{(6 \text{ keV/d})(8.398 \times 10^5 \text{ d/min})(1.6 \times 10^{-16} \text{ J/keV})}{54 \text{ kg}}$$

$$\dot{D}_0 = 1.493 \times 10^{-11} \text{ J/min·kg} = 1.493 \times 10^{-11} \text{ Gy/min}$$

The total dose she might receive (infinite time for elimination) is

$$D = \frac{\dot{D}_0}{\lambda_c}$$

Because the biological half-life of ^{3}H is much smaller than the half-life of decay, $t_{\frac{1}{2}c} \approx t_{\frac{1}{2}b}$ and $\lambda_c \approx \lambda_b$.

$$t_{\frac{1}{2}c} = \frac{(t_{\frac{1}{2}b})(t_{\frac{1}{2}r})}{t_{\frac{1}{2}b} + t_{\frac{1}{2}r}}$$

$$= \frac{(12 \text{ da})(12.26 \text{ yr})(365 \text{ da/yr})}{12 \text{ da} + (12.26 \text{ yr})(365 \text{ da/yr})} \approx 12 \text{ da}$$

$$\lambda_c = \frac{\ln 2}{(12 \text{ da})(1440 \text{ min/da})} = 4.011 \times 10^{-5}/\text{min}$$

$$D = \frac{\dot{D}_0}{\lambda_c} = \frac{1.493 \times 10^{-11} \text{ Gy/min}}{4.011 \times 10^{5}/\text{min}}$$

$$D = 3.72 \times 10^{-7} \text{ Gy} = 0.372 \text{ }\mu\text{Gy or } 37.2 \text{ }\mu\text{rad}$$

The dose equivalent is easily calculated since QF and DF $= 1$ for ^{3}H. Recall Equation 14-8.

$$H = (D)(QF)(DF) \tag{14-8}$$

$$H = (0.372 \text{ }\mu\text{Gy})(1)(1) = 0.372 \text{ }\mu\text{Sv}$$

$$H = (37.2 \text{ }\mu\text{rad})(1)(1) = 37.2 \text{ }\mu\text{rem}$$

Since the tritium, and hence dose, is uniformly distributed in the body the effective dose equivalent (H_E) is 0.372 μSv. Although a tritium concentration of 26 dpm/ml in a person's body water sounds alarming, it represents a very small dose. For the sake of comparison, the lifetime dose the technician would receive as a result of the hypothetical "tritium accident" is about the same as that a person would receive by flying at 39,000 ft for 4.5 min.

Equation 15-12 is a general equation for the total dose from a radionuclide in body fluids.

$$D = 2 \times 10^{-10} \text{ Gy} \cdot \text{SA} \cdot E_{av} \cdot t_{\frac{1}{2}c} \tag{15-12}$$

It is assumed that the emitter is uniformly distributed and that the body is 60% water. The dose is given in units of Grays, when SA is the specific activity of the fluid in dpm/ml, E_{av} is the average energy absorbed per disintegration in keV and $t_{\frac{1}{2}c}$ is the effective or combined half-life for elimination in days. The constant, 2×10^{-10}, was obtained by slight rounding of the conversion factors. Using the same units described for Equation (15-12), the equation for a dose expressed in rads is

$$D = 2 \times 10^{-8} \text{ rad} \cdot \text{SA} \cdot E_{av} \cdot t_{\frac{1}{2}c} \tag{15-13}$$

Calculations Involving Internal Sources of X- and γ-Rays

The calculations of doses to body organs from an internal source of γ-rays are not as straightforward as for α- and β-particles because γ-rays are incompletely absorbed. If the fraction of rays absorbed can be estimated, the remainder of the calculations are identical to those presented for β-emitters. The mathematics of absorption involve models and are beyond the scope of this book, but insight into the problem may be gained by consulting the texts by Cember (1983) and Shapiro (1981).

OCCUPATIONAL HEALTH RISKS, A PERSPECTIVE

The information presented in this and the preceding chapter may leave the reader without a firm conclusion concerning the reasonableness of such things as established exposure limits, maximum permissible concentrations, and national policies concerning man-made sources. Relative to other occupations, just how safe is it to work in a field involving possible radiation exposures?

Fatalities for those engaged in the "nuclear or radiation" industry were not given as a separate category in Table 15-1, but the 40-year average is approximately 1.0 per year per 10,000 workers. Effectively, this is the same value as the "all industries'" average and well below that of the mining, construction, agriculture, and transportation industries. The nuclear industry certainly appears to be one of the safest among the occupational sectors.

Individuals face other health risks such as driving an automobile, smoking, home accidents, and medical X-rays, some of which involve personal choices. How does the risk of radiation exposure compare to those risks? Table 15-6 provides a basis for some comparisons. Note that the industry average dose for occupational workers is 3.4 mSv/yr, which is about one-fifteenth the annual dose limit *and one-third the dose that causes a 1-day loss of life expectancy.*

Life-style affects the amount of radiation a person is subjected to also. Living in a brick house in Denver, Colorado as opposed to a wooden, sea-

TABLE 15-6 Estimated Loss of Life Expectancy from Health Risks

Health risk	Estimates of days of life expectancy lost, average
Smoking 20 cigarettes/day	2370 (6.5 years)
Overweight (by 20%)	985 (2.7 years)
All accidents combined	435 (1.2 years)
Auto accidents	200
Alcohol consumption (U.S. average)	130
Home accidents	95
Drowning	41
Safest jobs (such as teaching)	30
Natural background radiation, calculated	8
Medical X-rays (U.S. average), calculated	6
All catastrophes (earthquake, etc.)	3.5
10 mSv (1 rem) occupational radiation dose, calculated (industry average is 3.4 mSv or 0.34 rem/yr)	1
10 mSv/yr for 30 years, calculated	30
50 mSv/yr for 30 years, calculated	150

Source: From NRC (1980).

side cottage increases one's background dose by about 0.6 mSv per year, and a transatlantic airplane flight yields a dose around 40 µSv per round trip.

It certainly appears that a worker dealing with radioactive materials does not face an unusual occupational hazard, and comparing that hazard to other personal choices, the risks seem insignificant. Nevertheless, the value of the occupation must be weighed against the risks, and fortunately individual choices can be exercised.

PROBLEMS

1. What is the approximate exposure rate at a distance of 20 cm from a 10 mCi source of ^{125}I in units of R/hr and X-unit/da?

2. A transparent lead-impregnated plastic shield is available for working with sources, such as the one described in problem 1. The shield has a half-value layer of 0.26 cm for ^{125}I and is 1.5 cm thick. What is the exposure rate from the source in problem 1 when such a shield is utilized, assuming the distance from the source is not altered?

3. A 1.5 mCi source of ^{60}Co is stored in a room to which only occupational workers have access, but in an adjoining room, separated from the first room by an 8-in. concrete wall, several secretaries work. Assume the maximum occupancy time for the secretaries is 2080 hr/yr and the HVL for ^{60}Co and concrete is 6.2 cm. In units of Gy, calculate maximum annual absorbed dose a secretary could receive when the source is 2 m from the wall by estimating the dose rate at the surface of the wall in the secretaries' room. Recall from Chapter 14, 1 X-unit $\approx$ 34 Gy.

4. Assuming that the exposure is a whole-body exposure and the quality and distribution factors are unity, determine if the limits for nonoccupational exposures (maximum possible exposure to the secretaries) would be exceeded by the scenario given in problem 3.

5. Considering the situation described in problems 3 and 4 further; an inspector with a firm belief in the ALARA concept undoubtedly would recommend methods to reduce exposures, not only to the secretaries but to the occupational workers as well. Suggest some practical actions that could be used to reduce exposures to the secretaries.

6. Calculate the annual effective dose equivalent humans receive from the naturally occurring ^{14}C in their bodies. Assume that carbon constitutes about 18.1% of a person's live weight, the specific activity of the carbon is constant at 13.6 dpm/g and the distribution factor for ^{14}C is 1.0.

7. In equilibrium with the environment, a 70-kg man contains about 0.12 μCi of naturally occurring ^{40}K, which decays by several modes, β$^-$, β$^+$, and EC, with a half-life of 1.25×10^9 yr. However, the average energy dissipated in all kinds of body tissues is about 563 keV per disintegration. Calculate the annual absorbed dose (whole body) such a man would receive from his internal ^{40}K in units of Gy and mrad.

8. Approximately 30% of the iodine in a human body is located in the thyroid gland, which in a 70-kg man weighs about 20 g. The biological half-life ($t_{1/2b}$) of iodine is 138 da, and ^{131}I emits both β$^-$- and γ-rays. If a total of 9.0 μCi of ^{131}I were injected into a 70-kg man, what would be the resulting lifetime dose equivalent to (a) his thyroid gland, (b) the remainder of his body?

9. Using the information and answer from problem 8 and Table 14-5, calculate the lifetime effective dose equivalent (H_E) from the ^{131}I. Assume 70% of the ^{131}I is evenly distributed in the nonthyroid tissue and the weighting factor for that tissue is $1 - W_{\text{thyroid}}$.

10. Calculate the total (lifetime) dose equivalent for a person who has a specific activity of 1625 dpm/ml of urine and the activity is tritium. Assume 60% of the body's weight is due to water.

REFERENCES

Brodsky, A. (1978). *Handbook of Radiation Protection and Measurements,* Vol. I. CRC Press, Boca Raton, FL.

Cember, H. (1983). *Introduction to Health Physics,* 2nd ed. Pergamon Press, New York.

Libby, W. F. (1955). *Radiocarbon Dating.* University of Chicago Press, Chicago.

NCRP (1976). *Structural Shielding Design and Evaluation for Medical Use of X Rays and Gamma Rays of Energies up to 10 MeV.* NCRP Report No. 49, National Council on Radiation Protection and Measurements, Bethesda, MD.

NCRP (1987a). *Recommendations on Limits for Exposure to Ionizing Radiation.* NCRP Report No. 91, National Council on Radiation Protection and Measurements, Bethesda, MD.

NCRP (1987b). *Ionizing Radiation Exposure of the Population of the United States.* NCRP Report No. 93, National Council on Radiation Protection and Measurements, Bethesda, MD.

NCRP (1987c). *Exposure of the Population in the United States and Canada from Natural Background Radiation.* NCRP Report No. 94, National Council on Radiation Protection and Measurements, Bethesda, MD.

NCRP (1987d). *Radiation Exposure of the U.S. Population from Consumer Prod-*

ucts and Miscellaneous Sources. NCRP Report No. 95, National Council on Radiation Protection and Measurements, Bethesda, MD.

NRC (1980). *Task OH 902-1.* Nuclear Regulatory Commission, Washington, DC.

Shapiro, J. (1981). *Radiation Protection,* 2nd ed. Harvard University Press, Cambridge, MA.

Wang, Y. (1969). *Handbook of Radioactive Nuclides.* CRC Press, Boca Raton, FL.

CHAPTER 16

Practical Aspects of Radiation Protection

Radiological safety procedures are designed to accomplish two broad objectives: (1) to prevent untrained persons, the general public including non-occupational co-workers, from coming in contact with sources of ionizing radiation and (2) to prevent occupational workers from receiving exposures that are unsafe or above that necessary to perform their job functions. Thus, the purposes of many procedures are to warn persons of possible hazards, to provide security for radiation sources, to contain radioactivity, and to prevent possible spread of radioactive contamination.

THE RADIATION SYMBOL AND POSTING OF AREAS

A universal symbol is used to identify sources of radiation and areas in which such sources may exist. As shown in Figure 16-1, it consists of a three-bladed figure, colored magenta, on a yellow background. The size of the symbol varies with its use, but is always a conspicuous component of warning signs and labels.

High Radiation Area

Rooms where an individual may receive a dose of 1.0 mSv (100 mrem) in 1 hr must be posted with a warning sign stating "Caution" or "Danger High Radiation Area" along with the radiation symbol. Access to such areas must be controlled by locks and keys, and visual or audible alarms may be used as a signal when persons enter the area.

FIG. 16-1   The universal symbol for radiation hazards.

Radiation Area

Areas where a person may receive a dose of 0.05 mSv (5 mrem) in 1 hr or a dose of 1.0 mSv (100 mrem) in five consecutive days must be posted with a sign stating "Caution Radiation Area."

Airborne Radioactivity Area

Areas where the concentration of airborne radioactive materials exceeds certain levels must be posted with a sign stating "Caution Airborne Radioactive Area." The levels applicable depend upon the radionuclide present and may be determined from tables that list Maximum Permissible Concentrations (MPC) of radionuclides in air. Such values are normally available in the regulations of licensing agencies, and an older version is given by Wang (1969). An abbreviated table was presented as Table 15-5.

Radioactive Materials

The most common posting for radiotracer laboratories is a warning about the presence of radioactive materials. Rooms where any radioactive material, other than natural uranium or thorium, is used or stored must be posted with a sign stating "Caution Radioactive Materials" (see Figure 16-1) unless the amounts of radioactivity are below certain levels. The amount of radioactivity requiring a sign depends on the radionuclide in question, but it is 10 times the amount given as an exempt quantity for that radionuclide. Tables provided by regulatory agencies have values for all the radionuclides, but a few values were given in Table 1-1. For ^{32}P, the exempt quantity limit is 10 μCi. Therefore, a room where 100 μCi of ^{32}P is present must be posted with a "caution radioactive materials" sign.

In many laboratories sources of several different radionuclides are present and the quantities vary from time to time. Perhaps a good policy is, if you have a specific license or a user's permit under a broad scope license, then post the areas where the materials are stored and used.

Containers of Sources

Most containers of radioactive materials must have a label attached. An example of an appropriate label is shown in Figure 16-2. Besides the usual radiation symbol and the words "Caution Radioactive Materials," there should be space to clearly indicate the quantity of radioactivity, the radionuclide, and the date. Whether labels are required depends on the radionuclide in the container. For a particular radionuclide, a label is required if the amount of radioactivity is above its "exempt quantity" level (Table 1-1) and its concentration exceeds the MPC level for a restricted area (Table 15-7). For example, a vial containing over 100 μCi of ^{14}C in a volume less than 5000 ml would require a label.

Containers in transport are packaged and labeled differently, according to the regulations of the U.S. Department of Transportation. Such labels are frequently referred to as DOT labels and one is shown in Figure 16-3.

CAUTION
RADIOACTIVE MATERIAL
Isotope ——————————
Amount ——————————
Date ——————————

FIG. 16-2   Tape for labeling source containers.

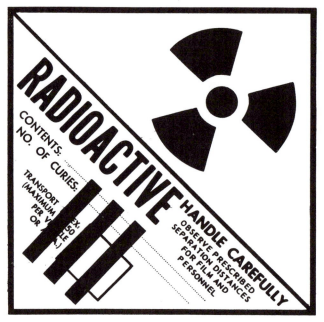

FIG. 16-3   A "DOT" label required by the Department of Transportation for shipment of radioactive materials.

Security

To prevent untrained individuals from coming in contact with radioactive materials or accidentally operating radiation-producing equipment, laboratories or storage areas for materials should be under lock and key except when trained personnel (occupational workers) are present.

SOURCES

Frequently, sources of radioactive materials are classified as "sealed" or "open" sources. Sealed sources are those in which the radioactivity is in an impervious container and mechanically protected from crushing or destruction. The radiation emanating from such sources is used, but the actual radioactive material is not. Examples of sealed sources are the metallic capsules of ^{137}Cs that are used as external standards in liquid scintillation counters. Beta-ionization detectors used on gas chromatographs also contain a sealed source.

Leak testing of sealed sources should be conducted periodically, usually at 6-month intervals, and is performed by wiping the outside of the source with absorbent paper or cotton, and counting the wiping material with an appropriate counter.

Open sources are those from which radioactive materials may be withdrawn. Radioactive materials in glass containers, whether capped or flame-sealed, are considered to be open sources.

PERSONNEL MONITORING

A number of methods are used to aid occupational workers in keeping radiation exposures as low as reasonably achievable and in keeping records of actual exposures.

Warning Devices

Areas where exposure levels may be quite high or may change unexpectedly may be equipped with monitors that activate sound and/or light warning devices. G-M or ionization chambers are frequently used as detectors for such devices.

Survey Meters

Some survey meters have an auditory output that emits a clicking sound or howl, and may serve as a warning device. However, the primary use of survey meters is to locate small sources of radiation that represent spills or contamination.

Survey meters are inexpensive rate meters that read in units, such as cpm, μGy/hr, and mR/hr, and most have multiple range switches for selecting sensitivities appropriate for various tasks. Portable battery-operated meters, such as those shown in Figure 16-4 are popular. However, alternating current-operated meters are available and each has a long flexible cable between the meter and its probe (detector) so that the probe can be used at a considerable distance from the meter.

All types of radiation, including neutrons, may be detected with survey

FIG. 16-4 Survey meters with different types of probes. Left to right: end-window G-M, pancake G-M, and solid scintillation probes.

meters depending on the type of detection probe employed. The most common probe for laboratory work is a rugged end-window G-M tube that has a relatively thin window so that weak β-particles and α-particles may be detected. Although detection efficiency is low, a G-M probe will detect X- and γ-rays. The so-called "pancake" probe is a G-M tube with a large diameter window and a short tube length that makes it quite sensitive in detecting low activity spots of contamination. Normally the windows of G-M tubes are protected against puncture by highly perforated covers such as plastic or wire grids. Solid scintillation crystal probes are preferred over G-M or proportional tubes when only X- and γ-rays are involved because of their greater sensitivities toward that type of radiation.

Whenever work with open sources is being performed, a survey meter with an appropriate probe should be available in the immediate vicinity, in case of an accident and to check for contamination.

Pocket Ionization Chambers

A pocket ionization chamber (or pocket dosimeter) and its charging device is shown in Figure 16-5. Such devices usually are about the size of a marking pen and are carried in a pocket by an occupational worker who might be exposed to relatively high levels of external sources of X- or γ-radiation.

The operating principle of these devices is similar to that of a Lauritsen electroscope, described in Chapter 6. At one end of the chamber is a lens with an exposure scale, perhaps with units of mR or μGy, and at the other end are contact points for charging the chamber. Translucent insulating material is used between the contact points so that light may enter, and allow a person to read the position of the quartz fiber relative to the scale.

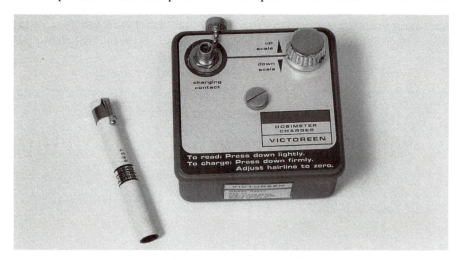

FIG. 16-5 A pocket dosimeter that operates as an electroscope, and its charger.

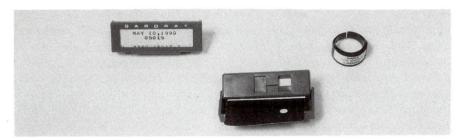

FIG. 16-6   Film badges for personnel monitoring. The "body badge" on the left is shown open in the center; note the different shielding sections. A "hand or ring badge" is shown on the right.

Beginning a work period, a person charges the dosimeter, setting the fiber to zero on the scale, and places it in his or her pocket. After completion of work the dosimeter is read, and the exposure is recorded in a log book. These pocket dosimeters are not as sensitive as some of the other types of personnel monitors. For that reason they are used primarily when the possibility of a fairly high exposure in a short period of time exists.

Film Monitors

Film badges utilize photographic emulsions as detectors. A film pack, film covered by paper to exclude light, is inserted in a plastic holder that is clipped to clothing and worn during working hours for a period of a week or a month (Figure 16-6). Film packs attached to rings are used to measure exposure to hands, while the body badges are used to estimate whole body exposures. The optical density of the developed film is related to the exposure levels. Absorption filters may be incorporated into the badge holder so that certain areas of the film are covered by different thicknesses of absorbing materials that permit estimation of the energies of the rays and determination of the type of radiation involved. These absorption filters also extend the measurable exposure range for film badges.

Generally, commercial laboratories supply, develop, read the films, and issue a monthly report to clients. The dose range most commonly used is about 30–1000 mrem (0.3–10 mSv).

Film badges are excellent monitors for X-, γ-, and high energy β-rays but are not effective for detecting low energy β-particles, α-particles, and neutrons.

Thermoluminescent Dosimeters

Thermoluminescent dosimeters are referred to as TLDs and are used in a manner similar to film badges. The energy levels of certain materials, such as LiF activated with Mg and Ti, may be raised to metastable states by the absorption of radiation, and the molecules remain in their metastable states unless they are heated strongly. On heating, 180–25-°C, the molecules release their excess energies as photons of light, the intensity of which is measured by an M-P tube. Generally, TLDs are small packets of the solid material that are worn in a holder. At weekly or monthly intervals the packets are replaced and sent to a commercial laboratory for reading. As far as the types of radiation detected, TLDs are similar to film monitors, however, they have a greater dose detection range, about 10^{-4} to 1 Gy or 0.01 to 100 rad.

The different personnel monitoring devices discussed above serve different functions, consequently the selection of a device(s) depends on the types and levels of radiation a worker is likely to experience. Essentially, every laboratory should have a survey meter to check for contamination and some type of dosimeter, either a pocket ionization chamber, film badge, or TLD if X- or γ-rays are involved. However, the purpose of these devices is to detect and measure sources of radiation external to the worker's body.

DETECTION OF INTERNAL SOURCES OF RADIATION

Weak β-emitters and α-emitters do not constitute a serious radiological hazard unless they are taken into the body. Therefore, if such radionuclides are used, methods are needed to detect and determine body burdens when they exist.

Bioassays

Most radionuclides are excreted in urine, therefore their presence can be detected by counting urine samples. Other body fluids, such as blood and saliva, may be better for detecting certan isotopes, but these are less easily obtained. Breath samples, nose wipes, and fecal samples may also be used. Internal concentrations of ^{131}I or ^{125}I can be estimated by monitoring the thyroid gland with a sensitive external detector.

Persons working with quantities of tritium in excess of 100 mCi should monitor their urine, and thyroid monitoring should be done when working with more than 10 mCi of volatile (unbound) ^{131}I or ^{125}I.

DETECTION OF CONTAMINATION

Contamination of working areas, and possibly beyond, represents a potential hazard to the public as well as to occupational workers. Contamination may also invalidate laboratory results, therefore emphasis is placed on methods to detect contamination so that corrective measures can be applied.

Survey Meters

Survey meters are useful for surveying working surfaces to detect possible spills, to check equipment and one's hands or clothing for contamination. Usually, the probe is held by hand and moved slowly and closely over the surfaces to be surveyed. Rate meters may have a relatively slow response time and geometric factors are not ideal for detecting a small spot of contamination, therefore considerable patience is required to adequately survey an area such as a lab bench.

Wipe Tests

If radioactive materials are present on a surface, wiping that surface with a filter paper disc or a cotton-tipped stick will remove some of the radioactivity. Assaying the wiping material will indicate if contamination is present. For example, if one is working with tritium, a 2-cm-diameter filter paper disc may be used to wipe an approximate area of 100 cm^2 in a location where contamination might be suspected. The disc is placed in a liquid scintillation vial, scintillaton fluid is added, and the sample counted for reasonably long time (10 min). Such measurements are not designed to be quantitative, but the level of activity is related to the extent of *removable* contamination. Sample counting rates less than 1.5 to 2 times the background counting rate generally are considered to be free of contamination. Of course, the types of radioactive materials that might be present govern the type(s) of counting instrument that should be used to assay wipe samples.

Occupational workers are expected to conduct wipe tests after experiments involving open sources and also at periodical intervals, weekly or monthly, to check for unsuspected contamination. Local regulations vary with respect to the frequency of wipe test surveys, but documentation of such surveys at specified intervals is required. A high frequency of surveys is considered a good practice.

GENERAL LABORATORY RULES FOR WORKERS

The rules for safe handling of radionuclides may vary among institutions, but the variation is largely a matter of format rather than intent. Therefore, the following list of rules is representative of rules a worker is likely to encounter.

Personal cleanliness and use of careful techniques are the primary means of preventing contamination and in protecting persons against inhalation, absorption, or ingestion of radionuclides. In order to minimize contamination and prevent entrance of radionuclides into the body, the following rules should be observed in laboratories where unsealed sources are used.

1. Eating, drinking, smoking, food preparation, food storage, and application of cosmetics shall not be permitted in any laboratories where radioactive materials are used or stored.

2. Storage of food and beverages is not permitted in the same storage location (refrigerator, freezer, etc) as radioactive materials.

3. Protective gloves shall be worn when handling radionuclides.

4. Pipetting of radioactive solutions by mouth shall not be permitted no matter what activity is involved. Remote devices are available and shall be used for such applications.

5. Containers for radioactive samples shall have only one distinctive label indicating the nuclide(s), amount(s), and date(s).

6. No experiment with radionuclides should be undertaken until trial runs, complete in every detail, are made with nonradioactive materials. Such trials should be repeated until the procedure is reproducible and improvements have been incorporated as needed.

7. Any work with radionuclides susceptible to atmospheric distribution (e.g., vaporizing, aerosol producing, spillage, dusting, effervescence of solution or other releases of radioactive gas) shall be confined to suitable hood or glove box.

8. Personnel shall not be permitted to work with radionuclides if there are open cuts or abrasions on the body (e.g., fingers, hands, or arms). Extreme caution must be taken to avoid cuts or puncture wounds, especially when working with materials of high activity or high hazard.

9. Care must be exercised when using organic solvents to avoid skin contact with radioactive materials. (Solvents may make the skin more permeable and many are biohazards.) Radioactive iodine (nonbound forms) has the ability to permeate polyvinyl, latex, and rubber gloves. It permeates polyethylene gloves at a much slower rate. Therefore, when handling radio iodine, 2 pairs of gloves (preferably polyethylene) shall be worn with the outer pair changed after each handling and hands surveyed at the end of the handling.

10. Monitoring of hands, feet, and clothing is recommended when using radionuclides and will be required when large amounts of radionuclides are being used. Protective garments should be left in the laboratory when work is completed or until monitored and found free of contamination.

Contamination of laboratory facilities represent a hazard to personnel and jeopardizes experimental results. The following rules are designed to avoid laboratory contamination.

1. Contaminated equipment, or equipment that has been used and is suspected of being contaminated, shall be isolated in designated areas in the laboratory or in suitable storage spaces until it can be wipe-tested to determine the contamination level, and equipment shall be decontaminated as soon as possible. Contaminated equipment must bear the "Caution Radioactive Material" label.

2. One sink in each laboratory shall be designated for washing contaminated glassware and equipment.

3. Tools, equipment, and apparatus when used in handling radioactive material, should be placed in nonporous metal trays or pans that are lined with plastic-backed absorbent (disposable) paper. This paper should be surveyed and changed frequently.

4. Any working surface where radioactive materials are used shall be covered with plastic-backed absorbent paper (disposable) or polyethylene sheets and be appropriately labeled. This paper or polyethylene sheet should be surveyed and changed frequently.

5. Auxiliary containers, blotters, and covers shall always be used where danger of spills and contamination of personnel exist.

6. Care should be taken that equipment not immediately necessary to the operations being performed is not brought into the working area.

7. Equipment and tools shall be routinely surveyed following their use. No equipment shall be returned to stock unless it is known to be completely free of contamination inside and out.

8. Contamination shall not be allowed to remain on working surfaces unless appropriately shielded.

ACCOUNTABILITY RECORDS

A license holder or authorized user of radioactive materials must be able to account for all radioactive materials he or she has received, used, and disposed. Consequently, records must be maintained that show a complete description of the source, isotope, chemical form, quantity, specific activity, and lot number. As the material is used, each withdrawal and deposition of resulting wastes are recorded. Eventually, all material must be accounted for by disposal methods. Usually, monthly or quarterly accountability reports showing all uses, disposals, and amounts on hand are required.

RECEIPT OF RADIOACTIVE MATERIALS

When a package containing radioactive materials is received the package should be inspected visually, surveyed with a meter, and wipe tested to determine if any leakage occurred during shipment. If no contamination is detected, the package should be opened carefully, the source container wipe tested, and its contents verified and duly recorded. If leakage or apparent breakage has occurred, the supplier should be notified and the package including packing material should be disposed of properly, after a consultation with the radiation safety officer.

DISPOSAL METHODS

Acceptable methods for disposal of radioactive materials vary somewhat among radiological safety programs at different institutions, but possible methods include the following:

1. Radioactive decay
2. Transfer to persons or agencies licensed to accept such material
3. Release into the sanitary sewer system
4. Release as gases, including incineration
5. Burial

Disposal by Decay

A radionuclide that has a relatively short half-life may be disposed of easily by storing wastes until the radioactivity has decayed. As a general rule, waste materials may be disposed of by normal methods if the materials have decayed for 10 half-lives. However, it is important to have wastes segregated according to the radionuclides present and be properly labeled with the quantity of radioactivity and date when placed in storage so that decayed material can be unquestionably identified. As an added precaution the packages should be surveyed before final disposal and all radioactive labels removed or obliterated. Although a good method for disposal of ^{131}I and ^{32}P, decay is not practical for isotopes with a long half-life such as ^{3}H and ^{14}C.

Disposal by Transfer

Unused portions of sources may be transferred to other authorized users if the transfer is proper in regards to possession limits and accountability records. However, true wastes are more difficult to transfer because the accepting agency must then dispose of the wastes. There are commercial firms that handle radioactive wastes and dispose of them, by burial primarily. Because shipment is involved, strict attention must be paid to preventing leakage of materials from containers and to follow proper labeling procedures. Effectively, liquids must be converted to solids, and this may be accomplished by using absorbents, such as sawdust and bentonite.

Disposal by Release into a Sewer System

Justification for flushing radioactive materials down a sink drain is based on the MPC levels adopted by NRC. MPC values for a few radionuclides were given in Table 15-5 and a complete list should be obtained from licensing agencies.

To an individual concerned about the environment, the fact that even low concentrations of radioactivity might be released deliberately may appear appalling. However, everything disposed of in any manner contributes to the environment. It may be modified by organisms in the environment, generally will be diluted greatly, but there are no methods certain to contain any material forever. For some radionuclides, rapid environmental dilution of a concentrated source of radioactivity is an excellent way to reduce the possibility that humans might ingest hazardous quantities of those radionuclides.

In Chapter 15 the annual absorbed dose for a person whose body water contained a concentration of tritium equal to the MPC level was calculated to be 2.02 mGy (0.202 rad). This is less than the limit for an incidental nonoccupational exposure. Nevertheless, the dose is significant and not one that should be tolerated without a significant offsetting benefit. Both ^{14}C and ^{3}H occur naturally, about 14 dpm per gram of carbon and 1 dpm per liter of water. Since vast quantities of carbon and water are in our biosphere, the current release rates of these isotopes could be increased many orders of magnitude without causing a measurable difference in the natural specific activities if adequate mixing occurs. Therefore, rapid dilution and dispersion is a logical method for disposal of ^{14}C and ^{3}H.

Because of their propensity to become diluted, soluble forms of radionuclides may be released in higher concentrations than insoluble or particulate forms. Also, higher concentrations are permitted for release in water than in air.

The point of release is an important factor in safety. An unrestricted area is one to which the public has immediate access. A restricted area is one to which the public does not have access and there is reasonable certainty that the radionuclide will be diluted to concentrations below the values given for an unrestricted area before the public can come in contact with the radionuclide. In Table 15-5 the MPC levels for tritium are listed as 1×10^{-1} and 1×10^{-3} μCi/ml for restricted and unrestricted areas, respectively. For example, suppose tritium is flushed down a sink drain and into a city's sewer system at a concentration of $1 \times 10^{-1} \mu Ci/ml$. It surely will be diluted by a factor of 33 (to $3 \times 10^{-3} \mu Ci/ml$) before anyone could possibly come in contact with the contaminated water.

When releasing radioactivity in the sewer system it does not necessarily need to be diluted to MPC levels prior to release. Instead the amount of material that can be released depends on the water usage of the institution. Suppose in a building water is used at a rate of 2500 gal/day. In a month, 2.84×10^8 ml of water would enter the city's sewer system and since the sewer system is a restricted area, the appropriate MPC level for tritium is $1 \times 10^{-1} \mu Ci/ml$.

$$(1 \times 10^{-1} \mu Ci/ml)(2.84 \times 10^8 \text{ ml}) = 2.84 \times 10^7 \mu Ci$$

Therefore, a total of 28.4 Ci of ^3H could be released in a period of 1 month without the average concentration exceeding the MPC level for tritium.

Disposal as Gases

Gaseous radioactive materials are not utilized extensively in most types of biological research, but when working with gaseous products, some release is almost inevitable. The production of gaseous products or aerosols is somewhat unpredictable, therefore radioisotopic work is performed in a fume hood whenever possible. Releases into a fume hood are wastes and hence represent disposal. Safety considerations for fume hoods will be discussed later in this chapter.

If burning produces radioactive gaseous products, incineration may be a practical disposal method. Particularly for ^{14}C wastes, incineration may be feasible since the carbon may be completely oxidized (CO_2) and the volume of gases that passes through the incinerator may be sufficient to dilute the resulting $^{14}CO_2$ to appropriate MPC levels. Which MPC level for $^{14}CO_2$, restricted ($5 \times 10^{-5} \mu Ci/ml$) or unrestricted ($1 \times 10^{-6} \mu Ci/ml$), should be considered? If persons can be prevented from approaching the top of the incinerator's stack the restricted levels *may* apply. Gases move with wind currents, which generally increases gaseous dilution, but without wind currents, dilution still occurs by gaseous diffusion. Therefore, if it can be shown that diffusion will reduce the concentration of the stack's effluents to levels below that for an unrestricted area before reaching the closest unrestricted area, then restricted area levels may be used. Otherwise, unrestricted area MPC levels apply.

Example Problem 16-1.   Would it be feasible to incinerate the carcasses of 10 laboratory rats, each of which had been injected with 50 μCi of a ^{14}C-labeled drug? The excreta, bedding, and other wastes would be incinerated along with the carcasses.

The incinerator available has three natural gas burners that supply a total of 1,050,000 BTU/hr and can burn 175 lb of waste in 1 hr. Three air blowers supply 500 ft³/min of air to ensure complete combustion of carboneous wastes. Although the burners will contribute to the volume of gases that goes up the stack, we will ignore that contribution as an extra margin of safety and consider the volume of air supplied by the blowers only. The volume discharged from the stack in 1 hr would be

$$(3)(500 \text{ ft}^3/\text{min})(60 \text{ min/hr})(2.83 \times 10^4 \text{ ml/ft}^3) = 2.55 \times 10^9 \text{ ml/hr}$$

To avoid calculating air diffusion, we will use the unrestricted area MPC level for CO_2 at the top of the stack. The amount of ^{14}C that could be combusted in 1 hr is

$$(1 \times 10^{-6} \text{ } \mu\text{Ci/ml})(2.55 \times 10^9 \text{ ml/hr}) = 2.55 \times 10^3 \text{ } \mu\text{Ci/hr}$$
$$\text{or}$$
$$2.55 \text{ mCi/hr}$$

Since the total amount of radioactivity administered was only 0.5 mCi, incineration would be a good method for disposing of the carcasses without exceeding MPC levels. To help distribute the burning of the carcasses over the hour period it would be a good idea to add nonlabeled wastes such as paper and wood shavings. The burning rate (175 lb/hr) would be considered in determining the amount of unlabeled wastes to be mixed with the carcasses. Currently, the Environmental Protection Agency is proposing legislation that may reduce acceptable gaseous release rates markedly.

Disposal by Burial

From a radiological safety perspective, burial is a suitable method for waste disposal provided the geological features of the burial site are such that the buried radionuclides do not move through the soil or contaminate surface or ground water supplies. Suitable sites exist in most states, but for a license holder or user, wastes destined for burial are usually transferred to commercial firms that operate burial sites.

Recent sensitivity to the burial issue by the public and several state legislative bodies has severely, and unreasonably, curtailed burial as a means of disposal. Many states have passed laws prohibiting burial in their states and most prohibit the shipment of wastes into their states for burial purposes. As of 1990, the states of Washington, Nevada, and North Carolina are the only states that permit radioactive wastes to be shipped into their state for burial. Because of supply and demand, burial is an expensive disposal method.

Generally, wastes must be segregated according to type, must be in a solid form, and are shipped in steel drums. Regulations for shipment of wastes are determined by the U.S. Department of Transportation.

Mixed Wastes and Liquid Scintillation Samples

Mixed wastes are those that contain radioactivity plus other material classified as "hazardous" by the Environmental Protection Agency. Laboratory wastes that contain radioactivity and carcinogens or hazardous solvents generally fall in this category of wastes. Regulations concerning mixed wastes will be forthcoming, but currently there are not ratified methods for their disposal. About the only way of dealing with mixed wastes is to eliminate or reduce one of the noxious agents to innocuous levels.

Spent liquid scintillation samples may be classified as mixed wastes. However, because of MPC levels, *tritium-* and *carbon-14*-containing liquid scintillation samples may be excluded if the specific activities of their solutions are below 0.05 μCi/g, which translates to about 1×10^5 dpm/ml. Since counting samples with specific activities above this level are rare, normally ^{3}H- and ^{14}C-containing liquid scintillation vials can be disposed by methods appropriate for the type of solvents they contain; as hazardous solvent, such as toluene, or as normal refuse if their solvents are not hazardous. Consequently, liquid scintillation vial disposal problems are reduced dramatically when ^{3}H and ^{14}C are used with nonhazardous liquid scintillation solutions.

Unfortunately, similar rules do not apply to liquid scintillation samples containing ^{35}S, ^{32}P, ^{125}I, or ^{131}I. Nevertheless, all of these isotopes have short enough half-lives to permit the activity to decay away. Thus, it is feasible to store such samples for a period sufficient to reduce the radioactivity to background levels and then dispose of the solutions as dictated by the type of solvent involved.

Materials classified as biohazards, such as pathogenic bacteria and recombinant DNA, must be biologically inactivated before disposal. The common methods of inactivation include autoclaving and chemical treatment. If biohazardous materials also contain radioactivity, the wastes may be regarded as radioactive wastes only after inactivation.

LABORATORY FACILITIES

The requirements concerning laboratory construction are not stringent if moderate levels of radioactive materials are used. To the extent possible, porous construction materials should be avoided so that decontamination procedures are effective in removing radioactivity. For example, bare concrete floors are difficult to decontaminate, paint would reduce porosity, floor tile would be better, but seamless linoleum is preferred because a spill could be cleaned up more easily, and, if necessary, the floor covering could be removed and replaced. Bench tops should be of nonporous materials; stainless steel is preferred but expensive. However, large $(24 \times 36 \times 2$

in.) plastic or stainless steel trays lined with absorbent paper are recommended for working surfaces, and, if used properly, should prevent lab bench contamination. Walls should be painted with a strippable yet chemically stable paint.

Fume Hoods

A fume hood with a face velocity of at least 100 ft/min is necessary for most research laboratory facilities. The face velocity of a hood is the linear flow of air across the open face of the hood; consequently, 100 ft/min represents a flow of 100 ft³/min for each square foot of the hood's open face or 100 ft³/min·ft². Such a flow rate is considered to be sufficient to prevent any diffusion of gases from inside the hood to the outside. Care should be taken to avoid clutter in a fume hood because objects can create eddy currents that may be directed backward.

LABORATORY ACCIDENTS

Experiments should be well planned to avoid unexpected problems but accidents do occur. In case of an accident, several immediate actions should be considered. Suppose a flask containing radioactive liquid has fallen on the floor and broken, but no injuries resulted from flying glass. What action should be taken?

1. Notify everybody in close proximity that a spill has occurred.
2. Consider your personal safety. Has your clothing been seriously contaminated? If so, remove the clothing (a good reason for requiring lab coats) and if your skin has been contaminated, wash that area lightly.
3. Prevent the radioactivity from spreading. Make an absorbent dam around the liquid and begin placing absorbent paper or towels on the liquid to soak up as much of it as possible.
4. Keep others from the area. Close the laboratory door and post a makeshift sign warning people of the accident and against entering the laboratory.
5. Call the Radiation Safety Officer or a person experienced in dealing with radioactivity for help and advice.
6. Begin decontamination procedures as soon as possible.

If injuries occur, they take first priority. For example, suppose the accident resulted in a lacerated wrist. Administer first aid to the injury, and depending on the severity of the injury, seek immediate medical attention. If you are a bystander to an accident, render aid if the person is seriously injured.

The philosophy to be considered is that serious injuries (severe lacerations, fainting, etc.) generally represent a much greater health hazard than the possible radiation dose.

DECONTAMINATION PROCEDURES

Procedures appropriate for decontamination depend on the degree of contamination and type of radionuclide involved. If large areas or quantities of radiation are involved, a person should contact his or her Radiation Safety Officer or an experienced radiological health physicist.

Laboratory Spills

For small spills or contamination, the usual method is to absorb all liquid with absorbent paper (paper towels). These materials should be placed in plastic bags (double bagging is advisable) and a label attached indicating the estimated maximum activity, the radionuclide involved, and the date. Then the area should be washed with warm soapy water. Iodine-based disinfectants or solutions are quite effective in cleaning radioactive iodine contamination. The wash water may be poured into the sink where contaminated glassware is normally washed if MPC levels are not exceeded after dilution with the building's water. After the area has dried, wipe tests or surveys with an appropriate detector should be conducted to see if contamination persists. Some areas may require several washings to reduce the contamination to an acceptable level (1.5 to 2 times the background counting rate). Of course, protective clothing such as lab coats, gloves, and probably disposable booties should be worn during the decontamination process.

Laboratory Glassware

Soap and water washing are usually sufficient to decontaminate glassware but special cleaning products are available commercially. Strong cleaning solutions $K_2Cr_2O_7-H_2SO_4$, $KMnO_4-NaOH$, 10% EDTA for metal radionuclides, and 6 N HCl are also appropriate for decontaminating glassware.

Hands

After completion of work and removal of gloves, a person should wash their hands with mild soap and water then monitor them with a survey meter. If contamination persists, a second or third washing may be required using a heavy lather and a soft brush. Stiff brushes or implements that abrade the skin should be avoided; instead numerous light washings are recommended.

CONCLUSION

"Results oriented" research workers may find the rules and regulations concerning the use of radioactive materials an inconvenience on some occasions, and perhaps examples can be cited where the inconvenience of a particular rule does not appear to be warranted in terms of the hazard involved. Nevertheless, it is obvious that unsafe practices are more likely to occur when individuals are given latitude in making judgments concerning the hazardousness of given operations on the spur of the moment. Therefore, rules and regulations are strictly enforced in most institutions. At this point, the reader should be able to see the logic behind all the rules and regulations, and this should aid him or her in accepting them willingly, which not only reduces personal risks but helps protect the public at large.

PROBLEMS

Some equations in Chapter 15 as well as data in Tables 15-4 and 15-5 may be used to solve some of the following problems.

1. Assume local rules permit release of radionuclides into the sanitary sewer system when their concentrations, averaged over a week period, do not exceed applicable MPC levels. In a building that has an average water usage rate of 3000 gal/da, what is the maximum amount of soluble ^{22}Na radioactivity that could be released per week?

2. Assume the same local rules described in problem 1 apply, and that during a 1-year period a total of 5.1 Ci of tritium was purchased by all of the users in the building. Would you anticipate the need for restrictions in flushing ^3H wastes into the sewer system? Support your conclusions with calculations.

3. A person planning to iodinate a protein with 10 μCi of ^{125}I estimated that the loss of volatile iodine would not exceed 2% during a 4-hr working period. The hood to be employed has a face velocity of 125 ft/min and an opening of 8 ft^2. If the maximum release occurred, would the average concentration in the hood exhaust during the working period exceed the MPC level for an unrestricted area?

4. To synthesize a desirable tracer, $^3H_2$ is to be employed for the reduction of a double bond in a biological compound. The work is to be conducted in a fume hood that has a face velocity of 125 ft/min and an opening of 10 ft^2. Calculate the maximum amount of activity that could be released in the hood during a 24-hr period without the average concentration of the hood's exhaust exceeding the MPC level for a restricted area.

5. Suppose a person transporting a solution containing 10 μCi of [^{14}C]alanine accidentally dropped and broke the container in a parking

lot. Propose a method of disposing of the resulting "liquid waste" using calculations to support your recommendations.

6. If a person somehow absorbed the amount of radioactivity in 10 ml of a ^{14}C waste solution that contained the MPC level of activity for an unrestricted area, what would be the resulting lifetime dose to that person assuming body weight to be 70 kg?

REFERENCE

Wang, Y. (1969). *Handbook of Radioactive Nuclides*. CRC Press, Boca Raton, FL.

PART IV

Radiotracer Methodology

The use of radiotracers in research has become so commonplace that some of the underlying assumptions and advantages frequently are not given due consideration. In this part such topics will be discussed along with general approaches to specific types of tracer experiments.

CHAPTER 17

Use of Labeled Compounds in Tracer Experiments

ADVANTAGES OF RADIOTRACERS

With modern counters, the measurement of 1000 dpm of activity can be accomplished with a high degree of precision and accuracy, perhaps less than 2% error at the 95% confidence level when counting times in excess of 10 min are used. Since tritium counting efficiencies are among the lowest of the common biological tracers, a calculation concerning detection sensitivity is appropriate. In Chapter 4, examples of maximum specific activity calculations were shown, and for a carrier-free ^3H-labeled compound with one atom of ^3H per molecule, the specific activity (SA) in dpm/mol is

$$SA = \frac{\lambda N}{\text{quantity}} = \frac{\dfrac{(\ln 2)(6.022 \times 10^{23} \text{ atoms})}{(12.26 \text{ yr})(5.256 \times 10^5 \text{ min/yr})}}{1.0 \text{ mol}}$$

$$SA = 6.478 \times 10^{16} \text{ dpm/mol}$$

Assuming 10^3 dpm can be measured accurately, the corresponding quantity of the labeled compound is

$$\frac{10^3 \text{ dpm}}{6.478 \times 10^{16} \text{ dpm/mol}} = 1.54 \times 10^{-14} \text{ mol}$$

This calculation indicates that approximately 10^{-14} mol can be measured quite accurately, and this quantity is several orders of magnitude below the lower limits of most chemical methods of detection.

Another rather obvious advantage of radiotracer use is the capability of ascertaining interconversion of compounds that would be impossible by quantitative analyses alone. Stable isotopes can be used as tracers as well so this is not a unique advantage of radiotracers, but the sensitivity of detection, even with modern mass spectrometers, makes stable isotope tracing less sensitive in most situations.

Although these advantages of radiotracers have made tracer methodology essential in many types of research, there are pitfalls that all too frequently are not considered when experiments are planned. Some of the assumptions and deficiencies in tracer methodology will be delineated as we discuss factors that should be considered in selecting, preparing, using, and storing labeled compounds.

AVAILABILITY OF SUITABLE TRACERS AND LABELED COMPOUNDS

The primary elements in biological compounds are H, C, O, N, P, and S, but ions of many metals such as Ca, Na, K, and Fe are present as well. As mentioned in Chapter 4, highly suitable radionuclides of O and N do not exist, so most tracer experiments involving those elements must be performed with stable isotopes. Therefore, the primary radiotracers in biological research are ^3H, ^{14}C, ^{32}P, ^{35}S, ^{125}I, and ^{131}I. Phosphorus-33 has a longer half-life and softer β than ^{32}P, which make it better suited for many radiotracer experiments; however, its cost has limited its use. Iodine is not a common element in biological compounds, but its radionuclides are used to label native compounds, primarily proteins. Some other radionuclides, such as ^{51}Cr and ^{99m}Tc, are used similarly.

Radionuclides are produced in nuclear reactors and accelerators by bombarding appropriate stable nuclides with neutrons, protons, other charged particles, or rays to induce nuclear reactions. As an example, ^{14}C is produced from ^{14}N by an (n, p^+) reaction that is identical to the reaction responsible for the cosmological formation of ^{14}C.

$$^{14}N + n \longrightarrow p^+ + {}^{14}C$$

After the transmutation reaction, the radionuclide must be isolated by purification methods and converted to a chemical form suitable for organic synthesis. Frequently, the starting compounds are oxides of the elements, but in addition to $^{14}CO_2$, $^{14}CN^-$ is used extensively in producing ^{14}C-labeled compounds.

Tritium labeling is relatively easy because of tritium's facile incorporation into molecules with suitable functional groups. Alkene and alkynes are easily reduced with tritium gas ($^3H_2$), and sodium or potassium boro[^3H]hydride may be used to reduce carbonyl groups.

Another labeling method, referred to as *isotope exchange,* may be used when usual synthetic methods are too time or expense consuming. Different agents or catalysts are used to promote the exchange of ^3H for ^1H, but in the *Wilzbach technique,* the energy from decaying ^3H atoms is utilized to promote exchange reactions. The compound is placed in a sealed container with $^3H_2$, generally under pressure, and allowed to exchange for at least several hours or perhaps weeks. Because of radiolysis, the effect of radiation on organic compounds, a significant proportion of the compound may be altered, but frequently the yield of the chemically unaltered compound with exchanged atoms is sufficient for the method to be successful. Isotope exchange with ^3H invariably yields compounds with *labile tritium.* Hydrogen atoms attached to heteroatoms, such as N and O, generally are exchangeable with hydrogen atoms in polar solvents that have similar groups, for example $-OH$ or $-NH_2$. Tritium atoms incorporated at such positions are referred to as labile tritium. Labile tritium is easily removed by dissolution of the compound in a suitable solvent containing exchangeable hydrogen atoms, but several changes of solvent may be necessary. Depending on the chemistry of a particular compound, tritium bonded to certain carbon atoms may be labile as well, for example the α-hydrogens of malonic esters. Acidic and basic conditions may increase the lability of tritium at certain positions in some organic compounds.

Generally, the purity of ^3H compounds prepared by isotope exchange is inferior to that of labeled compounds prepared by organic synthesis, frequently because of relatively low yields and the abundance of altered, but chemically similar, side products. Another deficiency in this type of labeling method is that it invariably leads to randomly labeled compounds.

Several commercial firms stock a number of ^3H- and ^{14}C-labeled compounds that may be available with either isotope and with the label being in different positions. Of course, commonly used compounds labeled with other isotopes are available as well, but there are fewer choices. In addition to stocking common compounds, some companies provide custom synthesis services and isotope exchange labeling by one or several methods.

Biosynthetic labeling, performed by feeding a labeled precursor to an organism, may be used to prepare tracers of complex molecules, but generally the label cannot be incorporated at a specific position. Therefore, most biosynthetically prepared tracers are randomly or uniformly labeled.

For many compounds, a choice between isotopes may not exist, but for the majority of organic compounds ^3H and ^{14}C labels are available. An advantage of ^3H over ^{14}C is due to the shorter half-life of ^3H, which yields higher specific activities, and a disadvantage is that tritium-labeled compounds are more likely to lose their labels by metabolic reactions that are not anticipated to be involved in the planned experiments.

Labeling of Proteins and Polypeptides

Synthesis of specific proteins labeled with ^3H, ^{14}C, or ^{35}S generally is not practical because of the size of such molecules, but two general methods for labeling proteins are used routinely.

Conjugation. Previously labeled low-molecular-weight molecules may be conjugated to a protein, usually by a reaction involving a particular functional group of certain amino acid side chains. The functional groups that may be utilized include free amino, hydroxyl, carboxyl, and sulfhydryl groups. Various radionuclides may be used for labeling the small molecule including radioiodine. Of course, the chemical nature of a protein is changed by conjugation, but in many cases the biological activity of the protein is not reduced below levels of usefulness.

Although many conjugation reactions have been proven to be effective, the general approach to such methods is illustrated by the popular Bolton and Hunter (1973) method. Bolton–Hunter reagent, *N*-succinimidyl 3-(4-hydroxy-5-[^{125}I]iodophenyl) propionate, is produced by iodination of *N*-succinimidyl 3-(4-hydroxyphenyl) propionate using chloramine-T as the oxidizing agent (see Direct Iodination of Proteins below). Then the reagent is conjugated with the protein of interest by forming a secondary amide bond between the carbonyl of the propionate moiety and one of the protein's free amino groups, such as an ε-amino group of a lysine residue. Milder conditions can be maintained for this type of labeling reaction than are possible in the direct iodination methods.

Direct Iodination of Proteins. Most proteins contain tyrosine residues that are readily iodinated with ^{125}I or ^{131}I to form mono- or diiodotyrosine residues in which the iodine atoms are incorporated *ortho* to the phenolic group. As with conjugation methods, the protein is altered structurally, and if the iodinated residues are critically involved in the structure or function of the protein, this type of labeling is not acceptable. However, the tyrosine residues most accessible to iodination reagents generally are among those least involved in structure and function; therefore, limiting the extent of the iodination reaction so that only a fraction of the tyrosine residues is labeled usually yields proteins with adequate specific radioactivities and full or only partially diminished biological activities. Radioiodine may also be incorporated into histidine residues as well, but tyrosine is the primary residue involved in most cases. Both ^{125}I and ^{131}I may be used, but because of its longer half-life and less biological damaging radiation, ^{125}I has become more popular. Several methods for the direct iodination of proteins have been developed, and although ^{125}I and ^{131}I may be employed, ^{125}I will be used in examples.

Iodine Monochloride. In the iodine monochloride method (McFarlane, 1958) unlabeled I-Cl is equilibrated with ^{125}I$^-$ to generate ^{125}I-Cl, which reacts with the tyrosine residues. Simple mixing is sufficient to equilibrate the forms of iodine. The protein is added and allowed to react for a short period of time, only a few seconds to a minute, then the reaction is quenched with sodium metabisulfite that reduces all iodine to the iodide form. To aid in the removal of residual $^{125}I^-$, NaI or KI is added, and the protein is repurified, often by gel filtration methods. Of course, to prevent ordinary protein denaturation, appropriate buffers and temperatures must be employed.

In this method the incorporation of radioiodine can be controlled reasonably well by limiting the quantities of the halogen reagents, but the use of I-Cl permits some incorporation of stable iodine into the protein. Consequently, the specific activity of the labeled protein is not as high as it could be if only radioactive iodine were incorporated.

Chloramine-T. The chloramine-T method (Hunter and Greenwood, 1962) is a widely used method for direct iodination of proteins, and in this method, $Na^{125}I$ is oxidized *in situ* to produce the radioiodine ($^{125}I_2$) that reacts with the residues. Reaction time periods, methods for stopping the reaction, and subsequent purification steps are similar to those described for the monochloride method. In the chloramine-T method, the specific activity of the ^{125}I incorporated is the same as that of the $Na^{125}I$ supplied.

Solid State Oxidizing Reagents. Oxidizing agents used to generate $^{125}I_2$ from $^{125}I^-$ usually cause some denaturation of the protein being labeled, but the degree of denaturation is dependent on the length of exposure and the sensitivity of the protein. The use of water-insoluble oxidizing agents appears to offer an advantage, because the protein receives less exposure to the agent.

A water-insoluble reagent, first described by Fraker and Speck (1978), may be used, and a commercial product called Iodo-gen® is marketed by Pierce. Also available commercially are polystyrene beads to which chloramine-T residues are bound.

Lactoperoxidase. Although the presence of oxidizing agents is not avoided, this method appears to afford a more gentle reaction environment. Lactoperoxidase catalyzes the iodination of tyrosine residues, but requires hydrogen peroxide (Marchalonis, 1969). To avoid high concentrations, the H_2O_2 is either added incrementally or generated continuously *in situ* by glucose oxidase (Hubbard and Cohn, 1972). Compared to other methods, reaction times are quite long, minutes to hours, and the reaction may be terminated by the addition of cysteine or by dilution and purification of the labeled protein. Usually more effort is required in this method to purify the product than when other labeling methods are used. To simplify purification of the product, lactoperoxidase immobilized on beads (Koch and Haustein, 1981) may be employed.

Other. A number of oxidizing agents have been used for iodination purposes, but electrolytic iodination avoids the use of harsh reagents (Pennis and Rosa, 1969). Iodination is usually performed by individual investigators, and the equipment and procedures for the electrolytic method are such that the method is not a practical solution for occasional use.

Ward (1984) has reviewed methods for labeling cell surface proteins, and Bolton (1983) has provided an excellent discussion of radioiodination techniques.

Criteria for Selecting Label Positions

In tracer experiments the labeled compound is assumed to have chemical and physical properties identical to the analogous unlabeled compound. Theoretically, identical behavior is unattainable, but with judicious consideration of isotopes, labeling patterns, and experimental design, practical objections to this assumption can be overcome usually.

The Isotope Effect. The stabilities of single bonds are dependent on the masses of the atoms involved, more precisely, for given elements stability is directly proportional to the square root of the atoms' masses. Thus, a $^{12}C-^3H$ bond is about 1.7 times more stable (neglecting the potential for radioactive decay) than a $^{12}C-^1H$ bond at a given temperature. Consequently, reaction rates involving the rupture of such bonds occur more slowly when heavier isotopes replace lighter ones. This phenomenon is referred to as the *isotope effect*. Theoretically, rates for reactions involving ^3H and ^{14}C bonds are respectively about 58 and 93% of the rates involving ^1H and ^{12}C bonds. Since bond stability is related to the square roots of the atom's mass, the isotope effects for heavier elements are not as dramatic as that for ^3H. However, even tritium may not exhibit significant isotope effects when multistep reaction mechanisms are involved. If actual bond rupture or formation is not the rate-limiting step in a reaction sequence, the isotope effect may be negligible. Isotope effects have been exploited to deduce enzyme-catalyzed reaction mechanisms, but stable isotopes are commonly employed for such studies.

For most tracer experiments, the molecular location of the isotope should not be a position where it is likely to undergo a chemical or metabolic reaction. For example, if [^{14}C]phenylalanine were to be used in studies involving peptide bond formation, one should choose to have the ^{14}C in a position least likely to be involved in the reaction. Thus, having the label in the phenyl group instead of in the carboxyl group is desirable to avoid possible isotope effects.

Isotopes Are Traced. In determining desirable label positions it is important to remember the obvious, *only the radioactive atoms are traced*. This may be an important benefit when a series of metabolic reactions is being investigated, but a deficiency when a particular reaction is being studied and unrelated metabolic activities change the chemical nature of the tracer. Of course, the nature and purpose of the experimental system have an important bearing on the better locations of specific labels. For this reason, a random or unknown labeling pattern is never a desirable situation. In some experiments, a randomly labeled compound may be the only type available, and it can be useful if the resulting data are interpreted accordingly.

Another factor that can influence the correct interpretation of tracer results is the presence of two or more radioactive atoms in a single molecule. The disintegration of one atom changes the compound drastically because a ^{14}C atom is transformed to ^{14}N and ^{32}P to ^{32}S, ^{35}S to ^{35}Cl, ^{3}H to ^{3}He, and so on. In addition to the transmutation of elements, the recoil energy of the decaying atom is usually sufficient to break chemical bonds. Obviously, the chemical identity of the residual compounds, which may retain a label, is unknown, and unfortunately such labeled products are subject to subsequent tracing.

Despite the potential for tracing unknown compounds, multiple labeling positions are used for some polymers, such as DNA and RNA, to increase detection sensitivity. As long as limitations of multiple label sites are recognized, useful information may be gained under certain circumstances. Many molecular biological techniques are based on complementary binding between two strands of DNA or RNA with one strand being labeled in several positions. Although the disintegration of the first radioactive atom in a strand changes the chemical nature of the strand (the strand may be broken), the unaltered segment(s) of the strand should retain their capacities to bind specifically at complementary regions. However as additional atoms decay, the "native" regions become shorter, and the possibility of detecting short complementary regions instead of fully complementary strands increases. Consequently, radioactive probes with multiple label sites have useful lifetimes that are much shorter than their levels of radioactivity may indicate.

PURITY OF LABELED COMPOUNDS

Chemical purity refers to the proportion of material in a given chemical form, whereas *radiochemical purity* refers to the proportion of radioactivity in the stated form. Unlabeled contaminates are ignored in assessing radiochemical purity, but experimentally, neither type of purity should be ignored. Labeled comounds in different chemical forms or with labeling patterns that differ from the stated form constitute radiochemical impurities. *Radioisotopic* or more properly *radionuclide purity* refers to the proportion of the radioactivity due to the stated radionuclide. Generally, radionuclide purity is not a problem for radionuclides such as ^{3}H and ^{14}C, but may be a problem when several radionuclides of the same element exist and are produced by competing nuclear reactions. As an example, the presence of ^{32}P in a ^{33}P-labeled compound constitutes a radionuclide impurity.

Assessing Radiochemical Purity

Most often, radiochemical purity is assessed by chromatographic methods, including paper, thin-layer, gas, and low- and high-pressure liquid chromatography. Electrophoresis is useful, especially for macromolecules, and when suitable quantities exist, recrystallization to constant specific activity may be used. All of these methods involve attempts to separate radioactivity from an assumed chemical component, and failure to do so indicates the presence of a labeled compound.

Carriers for the assumed labeled compound as well as for suspected contaminates may be added prior to the separation step for two reasons. First, the detection of radioactivity is more sensitive in most cases than is the detection of components by chemical (e.g., staining reactions) and physical (e.g., spectrophotometry) methods, and the addition of carriers to labeled components reduces specific activities. Second, in virtually all methods, nonideal separation behavior occurs to a limited extent (e.g., nonspecific adsorption by a chromatographic support), and the fraction of an individual chemical component that behaves nonideally is reduced by increasing the quantity of that component subjected to the separation method. Of course, adding too much carrier can overload the separation system and have the opposite effect.

AUTORADIOLYSIS

The purity of components in an experimental system is important for obvious reasons, but particular attention is given to labeled compounds because they are less stable than their unlabeled counterparts. The instability of labeled compounds is due primarily to radiolysis, the alteration of chemical structure due to radiation exposure (see Chapter 14). Of course, the radiation exposure to a labeled compound is self-generated, hence the radiolytic decomposition process may be called *autoradiolysis*.

STORAGE OF LABELED COMPOUNDS

Although compounds have different susceptibilities to radiolysis, some general methods for reducing self-decomposition and prolonging storage-life are available.

1. Reduce the molar specific activity to as low a level possible for intended experiments.

2. Dilute or disperse the labeled compound with unlabeled material, such as solvent.

3. Store labeled compounds at as low a temperature as possible.

4. Add free radical scavengers to solutions if possible.

5. Reduce exposure to oxygen.

6. Maintain storage conditions that maximize normal chemical stability.

Generally, a combination of these methods is used, but because of differing chemical properties, a given method may be effective for one compound yet counterproductive for another. Therefore, the above storage conditions should not be considered applicable for all compounds.

The reduction of the molar specific activity is always effective in prolonging radiochemical (but not chemical) purity. Of course, the reason is that the proportion of radiation energy absorbed by the labeled and nonlabeled compounds is directly related to the relative quantities present. While carrier-free labeled compounds provide more latitude in experimental designs, it should be recognized that their shelf-lives are considerably shorter than analogous low specific activity compounds.

Dilution or dispersal of a labeled compound with other materials generally is effective in reducing radiolysis because a greater proportion of the radiation energy is absorbed by the diluent. However, certain diluents may promote secondary radiolytic reactions, and the production of free radicals is of principal concern. Water and alcohols yield hydrogen and hydroxyl ($\cdot$OH and $\cdot$H) free radicals readily, and chlorinated hydrocarbons, such as chloroform and methylene chloride, are prone to producing chlorine radicals. Aromatic solvents give lower yields of free radicals, and benzene is considered the best solvent for dilution of labeled compounds. Unfortunately, many biochemicals are not soluble in benzene or other aromatic solvents so water, aqueous-alcohol, and alcohol (ethanol primarily) are used as solvents when necessary. Although there are exceptions, most compounds receive some protection by dilution with water or alcohols despite their propensity to form free radicals. Solvents used for dilution should always be of the highest purity, and those having, or capable of forming, peroxides should be avoided.

Some solid diluents have been used, for example, in the past, labeled carbohydrates were impregnated in chromatography paper strips for storage and then eluted with a small volume of solvent immediately before use. Powdered cellulose, charcoal, and sand have been used as solid dispersal agents as well.

Lower temperatures enhance the chemical stability of compounds and reduce secondary radiolytic reaction rates as well. Therefore, storing labeled compounds at as low a temperature as possible is a good practice generally. However, deleterious effects may arise when solutions are frozen due to the possibility of concentrating the solute. As a solution begins to freeze, crystallization of the solvent occurs first and solute molecules tend to be excluded from the growing crystal. Consequently, the last volumes to be frozen generally have much higher concentrations of the solute. Thus, the beneficial effects of dilution are diminished, and the lower temperature may not compensate for the concentrating effects. Freezing methods that yield a homogeneous dispersion of labeled solute should be used when possible.

Free radical scavengers reduce secondary radiolytic reactions, but they must be compatible with the labeled compound and its intended use. Some scavengers that have been used effectively include sulfhydryl reagents, such as mercaptoethanol and cysteamine, ascorbic acid, and benzyl alcohol.

Because of oxygen's propensity to participate in free radical reactions, storage conditions that eliminate or reduce oxygen concentrations generally are effective in prolonging storage-life. Antioxidants may be added to labeled compounds if they are compatible. The percentage of impurities resulting from air oxidation of solids and liquids is dependent on the ratio of mass to exposed surface area, consequently, having a tiny quantity of material spread on the interior surface of a vial should be avoided. To reduce the concentration of dissolved oxygen in solutions, solvents or the solutions themselves may be purged with an inert gas (N_2 or Ar), and all opened vials should be flushed with an inert gas before re-sealing or capping for storage.

Conditions that promote chemical stability are effective in reducing autoradiolysis as well. As an example, the stability of a compound may be affected by the pH of the solution; nucleotides are more stable in appropriately buffered solutions.

Although radiolysis may not be involved, other storage conditions may be important for maintaining radiochemical and chemical purity. The possibility of microbiological degradation of organic compounds should always be considered in selecting appropriate storage conditions. Also all containers, including glass, contribute contaminants that may or may not accelerate chemical and radiolytic decomposition. Normally, the concentration of such contaminates is inconsequential when a reasonable mass of material is involved, but when minuscule quantities of labeled compounds are involved, the container contaminates may reach substantial relative concentrations.

For additional details about autoradiolysis and storage conditions the reader should obtain a copy of the review by Evans (1983).

Suppliers of labeled compounds are quite interested in, and knowledgeable about, conditions that prolong the shelf-life of radiochemicals. Samples are shipped under conditions they deem best; usually storage recommendations are included and requests for additional information are normally honored. Also included in most shipments of labeled compounds are statements or evidence (usually chromatographic) concerning product purity. Although such information is valuable, careful investigators recheck purity according to their own criteria. Also if a significant storage period is involved between uses, the purity of a labeled compound should be reassessed.

DETECTION SYSTEMS FOR CHROMATOGRAPHY AND ELECTROPHORETIC METHODS

In biological and medical research, many different methods are used to isolate or separate specific compounds, and radioactivity measurements are frequently necessary during or following separation procedures. Although ordinary counting methods are satisfactory for some procedures, specialized instruments or instrument attachments may be more convenient and yield more accurate results by many methods. A discussion of radioactivity-assay methods associated with separation methods follows.

Autoradiography

Paper and thin-layer chromatography sheets or plates as well as electrophoretic gels or strips may be subjected to autoradiography and staining reagents. Obviously, techniques that yield high resolution, both for separations and autoradiography, are desirable. Nevertheless, the range of activities detectable by these methods is limited by the relative blackening of autoradiographical film, and quantitative data regarding the proportions of labeled components are difficult to judge accurately. Densiometric scanning of developed autoradiograms may provide approximate quantitative data.

Counting Segments

Segments of a separation specimen may be assayed for radioactivity by counting methods, but it is important for the radioactive material to be eluted from the segments if the adsorbent or gel absorbs a significant proportion of the radiation. See the Heterogeneous Systems section in Chapter 10 for assay methods involving liquid scintillation counting of chromatographic or electrophoretic segments. If γ-emitters are involved, the assay of segments by solid scintillation is quite convenient, and self-absorption by the segments is negligible in most cases. Specimens may be segmented by cutting or scraping absorbent from plates, and segment size, which affect resolution capabilities, may be kept even or may vary for specific purposes. For example, thin-layer absorbent may be scraped from equal areas of a plate or from areas corresponding to the location of known compounds (stained spots). The later method may be used when it is desirable to determine the percentages of activity associated with particular compounds.

Scanning Methods

Planer specimens may be scanned for radioactivity, and radiochromatogram scanners are available commercially. Most scanners utilize gas-ionization detectors that operate in the G-M or proportional regions. Such a detector was described in Chapter 6. Other types of detectors may be employed when compatible electronics are provided. In older instrument designs, either the chromatogram or detector is moved systematically, and a rate meter monitors the radioactivity beneath the detector window. Depending on the scanner, one or two scanning directions may be employed, and the readout usually is in the form of a two- or three-dimensional graph. Self-absorption by the separation matrix compromises quantitative measurements, but a uniform thickness of the matrix permits reasonable relative assay data.

Gel and Blot Imaging Systems

Although multiwire proportional detectors are used for various imaging purposes, a major use is for analyzing planner electrophoretic gels, blots, or thin-layer plates. The visual presentation of data from such instruments is similar to autoradiograms, and when ^{32}P is the tracer, excellent records of gels are obtained in a matter of minutes.

In general, multiwire proportional detectors behave as a planer array of many individual detectors. With the aid of a computer, the counts registered by each "detector cell" is recorded and may be plotted to yield a densiometric image. As originally designed (Charpak et al., 1968; Charpak, 1970), the detector chamber consists of a grid of uniformly spaced anode wires that has two grids of cathode wires, running at right angles to the anode grid. Electrons produced by an ionizing event between the cathode planes are accelerated toward the nearest anode wire as described for a common proportional detector, and the pulse resulting from the collection of electrons yields a count. The anode wire producing that count locates the position of the ionizing event in the x-direction. A second signal is generated by a particular cathode wire when the molecular ions are collected, and since the cathode wires are perpendicular to the anode wires, the cathode wire yielding the pulse locates the ionizing event in the y-direction. Of course, from a top view of the grid system, the position of the ionizing event would be near the intersection of the anode and cathode wires yielding pulses. It seems that separate "electronic or signal channels" (preamplifier, amplifier, etc.) would be needed for each wire, however, wires may be grouped to reduce the number of signal channels necessary. Because the volume in which the ionizing event occurs is larger than the spaces outlined by the grid wires, pulses are normally produced in more than one set of anode and cathode wires. Having neighboring anode or cathode wires assigned to different groups permits the principal wire affected to be identified from the particular combination of channels responding. Therefore, grouping of wires is used in some instruments to reduce the number of signal channels, and hence electronic complexity.

The entire grid system is enclosed in a chamber with a thin plastic face, usually of Mylar, and a gas suitable for proportional detection flows into the chamber slowly to maintain a constant gas composition. Argon and methane (90 to 10), carbon dioxide, and freon mixtures have been used as proportional gases.

Although a number of effective instrumental designs have been demonstrated, two commercial instruments on the U.S. market will be described to illustrate some operating principles.

A diagram of the detector used for the AMBIS radioanalytic imaging system is shown in Figure 17-1. The detector has a 20 × 20 cm detector face, and a 952-detector array is created at the intersections of 34 anode wires and 28 cathode strips. A guard electrode, operated at a potential between that of the anode and cathode, replaces the lower cathode grid in the general detector design described above. Below an anode–cathode intersection is a drift chamber that confines entering β-particles, or other ionizing rays, to that particular anode–cathode intersection. The drift cham-

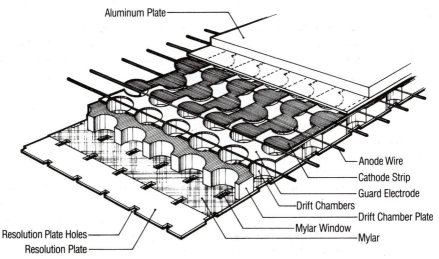

Aluminum Plate

Anode Wire
Cathode Strip
Guard Electrode
Drift Chambers
Drift Chamber Plate
Mylar Window
Mylar
Resolution Plate Holes
Resolution Plate

FIG. 17-1 A cross section of the multiwire proportional detector in an instrument produced by AMBIS Systems, Inc., San Diego, CA. (Reprinted from Nye, L., Colclough, J. M., Johnson, B. J., and Harrison, R. M. *American Laboratory* 7:18–27, 1989. Copyright 1988 by International Scientific Communications, Inc.)

bers are formed by holes drilled in a metal plate that is about 0.6 mm thick. An aluminized Mylar sheet below the drift chamber plate confines the proportional gas to the detector chamber. A resolution plate having regularly spaced holes that are smaller than the drift chamber's cross section forms windows for all detector cells, and interchangeable plates allow windows of different shapes and sizes to be employed so resolution and sensitivity may be optimized for various applications. It also provides mechanical strength to protect the thin Mylar film. The electrophoretic gel, blot, or chromatogram is pressed against the resolution plate.

A β-particle entering a drift chamber creates a trail of ion pairs and the electric field in the chamber causes them to be drawn toward the electrodes. As the electrons approach the anode they are accelerated sufficiently to initiate an electron avalanche and the collection of the electrons constitutes the anodic pulse. Collection of the molecular ions by the cathode represents the cathodic pulse, and the particular anode and cathode yielding pulses identifies the detector cell responding.

The AMBIS system incorporates a several-position scanning function also. After the counts from the various detector cells are recorded for a period of time, the specimen is lowered and moved to another slightly different position, then raised back up for another counting period. The appropriate number of moves in the x- and y-directions depends on the dimensions of the holes in the resolution plate and the desired degree of resolution, but the movements are controlled automatically by the computer according to options selected by the operator. Because of absorption of β-particles by the plastic window, ^{32}P is detected much more efficiently than ^{14}C, ^{35}S, and ^{125}I.

Another commercial instrument that utilizes a multiwire proportional detector is manufactured by Betagen Corporation. As shown in Figure 17-2, the detector chamber contains an anode–cathode grid in one plane. In addition, a second anode–cathode grid is located a slight distance away in a plane parallel to the first. A β-particle entering a responsive volume of the grid closest to the specimen produces a count, and, as described previously, the location of that volume can be deduced from the anode and cathode wires yielding simultaneous pulses. As the β-particle continues its flight, it interacts with a responsive volume in the second grid plane, producing pulses that delineate the location of the interaction in that plane. Although β-particle trajectories are not linear throughout their entire pathlengths, for a small distance the majority of the particles' paths are sufficiently close to being linear that trajectories can be used to locate the position of a source. A line between the points of a particle's interactions in the two grid planes, when projected back to the specimen, yields a reasonably accurate location of its origin. Because resolution is based on trajectories, the β-particles must have sufficient energy to traverse both grid

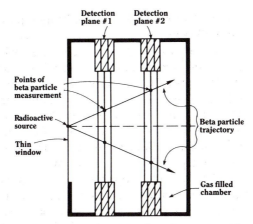

FIG. 17-2   A diagram indicating how two detection planes are used to obtain images from a ^{32}P-labeled blot. (Courtesy of the Betagen Corporation, Waltham, MA)

planes, therefore, this instrument is used for ^{32}P-labeled components primarily.

The principle of detection for the "PhosphorImager®," produced by Molecular Dynamics (Johnston et al., 1990), is similar to that involved in the operation of thermoluminescent dosimeters (TLDs), which were discussed in Chapter 16. Certain materials behave as storage phosphors that are activated by exposure to radiation. The activated complexes are trapped in a metastable energy state and will remain in that state for some time unless they are activated further, whereupon they revert to their ground-state energy levels by emitting a photon of light.

In the storage phosphor screen produced by Fuji Photo Film Company, fine crystals of a $BaFBr:Eu^{+2}$ complex are embedded in an organic binder that is coated on a rigid screen, and the screen is exposed to a planner specimen for a period of time in a manner similar to that for an X-ray film. Energy from an ionizing ray excites an individual Eu^{+2} ion by promoting one of its electrons to the conduction band, and the electron is trapped by a neighboring BaFBr to yield Eu^{+3} and $BaFBr^{-}$, which now is in an excited metastable energy state. This constitutes one molecular element in the latent image.

After exposure, the screen is scanned with an instrument that detects the latent image elements and their relative positions on the screen. The molecular events involved in the detection step follow. An excited $BaFBr^{-}$ complex (an element of the latent image) has an absorption band of around 600 nm, and upon absorbing a photon of such light, its extra electron is promoted back to the conduction band, which permits it to recombine with an Eu^{+3} ion, creating an *excited* $+2$ ion (Eu^{+2*}). The excited Eu ion subsequently loses its energy of excitation by emitting a photon of light with a wavelength of about 390 nm.

In the instrument, a beam from a helium–neon laser, which provides the light for excitation (emission maximum at 633 nm), is focused on an 88 μm^2-area of the screen. Those $BaFBr^-$ complexes in that area are excited, which leads to excited Eu ions ($Eu^{+2}*$) in the same area, and they lose their excitation energy by emitting photons at 390 nm. Since the emitted light is nondirectional, most of it can be separated from the laser's light by optics similar to that used in a fluorimeter. One or several elements in a linear bank of optical fibers collect emitted light and direct it to an MP-tube that yields an electrical signal proportional to the intensity of the light.

Imagine the optical fibers arranged as a flat fan. The length of the broad end of the fan, which is cut square rather than being rounded, extends across the screen, in the x-direction, and the gathered end forms a bundle of fibers that meets the face of the MP-tube. Initially, the laser beam is focused at one corner of the screen directly above one end of the optical-fiber fan; it then sweeps across the screen in the x-direction. The intensity of the emitted light relative to the laser beam's position is recorded digitally with a computer during the sweep. Then the fiber fan and laser beam are moved one step in the y-direction, and a sweep is made across that narrow strip. In this manner, the whole screen is scanned. The stored data may be presented in a variety of forms, but an autoradiogram format is the most popular.

Scanning the screen erases the latent image, but all vestiges of metastable molecules are deactivated by exposing the screen to light, after which the screen is ready for re-use with another specimen.

The primary advantage of this method of visualization appears to be that for equivalent images: the time required for screen exposure is much less than that required for photographic film. Having data stored permits further manipulations, and reasonably quantitative data for selective areas may be obtained. All types of ionizing radiation may be detected, but because of their applications in molecular biology, ^{32}P, ^{14}C, ^{35}S, and ^{125}I are used principally. Unfortunately, the screens are expensive, so specimens are exposed and scanned sequentially in most laboratories.

Gas Chromatographic Monitors

For very small quantities of labeled compounds, gas chromatography is a valuable separation method because of the very high sensitivities of commercially available detectors. Practically speaking, the response of a gas chromatography detector depends on the mass of organic compounds passing through the detector cell. Such cells do not detect radioactivity, although some of their operating principles are similar to those used in radioactivity detectors. Two common types of detectors behave as flowthrough gas ionization chambers, and electrometers are used to measure ion currents. Ionization is induced by a flame within the cell (flame ionization detector) or by β-particles emitted from a sealed source, such as ^{63}Ni, inside the cell (β-ray ionization detector). The flame ionization detector destroys the organic components of the column effluent, whereas β-ray ionization detectors cause slight damage.

Radioactivity in the effluent gases may be detected by an additional flowthrough cell. It too is a gas ionization detector, but generally it is operated in either the proportional or G-M region. A count rate meter monitors the detector's response, and a graphic readout, similar to that provided from the "mass" detector, is afforded. With flame ionization detectors, a stream splitter must be used so that parts of the column's effluent are directed to each of the two detectors. Stream splitters are also used with β-ray ionization detectors, but theoretically, higher sensitivities are possible when the two types of detectors are used in series rather than parallel. Although difficult, stream splitting can be used to trap gas chromatographic effluent components.

Low- and High-Pressure Liquid Chromatography Monitors

Effluents from liquid chromatography columns frequently are fractionated, and aliquots of each fraction may be assayed for radioactivity as well as for the presence of specific chemical compounds. Although radioactivity assays are simple for strong γ-emitters (solid scintillation counting) and permit recovery of labeled compounds, the assay of β-emitters is usually performed by liquid scintillation counting, and this requires sample preparation labor and materials. Furthermore, after mixing with liquid scintillation cocktail, fraction components are not recoverable. Of course, the use of smaller aliquots for radioactivity assays increases the percentage of recoverable components, but decreases assay sensitivity.

Complete instruments to monitor liquid chromatography effluents are available commercially, and they employ scintillation counting principles, flow cells, and MP-tubes as detectors. One or two MP-tubes may be used. Two-tube instruments have higher detection efficiencies and sensitivities, but coincidence counting is not used normally. Metering pumps, stream splitters, and associated electronics permit various monitoring configurations to be used, and in most instruments, different types of flow cells may be used interchangeably. There are three basic types of flow cells.

In one type of flow cell, column effluent is mixed with liquid scintillation cocktail, and the mixture flows through a small diameter plastic tube that has a high transmittancy of light. The space between the transparent faces of the cell is equivalent to the outside diameter of the plastic tubing, which is coiled inside the cell so that the diameter of the coil approximates the diameter of the face of the MP-tubes. Thus, each of two MP-tubes may "view" a cross section of the tubing coil. This type of flow cell offers the highest detection efficiency and sensitivity, particularly for weak emitters such as ^{3}H and ^{14}C. However, mixing of the effluent with liquid scintillation cocktail prevents recovery of separated components unless a stream splitter is employed. This system requires a pump for the cocktail and imposes some limitations on the chromatography solvents that may be employed because the effluent must form homogeneous, but not necessarily true, solutions with the cocktail.

A second type of cell involves a tube that is packed with a "solid scintillator" and has a high transmittancy of light. The effluent flows over the packing material, and rays emitted by effluent compounds interact with the solid scintillator beads producing scintillation events that are detected by the MP-tubes. Because the range of β-particles from ^{3}H is so small, the majority of the particles are absorbed by the solution rather than the scintillator beads, and this results in extremely low detection efficiencies. The situation is much better for ^{14}C and ^{35}S, and, as might be expected, ^{32}P can be detected quite efficiently.

In the past, anthracene crystals have been used as solid scintillator, but anthracene is subject to dissolution by some solvents. Europium activated calcium fluoride and organic scintillators embedded in plastic resins are used more commonly today. Bead sizes are in the range of 100–200 nm, and a low void volume in the packing material is desirable to increase detection efficiency because reducing the interstitial space size reduces β-particle absorption by the liquid phase. However, smaller spaces between solid scintillator beads result in higher back pressures. Consequently, the length of packed tubing is considerably shorter in this type of cell compared to the one described previously. Probably the most serious problems encountered are due to the exposure of the solid scintillator beads to a myriad of solvents and chemicals. Although quite inert, under some conditions the solid support may adsorb or absorb radioactivity from the effluent. This results in higher baselines (background) and possible memory effects. Also the beads may become coated with materials from the effluent, and this may cause lower detection efficiencies, if not a total loss.

The detection principles for the third type of flow cell are the same as that described for the second type, but the solid scintillator is incorporated into the tubing walls or cell faces instead of existing as packing material. Because of the thicknesses of the effluent streams and ranges of radioactive emission, this type of cell is used for hard β-emitters (^{32}P) and weak γ-emitters (^{125}I) primarily. Compared to packed cells, this physical arrangement is less prone to the problems caused by adsorption and absorption of materials from the column stream; also cleaning cells having residues is more feasible.

REFERENCES

Bolton, A. E. (1983). *Radioiodination Techniques*. Review 18, Amersham Corporation, Arlington Heights, IL.

Bolton, A. E., and Hunter, W. M. (1973). *Biochem. J.* 133:529.

Charpak, G. (1970). *Annu. Rev. Nucl. Sci.* 20:195.

Charpak, G., Bouchlier, R., Bressani, T., Favier, J., and Zupancic, C. (1968). *Nucl. Instrum. Methods* 62:262.

Evans, E. A. (1983). *Self Decomposition of Radiochemicals*. Review 16, Amersham Corporation, Arlington Heights, IL.

Fraker, P. J., and Speck, J. C. (1978). *Biochem. Biophys. Res. Commun.* 80:849.

Hubbard, A. L., and Cohn, Z. A. (1972). *J. Cell Biol.* 55:390.

Hunter, W. M., and Greenwood, F. C. (1962). *Nature* (London) 194:495.

Johnston, R. F., Pickett, S. C., and Barker, D. L. (1990). *Electrophoresis* 11:355.

Koch, N., and Haustein, D. (1981). *J. Immunol Methods* 41:163.

Marchalonis, J. J. (1969). *Biochem. J.* 113:199.

McFarlane, A. S. (1958). *Nature* (London) 182:53.

Pennis, F., and Rosa, U. (1969). *J. Nucl. Biol. Med.* 13:64.

Ward, G. M. (1984). In *Membranes, Detergents and Receptor Solubilization,* Vol. 1. J. C. Venter and L. C. Harrison, Eds., p. 109. Liss, New York.

CHAPTER 18

Radioimmunoassays
and Isotope Dilution

COMPETITIVE PROTEIN-BINDING ASSAYS

In *competitive protein-binding assays* the chemical principles operating involve the avidity of binding between a protein and a specific ligand, which frequently is a protein as well. *Radioimmunoassay* (RIA) methods belong in the general class of assays described above, but are more restrictive in that the ligand is an antigen and the binding protein is an antibody that has been produced immunologically by injecting the antigen into an animal. Antibodies are particularly good proteins for binding studies because they have high affinities and specificities for the particular antigens against which they were raised. The most frequent use of RIA methods is the quantitative determination of antigen concentrations in biological specimens, but RIA methods are also used to compare structural similarities of molecules that are closely related to a given antigen. The theoretical basis of competitive protein-binding assays is presented using RIA methods as examples.

THEORETICAL ASPECTS OF RIA

For our discussion, an asterisk will indicate that a particular molecular species is labeled with a radionuclide, most commonly ^{125}I, and Ag, Ab, and Ag–Ab represent an antigen, an antibody-binding site, and an antigen–antibody complex, respectively. Normally, antibodies have two antigen-binding sites per molecule, but there is no cooperativity between the binding sites. Thus mathematically, binding sites can be treated as separate entities. As usual, brackets denote concentrations. The reactions and corresponding affinity constants (K or K') for the binding of unlabeled and labeled antigens to antibody binding sites are

$$Ag + Ab \rightleftarrows Ag\text{–}Ab \qquad K = \frac{[Ag\text{–}Ab]}{[Ag][Ab]}$$

$$Ag^* + Ab \rightleftarrows Ag^*\text{–}Ab \qquad K' = \frac{[Ag^*\text{–}Ab]}{[Ag^*][Ab]}$$

The antibody (Ab) depicted in the above reactions would have been raised against a particular antigen (Ag, usually a protein) and should have a high specificity for that antigen, meaning that it binds little if anything else. In addition, the antibody should have a high affinity for the antigen, that is, the reaction equilibrium should lie far to the right. Affinity constants in the order of 10^{10} are not uncommon for antibody–antigen complexes. The reciprocal of an affinity constant is referred to as a dissociation constant, which may be used in competitive protein-binding assay studies.

Equilibrium RIA

Chemically, Ag^* is slightly different than Ag, but in useful RIA methods the difference is not great enough to cause significant differences in their affinities toward Ab, thus $K \approx K'$. If both Ag and Ag^*, with initial concentrations of $[Ag]_i$ and $[Ag^*]_i$, are mixed with a limited quantity of Ab, there will be competition between the two types of antigens for a limited number of binding sites ($[Ab]_i$), which normally are fully occupied at equilibrium. This is the phenomenon suggested by the phrase, competitive protein-binding.

$$
\begin{array}{ccccccccc}
& & & \textit{Initial} & & & & \textit{At Equilibrium} & \\
Ag & + & Ag^* & + & Ab & \rightarrow & Ag\text{–}Ab & + & Ag^*\text{–}Ab & + & Ag & + & Ag^* \\
[Ag]_i & & [Ag^*]_i & & [Ab]_i & & X & & Y & & [Ag]_i\text{–}X & & [Ag^*]_i\text{–}Y
\end{array}
$$

If the antibody binding sites are saturated, $X + Y = [Ab]_i$ and $X/Y = [Ag]_i/[Ag^*]_i$.1

In practice, a series of binding reactions are performed for which the initial concentrations of labeled antigen, $[Ag^*]_i$, and antibody binding sites, $[Ab]_i$, are held constant, but the initial concentration of the unlabeled antigen $[Ag]_i$ is varied. Normally, concentrations are selected so that when unlabeled antigen is absent ($[Ag]_i = 0$), about 50% of the labeled antigen is bound; thus $[Ab]_i \approx \frac{1}{2}[Ag^*]_i$. With such concentrations, when unlabeled antigen is present it competes with labeled antigen for the available binding sites. Since the amount of Ag^* bound at equilibrium is dependent on the initial concentration of unlabeled antigen $[Ag]_i$, measurements of the relative amounts of Ag^* in the bound and free forms are related to $[Ag]_i$.

Let B_0 = the radioactivity of the labeled antigen
in the bound form (Ag*–Ab) when $[Ag]_i = 0$.

B = the radioactivity of the bound form of the
labeled antigen (Ag*–Ab) when $[Ag]_i > 0$.

F = the radioactivity of the unbound form of the
labeled antigen (Ag*) when $[Ag]_i > 0$.

T = the total radioactivity of both forms of the
labeled antigen (Ag* + Ag*–Ab). Therefore, $T = B + F$.

In most quantitative assays, B_0, T and B or F are measured, but the actual concentrations of Ag*, Ag*–Ab, and Ab are never known. However, if the initial concentrations of labeled antigen and antibody ($[Ag^*]_i$ and $[Ab]_i$) are held constant for a given assay series, knowledge of the actual concentrations is not required because ratios may be employed, for which the proportionality constants relating radioactivity to concentration cancel. Thus,

$$\frac{B}{F} = \frac{[Ag^*-Ab]}{[Ag^*]} \quad \text{and} \quad \frac{B}{T} = \frac{[Ag^*-Ab]}{[Ag^*] + [Ag^*-Ab]} = \frac{[Ag^*-Ab]}{[Ag^*]_i}$$

Standard curves are prepared by adding reagents to a series of tubes in which $[Ag]_i$ is varied over a suitable range of known concentrations, including $[Ag]_i = 0$. Frequently, $[Ag]_i$ is referred to as the dose of antigen. After the reactions have reached equilibrium, the bound and free forms of the labeled antigen are separated and counted to obtain the desired assay parameters, B_0, T, B or F. Several ratios can be used for plotting, such as B/B_0, B/F, and B/T. The selected ratio is plotted against $[Ag]_i$ (dose), or perhaps $\log[Ag]_i$ (log dose), to yield a standard curve. A hypothetical plot is shown in Figure 18-1.

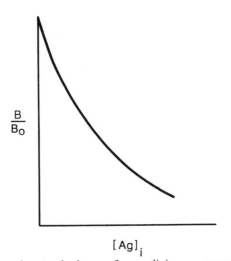

FIG. 18-1 A representative standard curve for a radioimmunoassay.

To determine the concentration of antigen in an experimental sample, the reaction is conducted in the same manner as that used for preparation of the standard curve (usually both operations are performed simultaneously), and the relevant ratio, e.g., B/B_0, is ascertained. The concentration corresponding to the experimental ratio is read from the standard curve. Most standard plots are curvilinear, but some may be fairly linear over limited concentration ranges. In an effort to obtain linear relationships, a logit function may be plotted against dose or log(dose). The logit–logarithm transformation is

$$\text{logit } (Y) = \ln\left(\frac{Y}{1 - Y}\right)$$

Consequently, the analogous transformation for the binding parameter, B/F, is

$$\text{logit}\left(\frac{B}{F}\right) = \ln\left(\frac{B}{F - B}\right)$$

In the preceding discussion several assumptions and conditions were alluded to, and perhaps now their necessity will be more obvious.

1. Ab is present in only one chemical form. Other antibodies may be present, but Ab is the only protein that binds Ag or Ag*.

2. Ab has a high specificity for Ag or Ag* and does not bind other antigens that may be present.

3. $K = K'$, that is, the binding characteristics of Ag and Ag* are not significantly different.

4. Sufficient time is allowed for each reaction to reach equilibrium.

5. The bound and free forms of the labeled antigen can be separated without disturbing the equilibrium.

6. There are no cooperative or allosteric effects in the binding reaction.

7. Radioactive measurements are proportional to the concentration of Ag* present regardless of its chemical form.

Sequential Saturation RIA

The previous list of assumptions and conditions applies to binding reactions that reach equilibrium, but another approach may be taken. A *sequential saturation* assay is performed by incubating an excess of antibody with limited quantities of nonlabeled antigen (from standards or test solutions). After equilibrium is reached, the labeled antigen is added and it binds to the remaining antibody sites. Obviously, the amount of labeled antigen bound is inversely proportional to the concentration of binding sites occupied by nonlabeled antigen. In this method a high avidity between the unlabeled antigen and antibody is necessary so that labeled antigen does not displace the nonlabeled antigen previously bound. In general, sequential saturation assays are more sensitive than equilibrium methods, but the range of concentrations suitable for assays is smaller.

Determination of Affinity Constants

The value of an affinity constant for a particular antigen and antibody is not needed to calculate RIA results, but an affinity constant may be important for studies involving other proteins that exhibit competitive binding behavior, such as membrane receptors and enzymes. Rather than introducing a different set of abbreviations, consider Ab to be a protein that binds a specific ligand denoted as Ag. Assuming that equilibrium methods are used and the affinity constants for binding labeled and nonlabeled ligand (Ag* and Ag) are equal ($K = K'$), the affinity constant for a competitive reaction is

$$K = \frac{[Ag^*\text{–}Ab] + [Ag\text{–}Ab]}{[Ab]([Ag^*] + [Ag])}$$

Since

$$\frac{B}{F} = \frac{[Ag^*\text{–}Ab]}{[Ag^*]} = \frac{[Ag^*\text{–}Ab] + [Ag\text{–}Ab]}{[Ag^*] + [Ag]}$$

$$K = \frac{B}{F} \cdot \frac{1}{[Ab]}$$

$$\frac{B}{F} = K[Ab]$$

$$[Ab] = [Ab]_i - ([Ag^*\text{–}Ab] + [Ag\text{–}Ab])$$

$$\text{Let } C = [Ag^*\text{–}Ab] + [Ag\text{–}Ab]$$

$$\frac{B}{F} = K[Ab]_i - KC \tag{18-1}$$

Equation 18-1 is a linear equation, and a plot of B/F versus C should yield a line with a slope of $-K$ and intercepts of $K[Ab]_i$ and $[Ab]_i$. However, the values of C must be known before such a plot can be prepared. Knowledge of the total initial concentration of antigen ($[Ag]_i + [Ag^*]_i$) and binding measurements permit C to be calculated as shown below.

$$\frac{B}{T} = \frac{[Ag\text{–}Ab^*]}{[Ag^*\text{–}Ab] + [Ag^*]} = \frac{[Ag^*\text{–}Ab] + [Ag\text{–}Ab]}{[Ag^*\text{–}Ab] + [Ag\text{–}Ab] + [Ag^*] + [Ag]}$$

$$\frac{B}{T} = \frac{C}{[Ag^*]_i + [Ag]_i}$$

$$C = \frac{B([Ag^*]_i + [Ag]_i)}{T}$$

Generally, $[Ag]_i$ is known or can be determined, therefore a standard curve such as that shown in Figure 18-2 may be prepared. From such a curve, the initial concentration of labeled ligand ($[Ag^*]_i$) may be determined by performing an additional assay where the unlabeled antigen is replaced ($[Ag]_i = 0$) by an additional aliquot of the labeled ligand. The concentration corresponding to the B/F ratio for that reaction is the increase in the initial concentration of the labeled ligand due to the added aliquot. Suppose the aliquot size was one-half the amount used for all of the other tubes and a concentration of R corresponds to the B/F ratio obtained for that assay. If $[Ag^*]_i$ is the initial concentration used for all other tubes,

$$\tfrac{1}{2}[Ag^*]_i = R \qquad \text{and} \qquad [Ag^*]_i = 2R$$

Knowing $[Ag^*]_i$ ($2R$ in this example), C can be calculated for each of the samples used to prepare the standard curve, in this illustration, those in Figure 18-2. Among those samples $[Ag]_i$ varies, but $[Ag^*]_i$ is constant.

Now, the same binding parameter, B/F, can be replotted against C yielding what is known as a Scatchard plot (Scatchard, 1949). Equation 18-1 is illustrated in the Scatchard plot shown in Figure 18-3. If the data yield a curve, cooperative binding effects are indicated.

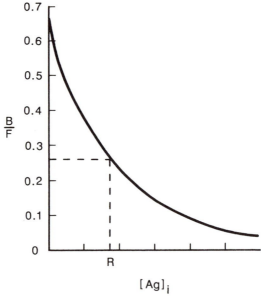

FIG. 18-2 A standard curve relating the ratio of bound and free antigen to the initial concentration of unlabeled antigen.

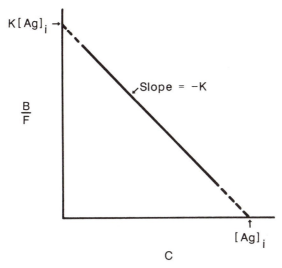

FIG. 18-3   The general features of a Scatchard plot. C is the initial concentration of labeled and unlabled antigen.

PRACTICAL ASPECTS OF RIA

For a few compounds that are assayed routinely by numerous laboratories, such as clinical diagnostic labs, RIA kits have been developed and are marketed commercially. A well-trained technician can perform kit assays effectively, but many research scientists wishing to employ RIA methods are faced with the task of producing most of the reagents necessary to conduct an assay. Because of the labor and experience needed to produce suitable materials, such an undertaking should not be considered lightly. Therefore, persons contemplating the use of RIA methods (except the kit methods) should consult the primary literature and individuals experienced in various techniques. An overview of steps involved follow.

Preparation of Antigen

The compound or substance to be assayed must be isolated in pure form, and it must be capable of eliciting an immunological response in an animal. Virtually all proteins, even small polypeptides, are effective antigens if they are foreign to an animal, hence the most common antigen is a protein. Small organic molecules do not evoke immunological responses, but sometimes an effective assay system can be produced by conjugating a small molecule, called a *hapten*, to a polypeptide. Cyclic nucleotides, thyroxine, prostaglandins, and certain "street drugs" are examples of haptens that have been used to develop sensitive analytical RIA methods for such substances. Of course, the hapten must be the *antigenic determinate* or *epitope,* that is, the hapten must be the group recognized by the antibody.

Immunization of Animals

After purification, the antigen is dispersed in an adjuvant, such as Freunds' adjuvant, and injected into a suitable animal. Adjuvants may contain emulsified lanolin, paraffin, and killed bacteria, and they stimulate the production of antibodies. Multiple injections, perhaps over a period of weeks, are administered.

Various animals may be used, but rabbits are popular because they generally produce antisera (blood serum containing antibodies) with high titers. *Titer* refers to the capacity of a serum to bind a given antigen, hence it is a measure of the affinity of antibody for antigen and the antibody's concentration. Blood is relatively easily withdrawn from rabbit ear veins, which makes them a convenient animal. Other small animals are used, but the rodents are not particularly responsive, immunologically. For larger quantities of antisera or when suitable housing facilities exist, domestic animals, such as goats, sheep, and donkeys, may be used.

Antibody Collection

Once immunized, an animal will continue to produce antibodies over a period of months, consequently blood is usually collected without sacrificing the animal so that a source of antiserum is available for an extended period of time. After several months or years, booster immunizations may be required to maintain high serum titers.

The blood from an immunized animal is collected, and after coagulation, the serum is isolated by centrifugation, yielding antiserum. The antiserum contains the antibodies, which may be called immunoglobulins, and the serum can be used directly or subjected to further fractionation to obtain the *immunoglobulin G* fraction, called *IgG*.

Polyclonal antibodies can be isolated by affinity chromatography in which the antigen or its epitope is linked to the chromatographic support. The population of polyclonal antibodies is not homogeneous because the cells responsible for antibody production in an animal are not identical. Furthermore, if an antigen is linked to the chromatographic support, antibodies that recognize different epitopes in the antigen may be included in the population. Nevertheless, all antibodies in the population should bind specifically to the particular antigen or epitope used in the affinity chromatography step.

Monoclonal antibodies are not produced by serum fractionation, instead they are isolated from cell cultures that are derived from a single antibody-producing cell. Consequently, the population of a species of monoclonal antibodies is homogeneous. Molecular biological cloning techniques offer an alternative for producing a large and continuous supply of a particular antibody, and many commercial sources of antibodies, such as those used in assay kits, are of the monoclonal variety.

Normally, serum is not fractionated extensively, although lipids, lipoproteins, and some nonimmunoglobulin proteins may be removed before use. The titer of a given serum sample must be determined so that the serum can be diluted appropriately before assay. Dilutions frequently are in the range of 10^{-3} to 10^{-6} depending on titer and desired sensitivity range for the intended assay.

Production of Labeled Antigen

For protein antigens, the most common labeling method is radioiodination with ^{125}I. Nevertheless, other radionuclides may be used if a labeled group can be conjugated to the antigen without destroying its affinity for the appropriate antibody. In some methods, antibodies may be labeled in place of antigens.

Counting Methods

Solid scintillation is the most popular counting method for RIAs involving ^{125}I and ^{131}I as labels. Generally, assay protocols are designed so that all counting is performed with samples in their respective assay tubes. This eliminates some pipetting steps, which reduces the possibilities of noncounting assay errors and labor. Most samples to be counted contain solids or liquids that can cause self-absorption problems in some counting methods. However, for counting X- or γ-emitters by solid scintillation, self-absorption is slight, or at least reasonably uniform, among similarly prepared samples.

Liquid scintillation may yield higher counting efficiencies, particularly for ^{125}I, but the use of higher activities can compensate for low counting efficiencies when statistical errors are considered. Nevertheless, liquid scintillation can be used for assays, but generally is not when a solid scintillation counter is available.

Separation of Bound and Free Forms of the Antigen

After individual reactions are complete, the antigen and antibody–antigen complex must be separated to determine the radioactivity in the bound (B) and free (F) forms. As an alternative the total radioactivity (T) may be counted in place of either B or F.

A number of methods have been used for separations.

1. Nonspecific precipitation. Because of size differentials, antibody–antigen complexes can be separated from free antigen by high-speed centrifugation in some cases. Usually, such separations yield fractions that still contain a significant amount of the other component. Salting out with $(NH_4)_2SO_4$ or solvent (ethanol) precipitation may permit separation in some cases.

2. Immunoprecipitation. This method requires an additional immunization step. First, antiserum against the antigen is produced by one species, for example, rabbit. Rabbit IgG is used to immunize a second species, perhaps goat, which yields antibodies against rabbit IgG (antirabbit serum). The assays are conducted as usual using the rabbit IgG, but on completion, goat serum, which binds rabbit antibodies, is added. Then the large antibody–antibody–antigen complex is separated from the antigen by centrifugation. Immunoprecipitation may be enhanced by the addition of polyethylene glycol before centrifugation.

3. Solid phase adsorption. Some adsorbents such as charcoal, talc, silica gel, as well as ion-exchange resins may adsorb free antigen preferentially and permit separation. However, a centrifugation step is usually involved.

4. Gel filtration (permeation) chromatography. Because of size differences, antigens and their antibody complexes can be separated by gel permeation chromatography. Unfortunately, this is a time consuming method when many assay samples are involved.

5. Electrophoresis. The electrophoretic mobilities of antigens and antibody complexes differ sufficiently for this method to be effective, but it is a tedious method if many quantitative assays are desired.

6. Immobilized antibodies. The most widely used technique for separating bound and free forms of antigens is by the use of immobilized antibodies, and there are many variations in the way separations are achieved.

Antibodies can be bound to a variety of plastics by treatment with relatively mild fixatives, such as glutaraldehyde and ethylchloroformate, without destroying their affinities toward their respective antigens. Both covalent bonds and adsorption have been used to bind antigens or antibodies to solid plastic supports.

In one of the earlier applications, antibody was bound to the interior of test tubes, and it remained immobilized during incubations with standard and test solutions of antigens. Separation was achieved by decanting the incubation mixture. Counting the rinsed tube yielded B, and T was obtained by counting the tube prior to decantation. One drawback to this method is that the number of test and standard samples in an assay series had to be determined in advance so an appropriate number of tubes with identical properties could be prepared.

Plastic beads have become more popular as a solid support because with little additional expense, a large batch of beads can be prepared. The beads are small enough to remain in suspension during pipetting operations yet easily sedimented by gentle centrifugation. Thus, the antigen test solutions, with a constant amount of labeled antigen, are incubated with the immobilized antibody (beads). Counting a tube in this condition yields T, then the tube is centrifuged, the supernatant decanted and in a similar manner the beads are rinsed. The amount of radioactivity bound to the rinsed beads represents B.

The double antibody solid phase method has become quite popular because it allows the normal competitive reaction to be conducted with all forms in solution. Suppose the primary antibody has been raised in rabbits and a secondary antibody in goats (antirabbit IgG). The completely soluble reaction mixture contains the antigen (test or standard solution), labeled antigen and rabbit IgG. At the completion of the incubation, one or several tubes are counted to ascertain T. Then goat IgG bound to beads is added and this complexes all of the rabbit antibody molecules including those binding antigen molecules. After separation and rinsing, the beads are counted, yielding B. Immobilized secondary antibodies for a few common animal species are available commercially.

The "sandwich" solid phase method is useful when the antigen is large and has more than one epitope. That is, the antigen has two regions to which different antibody species in a polyclonal population may bind. First, unlabeled antibody is bound to plastic beads, and incubations are performed with unlabeled antigen (standards or test samples) only. The antigen molecules bind to the immobilized antibody molecules, which are present in excess. Then soluble labeled antibody is added and it binds to the free epitope on the antigen forming a sandwich (immobilized antibody–antigen–antibody). The beads are separated and washed as usual, then counted. The amount of labeled antibody bound is a measure of the amount of antigen bound to the beads (B).

Another variation is a technique called IRMA (immunoradiometric assay). A test or standard antigen sample is incubated with labeled antibody, which is in excess. Then antigen covalently attached to a solid support is added. Any of the labeled antibody molecules that have a free binding site bind to immobilized antigen. After separation and washing, the solid support is counted and the amount of radioactivity is inversely proportional to the amount of antigen originally present in the test or standard sample.

In all RIA methods, blanks must be analyzed because there is some nonspecific binding of antigens or antibodies to other reactants and to the vessels themselves.

Data Reduction

When large numbers of test samples are to be analyzed by RIA methods, computer programs can eliminate the need for plotting standard curves, calculating concentrations, as well as performing logit transformations, linear regression, and curve fitting. The "spline fit" and "four parameter fit" are two examples of programs that perform such operations.

Additional Information

Some practical references concerning RIA methods include Collins (1985), Odell and Daughaday (1971), Kirkham and Hunter (1971), and Abraham (1977).

ISOTOPE DILUTION

A radioactivity measurement is a quantitative measurement of the amount of the tracer present, but certain methods may be used to determine the quantity of a nonlabeled compound in a mixture as well. One such group of methods is called *isotope dilution,* and they are valuable because in complex mixtures it is difficult, if not impossible, to isolate a pure compound quantitatively. For these illustrations, specific activities are assumed to be in dpm or dps per mol, but radioactivity per unit weight could be used as well in many cases.

Simple Isotope Dilution

By simple isotope dilution the amount of a given compound in a complex mixture can be ascertained if (1) a portion of the compound (with added tracer) can be isolated in pure form, and (2) there is a sufficient quantity of the pure compound to permit a specific activity measurement. In this method, specific activities must be known.

A known quantity (M_k) of a labeled compound having a known specific activity (S_k) is added to a mixture containing an unknown amount (M_u) of the same compound. After the labeled and nonlabeled forms are admixed, a portion of the compound is isolated in a chemically pure form. The fraction of the mass recovered is not important as long as the quantity is sufficient for a specific activity measurement (S_m). Since the total amount of radioactivity present is constant,

$$S_k \cdot M_k = S_m(M_k + M_u)$$

$$M_u = \frac{M_k(S_k - S_m)}{S_m} \qquad (18\text{-}2)$$

Reverse Isotope Dilution

Reverse isotope dilution is used to determine the quantity of a labeled compound when its specific activity is unknown, but the total activity present is known.

Suppose a person wishes to determine the metabolic pool size of a given intermediate called compound B. A commercial supply of ^{14}C-compound B is not available, but a small quantity can be produced biosynthetically. Although ^{14}C-compound B can be produced radiochemically pure (no ^{14}C in any compound except compound B), it cannot be isolated in a chemically pure form without adding carrier (unlabeled compound B). Therefore, a known amount of activity (A_k) of the biosynthetically prepared compound B is added to the system to be equilibrated with the pool, then a known quantity (M_k) of carrier is added. After isolation of pure compound B, its specific activity (S_m) is measured. Assuming adequate mixing,

$$A_k = S_m(M_u + M_k)$$

$$M_u = \frac{A_k - (S_m \cdot M_k)}{S_m} \tag{18-3}$$

Inverse Isotope Dilution

Inverse isotope dilution may be used to determine the quantity of a labeled compound when its specific activity is known, but the total activity present is unknown.

Suppose a labeled compound is metabolized by a system to produce several different compounds, and a person wishes to determine the resulting quantity (M_u) of one of the metabolites, compound B. The specific activity (S_k) of the labeled compound supplied is known, and because of its labeling pattern, it is safe to assume that the specific activity of compound B is the same. Therefore, a known quantity (M_k) of carrier compound B is added to the system; then after purification the specific activity (S_m) of compound B is measured. Since the amount of radioactivity due to compound B is constant,

$$S_k \cdot M_u = S_m(M_u + M_k)$$

$$M_u = \frac{S_m \cdot M_k}{S_k - S_m} \tag{18-4}$$

Derivative Dilution

A useful technique for assaying compounds involves the preparation of a labeled derivative by reacting the compound to be determined with a labeled derivatizing reagent. Many possibilities for suitable derivatives exist, but the preparation of acetate derivatives of alcohols will be used in the following examples. Acetic anhydride may be used to prepare acetate esters, but it is important to note that the specific activity of an acetate derivative is one-half that of the acetic anhydride used to prepare it (on a mole basis).

Let M_u be the quantity of alcohol to be determined and the specific activity of the [^{14}C]acetic anhydride be S_k. The reaction of the alcohol and acetic anhydride must be quantitative (complete derivatization). After the reaction, a relatively large, yet known, quantity (M_k) of unlabeled acetate derivative is added so that essentially the total mass of acetate derivative is due to M_k. That is, for all practical purposes, $M_k = M_k + M_u$. A quantity of the diluted acetate derivative is isolated in pure form and its specific activity (S_m) measured.

$$S_m \cdot M_k = \frac{1}{2} S_k \cdot M_u$$

$$M_u = \frac{2 S_m \cdot M_k}{S_k} \tag{18-5}$$

Double Isotope Dilution

In this method derivatization is used as before, but dual labels are employed. The second label is used to ascertain recoveries for the purification steps, which eliminates the need to isolate enough material to measure specific activities.

First [^{14}C]acetic anhydride with a specific activity of S_{Ck} is reacted with an unknown quantity (M_u) of the alcohol. A quantitative reaction is assumed. Then a quantity of the corresponding acetate derivative labeled with ^{3}H is added, and the amount of activity (A_{Hk}) added is known. After purification and isolation, the ^{14}C activity (A_{Cm}) and ^{3}H activity (A_{Hm}) are measured. In the sample isolated, the moles of ^{14}C-labeled derivative present is $A_{Cm}/\frac{1}{2}S_{Ck}$, and the fraction of the ^{14}C derivative isolated is A_{Hm}/A_{Hk}, therefore,

$$M_u = \frac{A_{Cm}/\frac{1}{2}S_{Ck}}{A_{Hm}/A_{Hk}} = \frac{2 A_{Cm} \cdot A_{Hk}}{S_{Ck} \cdot A_{Hm}} \tag{18-6}$$

By determining a recovery fraction that accounts for losses due to an incomplete reaction as well as purification, the usefulness of the double isotope dilution method can be extended.

A known quantity of activity (A_{Hk}) in the form of ^{3}H-labeled alcohol, which has a specific activity of S_{Hk}, is added to the unknown; then the mixture is derivatized with [^{14}C]acetic anhydride that has a specific activity of S_{Ck}. This yields a ^{14}C specific activity of $\frac{1}{2}S_{Ck}$ for the resulting derivative. After isolation and purification, the ^{3}H and ^{14}C activities in the sample are measured yielding A_{Hm} and A_{Cm}, respectively. The overall recovery fraction is A_{Hm}/A_{Hk} and when divided into $A_{Cm}/\frac{1}{2}S_{Ck}$ yields the total moles of alcohol present before derivatization. Subtraction of the moles of [^{3}H]alcohol added yields the moles of alcohol in the original sample.

$$M_u = \frac{A_{Cm}/\frac{1}{2}S_{Ck}}{A_{Hm}/A_{Hk}} - \frac{A_{Hk}}{S_{Hk}}$$

$$M_u = \frac{2A_{Cm} \cdot A_{Hk}}{S_{Ck} \cdot A_{Hm}} - \frac{A_{Hk}}{S_{Hk}} \tag{18-7}$$

The isotope dilution examples given above illustrate some common variations in quantitative measurements, but radioactivity measurements may be coupled with measurements of other physical parameters to extend the range of quantitative methods. The utility of such experimental designs is largely limited by an investigator's ingenuity.

REFERENCES

Abraham, G. E., Ed. (1977). *Handbook of Radioimmunoassay.* Marcel Dekker, New York.

Collins, W. P., Ed. (1985). *Alternative Immunoassays.* Wiley, London.

Kirkham, K. E., and Hunter, W. M. (1971). *Radioimmunoassay Methods.* Churchill Livingstone, London.

Odell, W. D., and Daughaday, W. H. (1971). *Principles of Competitive Protein Binding Assays.* Lippincott, Philadelphia.

Scatchard, G. (1949). *Ann. N.Y. Acad. Sci.* 51:660.

APPENDIX A

Physical Constants and Conversion Factors

Physical Constants and Conversion Factors

Quantity	Symbol	Value (rounded)
Avogadro's number	N	6.022×10^{23}
Velocity of light (vacuum)	c	2.9979×10^{8} m/s
		2.9979×10^{10} cm/s
Elementary charge	e	1.6022×10^{-19} C
		4.8029×10^{-10} esu
Planck's constant	h	6.6262×10^{-34} J/Hz
		6.6262×10^{-27} erg·s
		4.1357×10^{-15} eV·s
Atomic mass unit	amu	$1.6605655 \times 10^{-27}$ kg
		931.5016 MeV
Electron rest mass	m_e	$0.9109534 \times 10^{-30}$ kg
		5.4858×10^{-4} amu
		0.5110041 MeV
Proton rest mass	m_{p^+}	1.672648×10^{-27} kg
		1.0072765 amu
Neutron rest mass	m_n	$1.6749543 \times 10^{-27}$ kg
		1.0086650 amu
Electron volt	eV	1.60219×10^{-19} J
		1.60219×10^{-12} erg
		3.829324×10^{-20} cal
Joule	J	6.24146×10^{18} eV
		10^{7} erg
		1.0 kg·m^2/s^2
Erg	erg	6.24146×10^{11} eV
		10^{-7} J
		1.0 dyne·cm or g·cm^2/s^2

(*Continued*)

Quantity	Symbol	Value (rounded)
Disintegrations per second	dps	1 Bq
Disintegrations per minute	dpm	1/60 Bq
Becquerel	Bq	1 dps
Curie	Ci	3.7×10^{10} dps
		2.22×10^{12} dpm
X-unit	X-unit	1 C/kg air
Roentgen	R	1 esu/cm^3 air
Gray	Gy	1 J/kg
		100 rad
Radiation absorbed dose	rad	100 erg/g
		10^{-2} Gy
Sievert	Sv	Gy·QF·DF
Roentgen equivalent man	rem	rad·QF·DF
Year	yr	3.1536×10^7 s
		5.256×10^5 min
		8.76×10^3 hr
Day	da	8.64×10^4 s
		1.44×10^3 min
Ångstrom	Å	10^{-10} m
		10^{-1} nm

Decay Characteristics of Selected Radionuclides

Nuclide	Half-life	Decay mode	Decay energy (MeV)	Particle energy (MeV)	γ- and X-rays
$^3_1$H	12.26 yr	β⁻	0.01861	0.01861	No
$^{14}_6$C	5730 yr	β⁻	0.15648	0.1565	No
$^{22}_{11}$Na	2.605 yr	β⁺ (90%) EC (10%)	2.842 —	0.545	Yes
$^{32}_{15}$P	14.28 da	β⁻	1.710	1.710	No
$^{35}_{16}$S	87.2 da	β⁻	0.1674	0.1674	No
$^{90}_{38}$Sr	29 yr	β⁻	0.546	0.546	No
$^{125}_{53}$I	59.9 da	EC	0.178	—	Yes
$^{131}_{53}$I	8.04 da	β⁻	0.971	0.606	Yes

APPENDIX B

Answers to Problems

CHAPTER 2

1. 0.511 MeV, or 511 keV
2. 5.599×10^{-4} amu, or 9.29765×10^{-31} kg
3. 10.6 keV
4. 939.57 MeV
5. $0.42794 \times c$, or 1.2829×10^8 m/s
6. 8.2657×10^4 eV, or 82.657 keV
7. 9.5373×10^{-13} m, or 9.5373×10^{-3} Å
8. MD = 0.137006 amu, BE = 127.621 MeV
   BE_{av} = 7.976 MeV/nucleon

CHAPTER 3

1. Q = 7.594 MeV, $E_{k\alpha}$ = 7.450 MeV, E_{kd} = 0.144 MeV
2. $^{42}_{21}Sc \longrightarrow\ ^{42}_{20}Ca^- + \beta^+ + \nu$, Q = 6.424 MeV
   E_{max} = 5.402 MeV
3. mass of $^{22}_{11}Na$ = 23.990961 amu
4. Fig. Appendix B, 3-4
5. Fig. Appendix B, 3-5

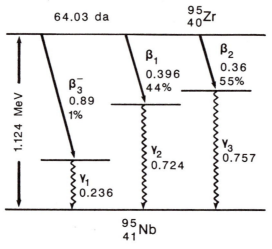

FIG. APPENDIX B, 3-4

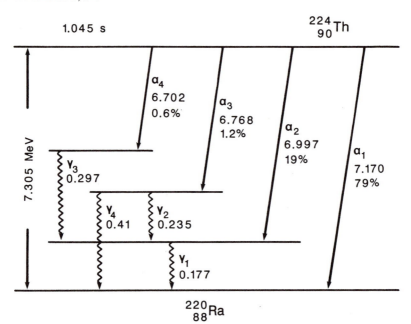

FIG. APPENDIX B, 3-5

CHAPTER 4

1. 1.85×10^5 Bq, 5×10^{-6} Ci, 5.0 µCi
2. 1.339×10^7 Bq, 13.39 MBq, 362 µCi
3. 79.6% are ^{125}I atoms

4. 54.8 yr

5. 6.207×10^{11} atoms

6. 1.191×10^4 dpm

7. 1.837×10^4 dpm

8. 4568 dpm

9. 1343 cpm

10. 2.352×10^8 dpm/mol

11. $\lambda = 0.1183$ da^{-1}, $t_{1/2} = 5.86$ da, $t = 8.45$ da

12. 9.34 yr

13. 73.9%

14. a. 2.338×10^5 dpm/ml, b. 32.33%

15. 46.57 μCi

16. 1.386×10^{14} dpm/mol, 2.31×10^6 MBq/mol, 373.4 μCi/mg

17. 0.1265 or 12.65%

18. 3.383×10^{17} Bq/mol, or 2.05×10^{17} dpm/g. Since ^{32}P decays to ^{32}S, the specific activity of the *phosphoric acid* in the sample remains constant, although its mass decreases according to the half-life of ^{32}P.

19. Dissolve 77.03 mg Na$_2$CO$_3$ plus 5.00 mg Na$_2$ ^{14}CO$_3$ in 1 liter.

20. a. 1.31 μCi, b. 1.688 μg/ml

CHAPTER 5

1. a. 0.84 cm, b. 4.82 cm

2. ^{32}P, 790.15 mg/cm^2; ^{14}C, 28.40 mg/cm^2; ^{3}H, 0.587 mg/cm^2

3. ^{32}P, 611 cm; ^{14}C, 22 cm; ^{3}H, 0.45 cm

4. The ranges of β-particles in glass are ^{32}P, 3.59 mm; ^{14}C, 0.13 mm; ^{3}H, 0.0027 mm. Therefore a 1.0 mm thickness would absorb all β-particles from ^{14}C and ^{3}H but not from ^{32}P.

5. 22.31% transmitted

6. 6.793 cm

7. 0.9 cm^2/g

8. 93.9% absorbed

9. 31.1% absorbed

10. a. Counting efficiency decreases with increasing quantum energies. b. Counting efficiency increases with increasing crystal size.

CHAPTER 6

1. 1.126×10^5 electrons

2. 1.0135×10^3 electrons

3. 6.107×10^{-8} min/ct or 3.66 μs/ct

4. 6.133×10^{-8} min/ct or 3.68 μs/ct

5. a. 60,222 cpm, b. 4,001 cpm

6. a. 0.0037, or 0.37%; b. 0.00025 or 0.025%

CHAPTER 13

1. a. 8 cpm, b. 1.2%, c. ≈ 0.78 cpm, d. 3.2%, e. 668 $\pm$ 8 cpm, f. 2.45%

2. $0.01 = \dfrac{2.58\sqrt{X}}{X}$, $X = 66,564$ counts

3. a. 2411 $\pm$ 25 cpm, b. 6739 $\pm$ 41 cpm, c. 9150 $\pm$ 94 cpm, d. 1.03%, e. $1.625 \times 10^7 \pm 1.94 \times 10^5$ (cpm)^2

4. 59 $\pm$ 3 dpm/ml

5. $s(\bar{X}) = 168$ cpm, $\bar{X} = 10,462 \pm 328$ cpm

CHAPTER 15

1. 1.75×10^{-2} R/hr and 1.08×10^{-4} X-unit/da

2. 3.21×10^{-4} R/hr and 1.98×10^{-6} X-unit/da

3. 7.66×10^{-4} Gy/yr

4. Since the quality and distribution factors are unity, the dose equivalent is 7.66×10^{-4} Sv/yr. Therefore the nonoccupational limit of 1×10^{-3} Sv/yr is not exceeded.

5. a. Build a lead brick shield around the source. Sufficient shielding could easily eliminate the need for any other actions.
   b. Rearrange the furniture in the secretaries' office so they spend more time at a greater distance from the wall.
   c. For storage, locate the source as far as possible from the wall adjoining the secretaries' office. Of course, all locations should be evaluated so that relocation does not result in even higher exposures to persons subject to exposure through the other walls.

6. 9.3 μSv/yr

7. 1.8×10^{-4} Gy/yr and 18 mrad/yr

8. a. 0.137 Sv, b. 9.15×10^{-5} Sv

9. 4.20 mSv

10. 23.4 μSv

CHAPTER 16

1. 79.4 mCi

2. Since water usage rates would permit the disposal of 7.9 Ci of ^{3}H per week, the disposal of 5.1 Ci during the year would not require scheduling of releases.

3. No. With the hood described, 0.54 μCi of ^{125}I could be released during a 4-hr period, and the maximum release anticipated is only 0.2 μCi.

4. 255 mCi

5. Since dilution of 10 μCi of soluble ^{14}C in 3.3 gallons of water would provide a concentration equivalent to the MPC level for an unrestricted area, pouring several buckets of water over the site should dilute the waste sufficiently to meet the "letter of the law." However, if a water hose, perhaps connected to a fire hydrant, could be obtained, flushing the site with copious quantitites of water and allowing it to drain via a storm sewer would be better.

6. 3.8×10^{-8} Gy or 3.8×10^{-6} rad

Index